W9-CNK-632

Springer Series in Statistics

Advisors:
P. Bickel, P. Diggle, S. Fienberg, K. Krickeberg,
I. Olkin, N. Wermuth, S. Zeger

Springer

New York
Berlin
Heidelberg
Barcelona
Hong Kong
London
Milan
Paris
Singapore
Tokyo

Springer Series in Statistics

Andersen/Borgan/Gill/Keiding: Statistical Models Based on Counting Processes.
Atkinson/Riani: Robust Diagnotstic Regression Analysis.
Berger: Statistical Decision Theory and Bayesian Analysis, 2nd edition.
Bolfarine/Zacks: Prediction Theory for Finite Populations.
Borg/Groenen: Modern Multidimensional Scaling: Theory and Applications
Brockwell/Davis: Time Series: Theory and Methods, 2nd edition.
Chan/Tong: Chaos: A Statistical Perspective.
Chen/Shao/Ibrahim: Monte Carlo Methods in Bayesian Computation.
David/Edwards: Annotated Readings in the History of Statistics.
Devroye/Lugosi: Combinatorial Methods in Density Estimation.
Efromovich: Nonparametric Curve Estimation: Methods, Theory, and Applications.
Eggermont/LaRiccia: Maximum Penalized Likelihood Estimation, Volume I:
    Density Estimation.
Fahrmeir/Tutz: Multivariate Statistical Modelling Based on Generalized Linear
    Models, 2nd edition.
Farebrother: Fitting Linear Relationships: A History of the Calculus of Observations
    1750-1900.
Federer: Statistical Design and Analysis for Intercropping Experiments, Volume I:
    Two Crops.
Federer: Statistical Design and Analysis for Intercropping Experiments, Volume II:
    Three or More Crops.
Fienberg/Hoaglin/Kruskal/Tanur (Eds.): A Statistical Model: Frederick Mosteller's
    Contributions to Statistics, Science and Public Policy.
Fisher/Sen: The Collected Works of Wassily Hoeffding.
Glaz/Naus/Wallenstein: Scan Statistics.
Good: Permutation Tests: A Practical Guide to Resampling Methods for Testing
    Hypotheses, 2nd edition.
Gouriéroux: ARCH Models and Financial Applications.
Grandell: Aspects of Risk Theory.
Haberman: Advanced Statistics, Volume I: Description of Populations.
Hall: The Bootstrap and Edgeworth Expansion.
Härdle: Smoothing Techniques: With Implementation in S.
Harrell: Regression Modeling Strategies: With Applications to Linear Models,
    Logistic Regression, and Survival Analysis
Hart: Nonparametric Smoothing and Lack-of-Fit Tests.
Hartigan: Bayes Theory.
Hastie/Tibshirani/Friedman: The Elements of Statistical Learning: Data Mining,
    Inference, and Prediction
Hedayat/Sloane/Stufken: Orthogonal Arrays: Theory and Applications.
Heyde: Quasi-Likelihood and its Application: A General Approach to Optimal
    Parameter Estimation.
Huet/Bouvier/Gruet/Jolivet: Statistical Tools for Nonlinear Regression: A Practical
    Guide with S-PLUS Examples.
Ibrahim/Chen/Sinha: Bayesian Survival Analysis.
Kolen/Brennan: Test Equating: Methods and Practices.

(continued after index)

Paul R. Rosenbaum

Observational Studies

Second Edition

 Springer

Paul R. Rosenbaum
Department of Statistics
The Wharton School
University of Pennsylvania
Philadelphia, PA 19104-6302
USA

Library of Congress Cataloging-in-Publication Data
Rosenbaum, Paul R.
    Observational studies / Paul R. Rosenbaum.—2nd ed.
      p. cm. — (Springer series in statistics)
    Includes bibliographical references and index.
    ISBN 0-387-98967-6 (acid-free paper)
    1. Experimental design.   2. Analysis of variance.   I. Title.   II. Series.
QA279.R67   2002
519.5′3—dc21   2001049264

Printed on acid-free paper.

Production managed by Jenny Wolkowicki; manufacturing supervised by Jeffrey Taub.
Photocomposed pages prepared from the author's LaTeX files.
Printed and bound by Edwards Brothers, Inc., Ann Arbor, MI.
Printed in the United States of America.

9 8 7 6 5 4 3 2 1

ISBN 0-387-98967-6   SPIN 10751865

Springer-Verlag   New York Berlin Heidelberg
A member of BertelsmannSpringer Science+Business Media GmbH

For Sarah, Hannah, and Aaron.

Preface

1. What the Book Is About: An Outline

An *observational study* is an empiric investigation of treatments, policies, or exposures and the effects they cause, but it differs from an experiment in that the investigator cannot control the assignment of treatments to subjects. Observational studies are common in most fields that study the effects of treatments or policies on people. Chapter 1 defines the subject more carefully, presents several observational studies, and briefly indicates some of the issues that structure the subject.

In an observational study, the investigator lacks experimental control; therefore, it is important to begin by discussing the contribution of experimental control to inference about treatment effects. The statistical theory of randomized experiments is reviewed in Chapter 2.

Analytical adjustments are widely used in observational studies in an effort to remove overt biases, that is, differences between treated and control groups, present before treatment, that are visible in the data at hand. Chapter 3 discusses the simplest of these adjustments, which do little more than compare subjects who appear comparable. Chapter 3 then examines the circumstances under which the adjustments succeed. Alas, these circumstances are not especially plausible, for they imply that the observational study differs from an experiment only in ways that have been recorded. If treated and control groups differed before treatment in ways not recorded, there would be a hidden bias. Chapter 4 discusses sensitivity analyses that ask how the findings of a study might be altered by hidden biases of various magnitudes. It turns out that observational studies vary markedly in their sensitivity to hidden bias. The degree of sensitivity to hidden bias is one important consideration in judging whether the treatment caused its ostensible effects or alternatively whether these seeming effects could merely reflect a hidden bias. Chapter 5 discusses models for treatment effects and illustrates their use in sensitivity analysis.

Although a sensitivity analysis indicates the degree to which conclusions could be altered by hidden biases of various magnitudes, it does not indicate whether hidden bias is present. Chapters 6 through 8 concern attempts to detect hidden biases using devices such as multiple control groups, multiple referent groups in a case-referent study, or known effects. Chapter 9 concerns coherence in observational studies, a concept that falls somewhere between attempts to detect hidden bias and sensitivity analyses.

Chapter 10 discusses methods and algorithms for matching and stratification for observed covariates. Chapters 3 and 10 both concern the control of overt biases; however, Chapter 3 assumes that matched pairs, sets, or strata are exactly homogeneous in the observed covariates. When there are many covariates each taking many values, the exact matching in Chapter 3 is not practical. In contrast, the methods and algorithms in Chapter 10 will often produce matched pairs, sets, or strata that balance many covariates simultaneously. The planning of observational studies is discussed in Chapter 11. Chapter 12 discusses the relationship between the design of an observational study and its intended audience.

2. Suggestions for the Reader

Chapter 1 is motivation and Chapter 2 is largely review. Chapter 4 depends on Chapter 3. Chapter 10 may be read immediately following Chapter 3. Other chapters depend strongly on Chapter 4 but only weakly on each other. Chapters 11 and 12 depend on all the previous chapters but may be read at any time.

It is not necessary, perhaps not wise, to read from cover to cover. This book discusses research design, scientific inference, concepts, methods, algorithms and technical results of statistics, together with many examples that serve varied purposes. Different topics will interest different readers, and some topics will be details for all readers. Several suggestions follow.

Chapters are organized by topic. For example, all of the material about randomized experiments appears in Chapter 2, but not all of this material is of equal importance, and parts can be skipped on the way to later chapters. Appendices and sections marked with an asterisk (*) provide some suggestions about what can be skipped.

Sections marked with an asterisk (*) may be skipped. Sections receive asterisks for widely varied reasons. An asterisk signals one thing only: the material in the section is not needed later in the book. A topic that is unconventional, or nonstandard, or technical, or a necessary but tiresome detail, or an unnecessary but interesting digression, is likely to receive an asterisk. Readers who skip these sections may suffer immediate loss, but without future peril.

Appendices discuss ideas that are primarily of interest to people who create statistical methods, rather than to people who use them. Appendices are leisurely and detailed, but they are aimed at a minority of readers. Sensitivity analyses—an important topic in the book—involve obtaining sharp bounds on certain probability distributions using aesthetically pleasing tools such as Holley's inequality and arrangement increasing functions. Topics of this sort are discussed in appendices.

A reader who wishes to understand all of the central concepts while minimizing the technical formalities should focus on matched pairs. It turns out that, for purely technical reasons, the case of matched pairs involves success-or-failure binary trials, so that the formalities are quite elementary. If a reader focuses on discussions of McNemar's test for paired binary outcomes and Wilcoxon's signed rank test for paired continuous outcomes, then the reader will encounter all of the important concepts, but will stay with probability distributions for binary trials. For example, such a reader might read only the first three sections of Chapter 4, thereby completely covering the case of matched pairs, before moving on to later chapters. Also, it is often possible to pick up concepts from the discussion of examples. Many chapters begin with an example, and many methods are illustrated with an example.

To be briefly introduced to the concepts underlying observational studies without formal mathematics, read the overview in Chapter 1, and read about randomized experiments in §2.1 of Chapter 2, adjustments for overt bias in §3.1, sensitivity analysis in §4.1, detecting hidden bias in §6.1, coherence in §9.1, multivariate matching and stratification in §10.1, planning an observational study in Chapter 11, and strategic issues in Chapter 12.

In the index, there is a list of examples under "examples," a list of mathematical symbols under "symbols," and a bits about software under "software."

The second edition of *Observational Studies* is about 50% longer than the first edition. There are two new chapters, namely, Chapter 5 about nonadditive models for treatment effects and Chapter 11 about planning observational studies. Chapter 9 about coherence has been completely rewritten. There are substantial additions to the other chapters, for example, §3.5 about covariance adjustment of rank tests using an estimated propensity score, §4.5 about sensitivity analysis based on asymptotic separability, and §6.5 about bias of known direction. There are many new examples and problems, and a few errors have been corrected.

3. Acknowledgments

To grow older as an academic is to accumulate intellectual debts, and to associate these debts with fond memories. I would like to acknowledge several people who have helped with this book and also collaborators on journal articles that are discussed here. My thanks go to Joe Gastwirth, Sam Gu, Daniel Heitjan, Bob Hornik, Abba Krieger, Marshall Joffe, Yunfei Paul Li, Bo Lu, Sue Marcus, Kewei Ming, Kate Propert, Don Rubin, Jeffrey Silber, and Elaine Zanutto.

This work was supported in part by the US National Science Foundation. The second edition was prepared while I was on sabbatical leave at the Center for Advanced Study in the Behavioral Sciences, and the Center's hospitality and support are gratefully acknowledged.

Chapter 11 is adapted from my article, "Choice as alternative to control in observational studies" which appeared with discussion in *Statistical Science*, 1999, **14**, 259—304. I am grateful to the editor, Leon Gleser, and the discussants, Charles Manski, James Robins, Thomas Cook, and William Shadish for their comments. Also, I am grateful to the Institute of Mathematical Statistics for their copyright policy which permits use by authors of publications in IMS journals, and for their specific encouragement to do so with this article.

Most of all, I'd like to thank Sarah, Hannah, Aaron, and Judy.

The Wharton School
University of Pennsylvania
October 2001   Paul R. Rosenbaum

Contents

1
Observational Studies

1.1 What Are Observational Studies?

William G. Cochran first presented "observational studies" as a topic defined by principles and methods of statistics. Cochran had been an author of the 1964 United States Surgeon General's Advisory Committee Report, *Smoking and Health*, which reviewed a vast literature and concluded: "Cigarette smoking is causally related to lung cancer in men; the magnitude of the effect of cigarette smoking far outweighs all other factors. The data for women, though less extensive, point in the same direction (p. 37)." Though there had been some experiments confined to laboratory animals, the direct evidence linking smoking with human health came from observational or nonexperimental studies.

In a later review, Cochran (1965) defined an observational study as an empiric investigation in which:

> ... the objective is to elucidate cause-and-effect relationships ... [in which] it is not feasible to use controlled experimentation, in the sense of being able to impose the procedures or treatments whose effects it is desired to discover, or to assign subjects at random to different procedures.

Features of this definition deserve emphasis. An observational study concerns treatments, interventions, or policies and the effects they cause, and in this respect it resembles an experiment. A study without a treatment is neither an experiment nor an observational study. Most public opinion polls, most forecasting efforts, most studies of fairness and discrimination, and many other important empirical studies are neither experiments nor observational studies.

In an experiment, the assignment of treatments to subjects is controlled by the experimenter, who ensures that subjects receiving different treatments are comparable. In an observational study, this control is absent for one of several reasons. It may be that the treatment, perhaps cigarette smoking or radon gas, is harmful and cannot be given to human subjects for experimental purposes. Or the treatment may be controlled by a political process that, perhaps quite appropriately, will not yield control merely for an experiment, as is true of much of macroeconomic and fiscal policy. Or the treatment may be beyond the legal reach of experimental manipulation even by a government, as is true of many management decisions in a private economy. Or experimental subjects may have such strong attachments to particular treatments that they refuse to cede control to an experimenter, as is sometimes true in areas ranging from diet and exercise to bilingual education. In each case, the investigator does not control the assignment of treatments and cannot ensure that similar subjects receive different treatments.

1.2   Some Observational Studies

It is encouraging to recall cases, such as *Smoking and Health*, in which observational studies established important truths, but an understanding of the key issues in observational studies begins elsewhere. Observational data have often led competent honest scientists to false and harmful conclusions, as was the case with Vitamin C as a treatment for advanced cancer.

Vitamin C and Treatment of Advanced Cancer: An Observational Study and an Experiment Compared

In 1976, in their article in the *Proceedings of the National Academy of Sciences*, Cameron and Pauling presented observational data concerning the use of vitamin C as a treatment for advanced cancer. They gave vitamin C to 100 patients believed to be terminally ill from advanced cancer and studied subsequent survival.

For each such patient, 10 historical controls were selected of the same age and gender, the same site of primary cancer, and the same histological tumor type. This method of selecting controls is called *matched sampling*—it consists of choosing controls one at a time to be similar to individual treated subjects in terms of characteristics measured prior to treatment. Used effectively, matched sampling often creates treated and control groups that are comparable in terms of the variables used in matching, though the groups may still differ in other ways, including ways that were not measured. Cameron and Pauling (1976, p. 3685) write: "Even though no formal process of randomization was carried out in the selection of our two groups, we believe that they come close to representing random subpopulations of the population of terminal cancer patients in the Vale of Leven Hospital." In a moment, we shall see whether this is so.

Patients receiving vitamin C were compared to controls in terms of time from "untreatability by standard therapies" to death. Cameron and Pauling found that, as a group, patients receiving vitamin C survived about four times longer than the controls. The difference was highly significant in a conventional statistical test, p-value < 0.0001, and so could not be attributed to "chance." Cameron and Pauling "conclude that there is strong evidence that treatment ... [with vitamin C] ... increases the survival time."

This study created interest in vitamin C as a treatment. In response, the Mayo Clinic (Moertel et al., 1985) conducted a careful randomized controlled experiment comparing vitamin C to placebo for patients with advanced cancer of the colon and rectum. In a *randomized experiment*, subjects are assigned to treatment or control on the basis of a chance mechanism, typically a random number generator, so it is only luck that determines who receives the treatment. They found no indication that vitamin C prolonged survival, with the placebo group surviving slightly but not significantly longer. Today, few scientists claim that vitamin C holds promise as a treatment for cancer.

What went wrong in Cameron and Pauling's observational study? Why were their findings so different from those of the randomized experiment? Could their mistake have been avoided in any way other than by conducting a true experiment?

Definite answers are not known, and in all likelihood will never be known. Evidently, the controls used in their observational study, though matched on several important variables, nonetheless differed from treated patients in some way that was important to survival.

The obvious difference between the experiment and the observational study was the random assignment of treatments. In the experiment, a single group of patients was divided into a treated and a control group using a random device. Bad luck could, in principle, make the treated and control groups differ in important ways, but it is not difficult to quantify the potential impact of bad luck and to distinguish it from an effect of the treatment. Common statistical tests and confidence intervals do precisely this. In fact, this is what it means to say that the difference could not reasonably be due to "chance." Chapter 2 discusses the link between statistical inference and random assignment of treatments.

In the observational study, subjects were not assigned to treatment or control by a random device created by an experimenter. The matched sampling ensured that the two groups were comparable in a few important ways, but beyond this, there was little to ensure comparability. If the groups were not comparable before treatment, if they differed in important ways, then the difference in survival might be no more than a reflection of these initial differences.

It is worse than this. In the observational study, the control group was formed from records of patients already dead, while the treated patients were alive at the start of the study. The argument was that the treated patients were terminally ill, that they would all be dead shortly, so the recent records of apparently similar patients, now dead, could reasonably be used to indicate the duration of survival absent treatment with vitamin C. Nonetheless, when the results were analyzed, some patients given vitamin C were still alive; that is, their survival times were censored. This might reflect dramatic effects of vitamin C, but it might instead reflect some imprecision in judgments about who is terminally ill and how long a patient is likely to survive, that is, imprecision about the initial prognosis of patients in the treated group. In contrast, in the control group, one can say with total confidence, without reservation or caveat, that the prognosis of a patient already dead is not good. In the experiment, all patients in both treated and control groups were initially alive.

It is worse still. While death is a relatively unambiguous event, the time from "untreatability by standard therapies" to death depends also on the time of "untreatability." In the observational study, treated patients were judged, at the start of treatment with vitamin C, to be untreatable by other therapies. For controls, a date of untreatability was determined from records. It is possible that these two different processes would produce the same number, but it is by no means certain. In contrast, in the experiment, the starting date in treated and control groups was defined in the same way for both groups, simply because the starting date was determined before a subject was assigned to treatment or control.

What do we conclude from the studies of vitamin C? First, observational studies and experiments can yield very different conclusions. When this happens, the experiments tend to be believed. Chapter 2 develops some of the reasons why this tendency is reasonable. Second, matching and similar adjustments in observational studies, though often useful, do not ensure that treated and control groups are comparable in all relevant ways. More than this, the groups may not be comparable and yet the data we have may fail to reveal this. This issue is discussed extensively in later chapters. Third, while a controlled experiment uses randomization and an observational study does not, experimental control also helps in other ways. Even if we cannot randomize, we wish to exert as much experimental control as is possible, for instance, using the same eligibility criteria for treated and control groups, and the same methods for determining measurements.

Observational studies are typically conducted when experimentation is not possible. Direct comparisons of experiments and observational studies are less common, vitamin C for cancer being an exception. Another direct comparison occurred in the Salk vaccine for polio, a story that is well told by Meier (1972). Others are discussed by Chalmers, Block, and Lee (1970), LaLonde (1986), Fraker and Maynard (1987), Zwick (1991), Friedlander and Robins (1995), and Dehejia and Wahba (1999).

Smoking and Heart Disease: An Elaborate Theory

Doll and Hill (1966) studied the mortality from heart disease of British doctors with various smoking behaviors. While dramatic associations are typically found between smoking and lung cancer, much weaker associations are found with heart disease. Still, since heart disease is a far more common cause of death, even modest increases in risk involve large numbers of deaths.

The first thing Doll and Hill did was to "adjust for age." The old are at greater risk of heart disease than the young. As a group, the smokers tended to be somewhat older than the nonsmokers, though of course there were many young smokers and many old nonsmokers. Compare smokers and nonsmokers directly, ignoring age, and you compare a somewhat older group to a somewhat younger group, so you expect a difference in coronary mortality even if smoking has no effect. In its essence, to "adjust for age" is to compare smokers and nonsmokers of the same age. Often results at different ages are combined into a single number called an age-adjusted mortality rate. Methods of adjustment and their properties are discussed in Chapters 3 and 10. For now, it suffices to say that differences in Doll and Hill's age-adjusted mortality rates cannot be attributed to differences in age, for they were formed by comparing smokers and nonsmokers of the same age. Adjustments of this sort, for age or other variables, are central to the analysis of observational data.

The second thing Doll and Hill did was to consider in detail what should be seen if, in fact, smoking causes coronary disease. Certainly, increased deaths among smokers are expected, but it is possible to be more specific. Light smokers should have mortality somewhere between that of nonsmokers and heavy smokers. People who quit smoking should also have risks between those of nonsmokers and heavy smokers, though it is not clear what to expect when comparing continuing light smokers to people who quit heavy smoking.

Why be specific? Why spell out in advance what a treatment effect should look like? The importance of highly specific theories has a long history, having been advocated in general by Sir Karl Popper (1959) and in observational studies by Sir Ronald Fisher, the inventor of randomized experiments, as quoted by Cochran (1965, §5):

> About 20 years ago, when asked in a meeting what can be done in observational studies to clarify the step from association to causation, Sir Ronald Fisher replied: 'Make your theories

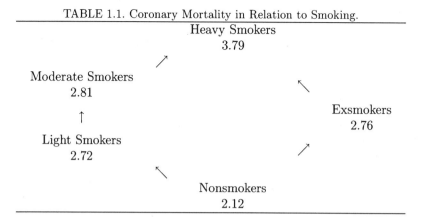

TABLE 1.1. Coronary Mortality in Relation to Smoking.

elaborate.' The reply puzzled me at first, since by Occam's razor, the advice usually given is to make theories as simple as is consistent with known data. What Sir Ronald meant, as subsequent discussion showed, was that when constructing a causal hypothesis one should envisage as many different consequences of its truth as possible, and plan observational studies to discover whether each of these consequences is found to hold.

... this multi-phasic attack is one of the most potent weapons in observational studies.

Chapters 6 through 9 consider this advice formally and in detail.

Table 1.1 gives Doll and Hill's six age-adjusted mortality rates for death from coronary disease not associated with any other specific disease. The rates are deaths per 1000 per year, so the value 3.79 means about 4 deaths in each 1000 doctors each year. The six groups are nonsmokers, exsmokers, and light smokers of 1 to 14 cigarettes, moderate smokers of 15 to 24 cigarettes, and heavy smokers of 25 or more cigarettes per day. Doll and Hill did not separate exsmokers by the amount they had previously smoked, though this would have been interesting and would have permitted more detailed predictions. Again, differences in age do not affect these mortality rates.

Table 1.1 confirms each expectation. Mortality increases with the quantity smoked. Quitters have lower mortality than heavy smokers but higher mortality than nonsmokers. Any alternative explanation, any claim that smoking is not a cause of coronary mortality, would need to explain the entire pattern in Table 1.1. Alternative explanations are not difficult to imagine, but the pattern in Table 1.1 restricts their number.

DES and Vaginal Cancer: Sensitivity to Bias

Cancer of the vagina is a rare condition, particularly in young women. In 1971, Herbst, Ulfelder, and Poskanzer published a report describing eight cases of vaginal cancer in women aged 15 to 22. They were particularly interested in the possibility that a drug, diethylstilbestrol or DES, given to pregnant women, might be a cause of vaginal cancer in their daughters. Each of the eight cases was matched to four *referents*, that is, to four women who did not develop vaginal cancer. These four referents were born within five days of the birth of the case at the same hospital, and on the same type of service, ward or private. There were then eight cases of vaginal cancer and 32 referents, and the study compared the use of DES by their mothers.

This sort of study is called a *case-referent study* or a *case-control study* or a *retrospective study*, no one terminology being universally accepted. In an experiment and in many observational studies, treated and control groups are followed forward in time to see how outcomes develop. In the current context, this would mean comparing two groups of women, a treated group whose mothers had received DES and a control group whose mothers had not. That sort of study is not practical because the outcome, vaginal cancer, is so rare—the treated and control groups would have to be enormous and continue for many years to yield eight cases of vaginal cancer. In a case-referent study, the groups compared are not defined by whether or not they received the treatment, but rather by whether or not they exhibit the outcome. The cases are compared to the referents to see if exposure to the treatment is more common among cases.

In general, the name "case-control" study is not ideal because the word "control" does not have its usual meaning of a person who did not receive the treatment. In fact, in most case-referent studies, many referents did receive the treatment. The name "retrospective" study is not ideal because there are observational studies in which data on entire treated and control groups are collected after treatments have been given and outcomes have appeared, that is, collected retrospectively, and yet the groups being compared are still treated and untreated groups. See MacMahon and Pugh (1970, pp. 41–46) for some detailed discussion of this terminology.

So the study compared eight cases of vaginal cancer to 32 matched referents to see if treatment with diethylstilbestrol was more common among mothers of the cases, and indeed it was. Among the mothers of the eight cases, seven had received DES during pregnancy. Among mothers of the 32 referents, none had received DES. The association between vaginal cancer and DES appears to be almost as strong as a relationship can be, though of course only eight cases have been observed. If a conventional test designed for use in a randomized experiment is used to compare cases and referents in terms of the frequency of exposure to DES, the difference is highly significant. However, experience with the first example, vitamin C and cancer, suggests caution here.

What should be concluded from the strong association observed between DES and vaginal cancer in eight cases and 32 matched referents? Unlike the case of vitamin C and cancer, it would be neither practical nor ethical to follow up with a randomized experiment. Could such a hypothetical experiment produce very different findings? That possibility can never be entirely ruled out. Still, it is possible to ask: How severe would the unseen problems in this study have to be to produce such a strong relationship if DES did not cause vaginal cancer? How far would the observational study have to depart from an experiment to produce such a relationship if DES were harmless? How does the small size of the case group, eight cases, affect these questions? Chapter 4 provides answers. As it turns out, only severe unseen problems and hidden biases, only dramatic departures from an experiment, could produce such a strong association in the absence of an effect of DES, the small sample size notwithstanding. In other words, this study is highly insensitive to hidden bias; its conclusions could be altered by dramatic biases, but not by small ones. This is by no means true of all observational studies. Chapter 4 concerns general methods for quantifying the sensitivity of findings to hidden biases, and it discusses the uses and limitations of sensitivity analyses.

Academic Achievement in Public and Catholic High Schools: Specific Responses to Specific Criticisms

A current controversy in the United States concerns the effectiveness of public or state-run schools, particularly as compared to existing privately operated schools. The 1985 paper by Hoffer, Greely, and Coleman is one of a series of observational studies of this question. They used data from the High School and Beyond Study (HSB), which includes a survey of US high-school students as sophomores with follow-up in their senior year. The HSB study provided standardized achievement test scores in several areas in sophomore and senior years, and included follow-up of students who dropped out of school, so as these things go, it is a rather complete and attractive source of data. Hoffer, Greely, and Coleman (1985) begin with a list of six objections made to their earlier studies, which had compared achievement test scores in public and Catholic schools, concluding that "... Catholic high schools are more effective than public high schools." As an illustration, objection #3 states: "Catholic schools seem to have an effect because they eliminate their disciplinary problems by expelling them from the school." The idea here is that Catholic schools eliminate difficult students while the public schools do not, so the students who remain in Catholic schools would be more likely to perform well even if there were no difference in the effectiveness of the two types of schools.

Criticism is enormously important to observational studies. The quality of the criticism offered in a particular field is intimately connected with the quality of the studies conducted in that field. Quality is not quantity, nor is harshness quality. What is scientifically plausible must be distinguished from what is just logically possible (Gastwirth, Krieger and Rosenbaum 1997). Cochran (1965, §5) argues that the first critic of an observational study should be its author:

> When summarizing the results of a study that shows an association consistent with the causal hypothesis, the investigator should always list and discuss all alternative explanations of his results (including different hypotheses and biases in the results) that occur to him. This advice may sound trite, but in practice is often neglected.

Criticisms of observational studies are of two kinds, the tangible and the dismissive, objection #3 being of the tangible kind. A tangible criticism is a specific and plausible alternative interpretation of the available data; indeed, a tangible criticism is itself a scientific theory, itself capable of empirical investigation. Bross (1960) writes:

> ... a critic who objects to a bias in the design or a failure to control some established factor is, in fact, raising a counter-hypothesis ... [and] has the responsibility for showing [it] is tenable. In doing so, he operates under the same ground rules as the proponent ... : When a critic has shown that his counterhypothesis is tenable, his job is done, while at this point the proponent's job is just beginning. A proponent's job is not finished as long as there is a tenable hypothesis that rivals the one he asserts.

On the second page of his *The Design of Experiments*, Fisher (1935) described dismissive criticism as he argued that a theory of experimental design is needed:

> This type of criticism is usually made by what I might call a heavyweight *authority*. Prolonged experience, or at least the long possession of a scientific reputation, is almost a pre-requisite for developing successfully this line of attack. Technical details are seldom in evidence. The authoritative assertion: "His *controls* are *totally* inadequate" must have temporarily discredited many a promising line of work; and such an authoritarian method of judgement must surely continue, human nature being what it is, so long as theoretical notions of the principles of experimental design are lacking ... .

Dismissive criticism rests on the authority of the critic and is so broad and vague that its claims cannot be studied empirically. Judging the weight of evidence is inseparable from judging the criticisms that have been or can be raised.

Concerning objection #3, Hoffer, Greely, and Coleman (1985) respond: "... the evidence from the HSB data, although indirect, does not support this objection. Among students who reported that they had been suspended during their sophomore year, those in the Catholic sector were more likely to be in the same school as seniors than those in the public sector (63 percent to 56 percent)." In other words, difficult students, or at any rate students who were suspended, remained in Catholic school more often, not less often, than in public schools. This response to objection #3, though not decisive, does gives one pause.

Successful criticism of an observational study points to ambiguity in evidence or argument, and then points to methods for removing the ambiguity. Efforts to resolve an ambiguity are sometimes undermined by efforts to win an argument. Popper (1994, p. 44) writes:

> Serious critical discussions are always difficult ... Many participants in a rational, that is, a critical, discussion find it particularly difficult to unlearn what their instincts seem to teach them (and what they are taught, incidently, by every debating society): that is, to win. For what they have to learn is that victory in debate is nothing, while even the slightest clarification of one's problem—even the smallest contribution made towards a clearer understanding of one's own position or that of one's opponent—is a great success. A discussion which you win but which fails to help you change or to clarify your mind at least a little should be regarded as a sheer loss.

1.3   Purpose of This Book

Scientific evidence is commonly and properly greeted with objections, skepticism, and doubt. Some objections come from those who simply do not like the conclusions, but setting aside such unscientific reactions, responsible scientists are responsibly skeptical. We look for failures of observation, gaps in reasoning, alternative interpretations. We compare new evidence with past evidence. This skepticism is itself scrutinized. Skepticism must be justified, defended. One needs "grounds for doubt," in Wittgenstein's (1969, §122) phrase. The grounds for doubt are themselves challenged. Objections bring forth counterobjections and more evidence. As time passes, arguments on one side or the other become strained, fewer scientists are willing to offer them, and the arguments on that side come increasingly from individuals who seem to have some stake in the outcome. In this way, questions are settled.

Scientific questions are not settled on a particular date by a single event, nor are they settled irrevocably. We speak of the weight of evidence. Eventually, the weight is such that critics can no longer lift it, or are too weary to try. Overwhelming evidence is evidence that overwhelms responsible critics.

Experiments are better than observational studies because there are fewer grounds for doubt. The ideal experiment would leave few grounds for doubt, and at times this ideal is nearly achieved, particularly in the laboratory. Experiments often settle questions faster.

Despite this, experiments are not feasible in some settings. At times, observational studies have produced overwhelming evidence, as compelling as any in science, but at other times, observational data have misled investigators to advocate harmful policies or ineffective treatments.

A statistical theory of observational studies is a framework and a set of tools that provide measures of the weight of evidence. The purpose of this book is to give an account of statistical principles and methods for the design and analysis of observational studies. An adequate account must relate observational studies to controlled experiments, showing how uncertainty about treatment effects is greater in the absence of randomization. Analytical adjustments are common in observational studies, and the account should indicate what adjustments can and cannot do. A large literature offers many devices to detect hidden biases in observational studies, for instance, the use of several control groups, and the account must show how such devices work and when they may be expected to succeed or fail. Even when it is not possible to reduce or dispel uncertainty, it is possible to be careful in discussing its magnitude. That is, even when it is not possible to remove bias through adjustment or to detect bias through careful design, it is nonetheless possible to give quantitative expression to the magnitude of uncertainties about bias, a technique called *sensitivity analysis*. The account must indicate what can and cannot be done with a sensitivity analysis.

1.4 Bibliographic Notes

Most scientific fields that study human populations conduct observational studies. Many fields have developed a literature on the design, conduct, and interpretation of observational studies, often with little reference to related work in other fields. It is not possible to do justice to these several literatures in a short bibliographic note. There follows a short and incomplete list of fine books that contain substantial general discussions of the methodology used for observational studies in epidemiology, public program evaluation, or the social sciences. A shared goal in these diverse works is evaluation of treatments, exposures, programs, or policies from nonexperimental data. The list is followed by references cited in Chapter 1.

Some Books and a Few Papers

Angrist, J. D. and Krueger, A. B. (1999) Empirical strategies in labor economics. In: *Handbook of Labor Economics*, O. Ashenfelter and D. Card, eds., Volume 3A, Chapter 23, New York: Elsevier.

Ashenfelter, O., ed. (2000) *Labor Economics*. New York: Worth.

Becker, H. S. (1997) *Tricks of the Trade*. Chicago: University of Chicago Press.

Blaug, M. (1980) *The Methodology of Economics*. New York: Cambridge University Press.

Breslow, N. and Day, N. (1980, 1987) *Statistical Methods in Cancer Research*, Volumes 1 and 2. Lyon, France: International Agency for Research on Cancer.

Campbell, D. T. (1988) *Methodology and Epistemology for Social Science: Selected Papers*. Chicago: University of Chicago Press, pp. 315—333.

Campbell, D. and Stanley, J. (1963) *Experimental and Quasi-Experimental Design for Research*. Chicago: Rand McNally.

Chamberlain, G. (1984) Panel data. In: *Handbook of Econometrics*, Chapter 22, Volume 2, Z. Griliches and M. D. Intriligator, eds., New York: Elsevier.

Cochran, W. G. (1965) The planning of observational studies of human populations (with discussion). *Journal of the Royal Statistical Society*, Series **A**, **128**, 134—155.

Cochran, W. (1983) *Planning and Analysis of Observational Studies*. New York: Wiley.

Cook, T. D. and Campbell, D. C. (1979) *Quasi-Experimentation*. Chicago: Rand McNally.

Cook, T. D., Campbell, D. T., and Peracchio, L. (1990) Quasi-experimentation. In: *Handbook of Industrial and Organizational Psychology*, M. Dunnette and L. Hough, eds., Palo Alto, CA: Consulting Psychologists Press, Chapter 9, pp. 491—576.

Cook, T. D. and Shadish, W. R. (1994) Social experiments: Some developments over the past fifteen years. *Annual Review of Psychology*, **45**, 545—580.

Cornfield, J., Haenszel, W., Hammond, E., Lilienfeld, A., Shimkin, M., and Wynder, E. (1959) Smoking and lung cancer: Recent evidence and a discussion of some questions. *Journal of the National Cancer Institute*, **22**, 173—203.

Cox, D. R. (1992) Causality: Some statistical aspects. *Journal of the Royal Statistical Society*, Series **A**, **155**, 291–301.

Elwood, J. M. (1988) *Causal Relationships in Medicine*. New York: Oxford University Press.

Emerson, R. M. (1981) Observational field work. *Annual Review of Sociology*, **7**, 351—378.

Freedman, D. (1997) From association to causation via regression. *Advances in Applied Mathematics*, **18**, 59—110.

Friedman, M. (1953) *Essays in Positive Economics*. Chicago: University of Chicago Press.

Gastwirth, J. (1988) *Statistical Reasoning in Law and Public Policy*. New York: Academic Press.

Gordis, L. (2000) *Epidemiology* (Second Edition) Philadelphia: Saunders.

Greenhouse, S. (1982) Jerome Cornfield's contributions to epidemiology. *Biometrics*, **28**, Supplement, 33–46.

Heckman, J. J. (2001) Micro data, heterogeneity, and the evaluation of public policy: The Nobel lecture. *Journal of Political Economy*, **109**, 673–748.

Hill, A. B. (1965) The environment and disease: Association or causation? *Proceedings of the Royal Society of Medicine*, **58**, 295—300.

Holland, P. (1986) Statistics and causal inference (with discussion). *Journal of the American Statistical Association*, **81**, 945–970.

Kelsey, J., Whittemore, A., Evans, A., and Thompson. W. (1996). *Methods in Observational Epidemiology*. New York: Oxford University Press.

Khoury, M. J., Cohen, B. H., and Beaty, T. H. (1993) *Fundamentals of Genetic Epidemiology*. New York: Oxford University Press.

Kish, L. (1987) *Statistical Design for Research*. New York: Wiley.

Lilienfeld, A. and Lilienfeld, D. E. (1980) *Foundations of Epidemiology*. New York: Oxford University Press.

Lilienfeld, D. E. and Stolley, P. D. (1994) *Foundations of Epidemiology*. New York: Oxford University Press.

Lipsey, M. W. and Cordray, D. S. (2000) Evaluation methods for social intervention. *Annual Review of Psychology*, **51**, 345—375.

Little, R. J. and Rubin, D. B. (2000) Causal effects in clinical and epidemiological studies via potential outcomes. *Annual Review of Public Health*, **21**, 121—145.

Maclure, M. and Mittleman, M. A. (2000) Should we use a case-crossover design? *Annual Review of Public Health*, **21**, 193—221.

MacMahon, B. and Pugh, T. (1970) *Epidemiology*. Boston: Little, Brown.

MacMahon, B. and Trichopoulos, D. (1996) *Epidemiology*. Boston: Little, Brown.

Manski, C. (1995) *Identification Problems in the Social Sciences*. Cambridge, MA: Harvard University Press.

Mantel, N. and Haenszel, W. (1959) Statistical aspects of retrospective studies of disease. *Journal of the National Cancer Institute*, **22**, 719–748.

Meyer, B. D. (1995) Natural and quasi-experiments in economics. *Journal of Business and Economic Statistics*, **13**, 151—161.

Meyer, M. and Fienberg, S., eds. (1992) *Assessing Evaluation Studies: The Case of Bilingual Education Strategies*. Washington, DC: National Academy Press.

Miettinen, O. (1985) *Theoretical Epidemiology*. New York: Wiley.

Pearl, J. (2000) *Causality: Models, Reasoning, Inference*. New York: Cambridge University Press.

Reichardt, C. S. (2000) A typology of strategies for ruling out threats to validity. In: *Research Design: Donald Campbell's Legacy*, L. Brickman, ed., Thousand Oaks, CA: Sage, Volume 2, pp., 89–115.

Reiter, J. (2000) Using statistics to determine causal relationships. *American Mathematical Monthly*, **107**, 24—32.

Robins, J. M. (1999) Association, causation, and marginal structural models. *Synthese*, **121**, 151—179.

Robins, J., Blevins, D., Ritter, G., and Wulfsohn, M. (1992) G-estimation of the effect of prophylaxis therapy for pneumocystis carinii pneumonia on the survival of AIDS patients. *Epidemiology,* **3**, 319—336.

Rosenthal, R. and Rosnow, R., eds. (1969) *Artifact in Behavioral Research*. New York: Academic.

Rosenzweig, M. R. and Wolpin, K. I. (2000) Natural "natural experiments" in economics. *Journal of Economic Literature*, **38**, 827—874.

Rosnow, R. L. and Rosenthal, R. (1997) *People Studying People: Artifacts and Ethics in Behavioral Research.* New York: W. H. Freeman.

Rossi, P., Freeman, H., and Lipsey, M. W. (1999) *Evaluation.* Beverly Hills, CA: Sage.

Rothman, K. and Greenland, S. (1998) *Modern Epidemiology.* Philadelphia: Lippincott-Raven.

Rubin, D. (1974) Estimating causal effects of treatments in randomized and nonrandomized studies. *Journal of Educational Psychology,* **66**, 688–701.

Schlesselman, J. (1982) *Case-Control Studies.* New York: Oxford University Press.

Schulte, P. A. and Perera, F. (1993) *Molecular Epidemiology: Principles and Practices.* New York: Academic.

Shadish, W. R., Cook, T. D., and Campbell, D. T. (2002) *Experimental and Quasi-Experimental Designs for Generalized Causal Inference.* Boston: Houghton-Mifflin.

Shafer, G. (1996) *The Art of Causal Conjecture.* Cambridge, MA: MIT Press.

Sobel, M. (1995) Causal inference in the social and behavioral sciences. In: *Handbook of Statistical Modelling for the Social and Behavioral Sciences,* G. Arminger, C. Clogg, and M. Sobel, eds., New York: Plenum, 1—38.

Steenland, K., ed. (1993) *Case Studies in Occupational Epidemiology.* New York: Oxford University Press.

Strom, B. (2000) *Pharmacoepidemiology.* New York: Wiley.

Suchman, E. (1967) *Evaluation Research.* New York: Sage.

Susser, M. (1973) *Causal Thinking in the Health Sciences: Concepts and Strategies in Epidemiology.* New York: Oxford University Press.

Susser, M. (1987) *Epidemiology, Health and Society: Selected Papers.* New York: Oxford University Press.

Tufte, E., ed. (1970) *The Quantitative Analysis of Social Problems.* Reading, MA: Addison-Wesley.

Weiss, C. (1997) *Evaluation.* Englewood Cliffs, NJ: Prentice-Hall.

Weiss, N. S. (1996) *Clinical Epidemiology.* New York: Oxford University Press.

Willett, W. (1998) *Nutritional Epidemiology*. New York: Oxford University Press.

Winship, C. and Morgan, S. L. (1999) The estimation of causal effects from observational data. *Annual Review of Sociology*, **25**, 659—706.

Zellner, A. (1968) *Readings in Economic Statistics and Econometrics*. Boston: Little, Brown.

1.5   References

Bross, I. D. J. (1960) Statistical criticism. *Cancer*, **13**, 394–400. Reprinted in: *The Quantitative Analysis of Social Problems*, E. Tufte, ed., Reading, MA: Addison-Wesley, pp. 97–108.

Cameron, E. and Pauling, L. (1976) Supplemental ascorbate in the supportive treatment of cancer: Prolongation of survival times in terminal human cancer. *Proceedings of the National Academy of Sciences (USA)*, **73**, 3685–3689.

Chalmers, T., Block, J., and Lee, S. (1970) Controlled studies in clinical cancer research. *New England Journal of Medicine*, **287**, 75–78.

Cochran, W.G. (1965) The planning of observational studies of human populations (with discussion). *Journal of the Royal Statistical Society*, Series A, **128**, 134–155. Reprinted in *Readings in Economic Statistics and Econometrics*, A. Zellner, ed., 1968, Boston: Little Brown, pp. 11–36.

Dehejia, R. H. and Wahba, S. (1999) Causal effects in nonexperimental studies: Reevaluating the evaluation of training programs. *Journal of the American Statistical Association*, **94**, 1053—1062.

Doll, R. and Hill, A. (1966) Mortality of British doctors in relation to smoking: Observations on coronary thrombosis. In: *Epidemiological Approaches to the Study of Cancer and Other Chronic Diseases*, W. Haenszel, ed., U.S. National Cancer Institute Monograph 19, Washington, DC: US Department of Health, Education, and Welfare, pp. 205–268.

Fisher, R.A. (1935, 1949) *The Design of Experiments*. Edinburgh: Oliver & Boyd.

Fraker, T. and Maynard, R. (1987) The adequacy of comparison group designs for evaluations of employment-related programs. *Journal of Human Resources*, **22**, 194–227.

Friedlander, D. and Robins, P. K. (1995) Evaluating program evaluations: New evidence on commonly used nonexperimental methods. *American Economic Review*, **85**, 923—937.

Gastwirth, J. L., Krieger, A. M., and Rosenbaum, P. R. (1997) Hypotheticals and hypotheses. *American Statistician*, **51**, 120—121.

Herbst, A., Ulfelder, H., and Poskanzer, D. (1971) Adenocarcinoma of the vagina: Association of maternal stilbestrol therapy with tumor appearance in young women. *New England Journal of Medicine*, **284**, 878–881.

Hoffer, T., Greeley, A., and Coleman, J. (1985) Achievement growth in public and Catholic schools. *Sociology of Education*, **58**, 74–97.

LaLonde, R. (1986) Evaluating the econometric evaluations of training programs with experimental data. *American Economic Review*, **76**, 604–620.

Meier, P. (1972) The biggest public health experiment ever: The 1954 field trial of the Salk poliomyelitis vaccine. In: *Statistics: A Guide to the Unknown*, J. Tanur, ed., San Francisco: Holden-Day, pp. 2–13.

Moertel, C., Fleming, T., Creagan, E., Rubin, J., O'Connell, M., and Ames, M. (1985) High-dose vitamin C versus placebo in the treatment of patients with advanced cancer who have had no prior chemotherapy: A randomized double-blind comparison. *New England Journal of Medicine*, **312**, 137–141.

Popper, K. (1959) *The Logic of Scientific Discovery*. New York: Harper & Row.

Popper, K. (1994) *The Myth of the Framework*. New York: Routledge.

United States Surgeon General's Advisory Committee Report (1964) *Smoking and Health*. Washington, DC: US Department of Health, Education and Welfare.

Wittgenstein, L. (1969) *On Certainty*. New York: Harper & Row.

Zwick, R. (1991) Effects of item order and context on estimation of NAEP reading proficiency. *Educational Measurement: Issues and Practice*, **3**, 10–16.

)

2
Randomized Experiments

2.1 Introduction and Example: A Randomized Clinical Trial

Observational studies and controlled experiments have the same goal, inference about treatment effects, but random assignment of treatments is present only in experiments. This chapter reviews the role of randomization in experiments, and so prepares for discussion of observational studies in later chapters. A theory of observational studies must have a clear view of the role of randomization, so it can have an equally clear view of the consequences of its absence. Sections 2.1 and 2.2 give two examples: a large controlled clinical trial, and then a small but famous example due to Sir Ronald Fisher, who is usually credited with the invention of randomization, which he called the "reasoned basis for inference" in experiments. Later sections discuss the meaning of this phrase, that is, the link between randomization and statistical methods. Most of the material in this chapter is quite old.

Randomized Trial of Coronary Surgery

The US Veterans Administration (Murphy et al. 1977) conducted a randomized controlled experiment comparing coronary artery bypass surgery with medical therapy as treatments for coronary artery disease. Bypass surgery is an attempt to repair the arteries that supply blood to the heart, arteries that have been narrowed by fatty deposits. In bypass surgery, a

TABLE 2.1. Base-Line Comparison of Coronary Patients in the Veterans Administration Randomized Trial.

Covariate	Medical %	Surgical %
New York Heart Association Class II & III	94.2	95.4
History of myocardial infarction (MI)	59.3	64.0
Definite or possible MI based on electrocardiogram	36.1	40.5
Duration of chest pain > 25 months	50.0	51.8
History of hypertension	30.0	27.6
History of congestive heart failure	8.4	5.2
History of cerebral vascular episode	3.2	2.1
History of diabetes	12.9	12.2
Cardiothoracic ratio > 0.49	10.4	12.2
Serum cholesterol > 249 mg/100 ml	31.6	20.6

bypass or bridge is formed around a blockage in a coronary artery. In contrast, medical therapy uses drugs to enhance the flow of blood through narrowed arteries. The study involved randomly assigning 596 patients at 13 Veterans Administration hospitals, of whom 286 received surgery and 310 received drug treatments. The random assignment of a treatment for each patient was determined by a central office after the patient had been admitted into the trial.

Table 2.1 is taken from their study. It compares the medical and surgical treatment groups in terms of 10 important characteristics of patients measured at "base-line," that is, prior to the start of treatment. A variable measured prior to the start of treatment is called a *covariate*. Similar tables appear in reports of most clinical trials.

Table 2.1 shows the two groups of patients were similar in many important ways prior to the start of treatment, so that comparable groups were being compared. When the percentages for medical and surgical are compared, the difference is not significant at the 0.05 level for nine of the variables in Table 2.1, but is significant for serum cholesterol. This is in line with what one would expect from 10 significance tests if the only differences were due to chance, that is, due to the choice of random numbers used in assigning treatments.

For us, Table 2.1 is important for two reasons. First, it is an example showing that randomization tends to produce relatively comparable or balanced treatment groups in large experiments. The second point is separate and more important. The 10 covariates in Table 2.1 were not used in assigning treatments. There was no deliberate balancing of these variables. Rather the balance we see was produced by the random assignment, which made no use of the variables themselves. This gives us some reason to hope and expect that other variables, not measured, are similarly balanced. Indeed, as shown shortly, statistical theory supports this expectation. Had the trial not used random assignment, had it instead assigned patients one at a time to balance these 10 covariates, then the balance might well have been better than in Table 2.1, but there would have been no basis for expecting other unmeasured variables to be similarly balanced.

The VA study compared survival in the two groups three years after treatment. Survival in the medical group was 87% and in the surgical group 88%, both with a standard error of 2%. The 1% difference in mortality was not significant. Evidently, when comparable groups of patients received medical and surgical treatment at the VA hospitals, the outcomes were quite similar.

The statement that randomization tends to balance covariates is at best imprecise; taken too literally, it is misleading. For instance, in Table 2.1, the groups do differ slightly in terms of serum cholesterol. Presumably there are other variables, not measured, exhibiting imbalances similar to if not greater than that for serum cholesterol. What is precisely true is that random assignment of treatments can produce some imbalances by chance, but common statistical methods, properly used, suffice to address the uncertainty introduced by these chance imbalances. To this subject, we now turn.

2.2   The Lady Tasting Tea

"A lady declares that by tasting a cup of tea made with milk she can discriminate whether the milk or the tea infusion was first added to the cup," or so begins the second chapter of Sir Ronald Fisher's (1935, 1949) book *The Design of Experiments*, which introduced the formal properties of randomization. This example is part of the tradition of statistics, and in addition it was well selected by Fisher to illustrate key points. He continues:

> Our experiment consists in mixing eight cups of tea, four in one way and four in the other, and presenting them to the subject for judgement in a random order. The subject has been told in advance of what the test will consist, namely that she

will be asked to taste eight cups, that these shall be four of each kind, and that they shall be presented to her in a random order, that is in an order not determined arbitrarily by human choice, but by the actual manipulation of the physical apparatus used in games of chance, cards, dice, roulettes, etc., or more expeditiously, from a published collection of random sampling numbers purporting to give the actual results of such a manipulation. Her task is to divide the 8 cups into two sets of 4, agreeing, if possible, with the treatments received.

Fisher then asks what would be expected if the Lady was "without any faculty of discrimination," that is, if she makes no changes at all in her judgments in response to changes in the order in which tea and milk are added to the cups. To change her judgments is to have some faculty of discrimination, however slight. So suppose for the moment that she cannot discriminate at all, that she gives the same judgments no matter which four cups receive milk first. Then it is only by accident or chance that she correctly identifies the four cups in which milk was added first. Since there are $\binom{8}{4} = 70$ possible divisions of the eight cups into two groups of four, and randomization has ensured that these are equally probable, the chance of this accident is $1/70$. In other words, the probability that the random ordering of the cups will yield perfect agreement with the Lady's fixed judgments is $1/70$. If the Lady correctly classified the cups, this probability, $0.014 = 1/70$, is the significance level for testing the null hypothesis that she is without the ability to discriminate.

Fisher goes on to describe randomization as the "reasoned basis" for inference and "the physical basis of the validity of the test"; indeed, these phrases appear in section headings and are clearly important to Fisher. He explains:

> We have now to examine the physical conditions of the experimental technique needed to justify the assumption that, if discrimination of the kind under test is absent, the result of the experiment will be wholly governed by the laws of chance ... .
>
> It is [not sufficient] to insist that "all the cups must be exactly alike" in every respect except that to be tested. For this is a totally impossible requirement in our example, and equally in all other forms of experimentation ... .
>
> The element in the experimental procedure which contains the essential safeguard is that the two modifications of the test beverage are to be prepared "in random order." This, in fact, is the only point in the experimental procedure in which the laws of chance, which are to be in exclusive control of our frequency distribution, have been explicitly introduced.

Fisher discusses this example for 15 pages, though its formal aspects are elementary and occupy only a part of a paragraph. He is determined to establish that randomization has justified or grounded a particular inference, formed its "reasoned basis," a basis that would be lacking had the same pattern of responses, the same data, been observed in the absence of randomization.

The example serves Fisher's purpose well. The Lady is not a sample from a population of Ladies, and even if one could imagine that she was, there is but one Lady in the experiment and the hypothesis concerns her alone. Her eight judgments are not independent observations, not least because the rules require a split into four and four. Later cups differ from earlier ones, for by cup number five, the Lady has surely tasted one with milk first and one with tea first. There is no way to construe, or perhaps misconstrue, the data from this experiment as a sample from a population, or as a series of independent and identical replicates. And yet, Fisher's inference is justified, because the only probability distribution used in the inference is the one created by the experimenter.

What are the key elements in Fisher's argument? First, experiments do not require, indeed cannot reasonably require, that experimental units be homogeneous, without variability in their responses. Homogeneous experimental units are not a realistic description of factory operations, hospital patients, agricultural fields. Second, experiments do not require, indeed, cannot reasonably require, that experimental units be a random sample from a population of units. Random samples of experimental units are not the reality of the industrial laboratory, the clinical trial, or the agricultural experiment. Third, for valid inferences about the effects of a treatment on the units included in an experiment, it is sufficient to require that treatments be allocated at random to experimental units—these units may be both heterogeneous in their responses and not a sample from a population. Fourth, probability enters the experiment only through the random assignment of treatments, a process controlled by the experimenter. A quantity that is not affected by the random assignment of treatments is a fixed quantity describing the units in the experiment.

The next section repeats Fisher's argument in more general terms.

2.3   Randomized Experiments

2.3.1   Units and Treatment Assignments

There are N units available for experimentation. A unit is an opportunity to apply or withhold the treatment. Often, a unit is a person who will receive either the treatment or the control as determined by the experimenter. However, it may happen that it is not possible to assign a treatment to a single person, so a group of people form a single unit, perhaps all children in a particular classroom or school. On the other hand, a single person may present several opportunities to apply different treatments, in which case each opportunity is a unit; see Problem 2. For instance, in §2.2, the one Lady yielded eight units.

The N units are divided into S strata or subclasses on the basis of covariates, that is, on the basis of characteristics measured prior to the assignment of treatments. The stratum to which a unit belongs is not affected by the treatment, since the strata are formed prior to treatment. There are n_s units in stratum s for $s = 1, \ldots, S$, so $N = \sum n_s$.

Write $Z_{si} = 1$ if the ith unit in stratum s receives the treatment and write $Z_{si} = 0$ if this unit receives the control. Write m_s for the number of treated units in stratum s, so $m_s = \sum_{i=1}^{n_s} Z_{si}$, and $0 \leq m_s \leq n_s$. Finally, write $\mathbf{Z}$ for the N-dimensional column vector containing the Z_{si} for all units in the lexical order; that is,

$$
\mathbf{Z} = \begin{bmatrix} Z_{11} \\ Z_{12} \\ \vdots \\ Z_{1,n_1} \\ Z_{21} \\ \vdots \\ Z_{S,n_S} \end{bmatrix} = \begin{bmatrix} \mathbf{Z}_1 \\ \vdots \\ \mathbf{Z}_S \end{bmatrix}, \qquad \text{where} \quad \mathbf{Z}_s = \begin{bmatrix} Z_{s1} \\ \vdots \\ Z_{s,n_s} \end{bmatrix}. \tag{2.1}
$$

This notation covers several common situations. If no covariates are used to divide the units, then there is a single stratum containing all units, so $S = 1$. If $n_s = 2$ and $m_s = 1$ for $s = 1, \ldots, S$, then there are S pairs of units matched on the basis of covariates, each pair containing one treated unit and one control. The situation in which $n_s \geq 2$ and $m_s = 1$ for $s = 1, \ldots, S$ is called matching with multiple controls. In this case there are S matched sets, each containing one treated unit and one or more controls.

The case of a single stratum, that is $S = 1$, is sufficiently common and important to justify slight modifications in notation. When there is only a single stratum the subscript s is dropped, so Z_i is written in place of Z_{1i}. The same convention applies to other quantities that have subscripts s and i.

2.3.2 Several Methods of Assigning Treatments at Random

In a *randomized experiment*, the experimenter determines the assignment of treatments to units, that is the value of $\mathbf{Z}$, using a known random mechanism such as a table of random numbers. To say that the mechanism is known is to say that the distribution of the random variable $\mathbf{Z}$ is known because it was created by the experimenter. One requirement is placed on this random mechanism, namely, that, before treatments are assigned, every unit has a nonzero chance of receiving both the treatment and the control, or formally that $0 < \text{prob}(Z_{si} = 1) < 1$ for $s = 1, \ldots, S$ and $i = 1, \ldots, n_s$. Write Ω_0 for the set containing all possible values of $\mathbf{Z}$, that is, all values of $\mathbf{Z}$ which are given nonzero probability by the mechanism.

In practice, many different random mechanisms have been used to determine $\mathbf{Z}$. The simplest assigns treatments independently to different units, taking $\text{prob}(Z_{si} = 1) = 1/2$ for all s, i. This method was used in the Veterans Administration experiment on coronary artery surgery in §2.1. In this case, Ω_0 is the set containing 2^N possible values of $\mathbf{Z}$, namely, all N-tuples of zeros and ones, and every assignment in Ω_0 has the same probability; that is, $\text{prob}(\mathbf{Z} = \mathbf{z}) = 1/2^N$ for all $\mathbf{z} \in \Omega_0$. The number of elements in a set A is written $|A|$, so in this case $|\Omega_0| = 2^N$. This mechanism has the peculiar property that there is a nonzero probability that all units will be assigned to the same treatment, though this probability is extremely small when N is moderately large. From a practical point of view, a more important problem with this mechanism arises when S is fairly large compared to N. In this case, the mechanism may give a high probability to the set of treatment assignments in which all units in some stratum receive the same treatment. If the strata were types of patients in a clinical trial, this would mean that all patients of some type received the same treatment. If the strata were schools in an educational experiment, it would mean that all children in some school received the same treatment. Other assignment mechanisms avoid this possibility.

The most commonly used assignment mechanism fixes the number m_s of treated subjects in stratum s. In other words, the only assignments $\mathbf{Z}$ with nonzero probability are those with m_s treated subjects in stratum s for $s = 1, \ldots, S$. If m_s is chosen sensibly, this avoids the problem mentioned in the previous paragraph. For instance, if n_s is required to be even and m_s is required to equal $n_s/2$ for each s, then half the units in each stratum receive the treatment and half receive the control, so the final treated and control groups are exactly balanced in the sense that they contain the same number of units from each stratum.

When m_s is fixed in this way, let Ω be the set containing the $K = \prod_{s=1}^{S} \binom{n_s}{m_s}$ possible treatment assignments $\mathbf{z} = \begin{bmatrix} \mathbf{z}_1 \\ \vdots \\ \mathbf{z}_S \end{bmatrix}$ in which $\mathbf{z}_s$ is an n_s-tuple with m_s ones and $n_s - m_s$ zeros for $s = 1, \ldots, S$. In the most common assignment mechanism, each of these K possible assignments is given the same probability, $\text{prob}(\mathbf{Z} = \mathbf{z}) = 1/K$ all $\mathbf{z} \in \Omega$. This type of randomized experiment, with equal probabilities and fixed m_s, will be called a *uniform randomized experiment*. When there is but a single stratum, $S = 1$, it has traditionally been called a *completely randomized experiment*, but when there are two or more strata, $S \geq 2$, it has been called a *randomized block experiment*. If the strata each contain two units, $n_s = 2$, and one receives the treatment, $m_s = 1$, then it has been called a *paired randomized experiment*.

As a small illustration, consider a uniform randomized experiment with two strata, $S = 2$, four units in the first stratum, $n_1 = 4$, and two in the second, $n_2 = 2$, and $N = n_1 + n_2 = 6$ units in total. Half of the units in each stratum receive the treatment, so $m_1 = 2$ and $m_2 = 1$. There are $K = 12$ possible treatment assignments $\mathbf{z} = (z_{11}, z_{12}, z_{13}, z_{14}, z_{21}, z_{22})^T$ contained in the set Ω, and each has probability $1/12$. So Ω is the following set of $K = 12$ vectors $\mathbf{z}$ of dimension $N = 6$ with binary coordinates such that $2 = z_{11} + z_{12} + z_{13} + z_{14}$ and $1 = z_{21} + z_{22}$.

$$\Omega = \left\{ \begin{matrix} \begin{bmatrix} 1 \\ 1 \\ 0 \\ 0 \\ 1 \\ 0 \end{bmatrix} & \begin{bmatrix} 1 \\ 0 \\ 1 \\ 0 \\ 1 \\ 0 \end{bmatrix} & \begin{bmatrix} 1 \\ 0 \\ 0 \\ 1 \\ 1 \\ 0 \end{bmatrix} & \begin{bmatrix} 0 \\ 1 \\ 1 \\ 0 \\ 1 \\ 0 \end{bmatrix} & \begin{bmatrix} 0 \\ 1 \\ 0 \\ 1 \\ 1 \\ 0 \end{bmatrix} & \begin{bmatrix} 0 \\ 0 \\ 1 \\ 1 \\ 1 \\ 0 \end{bmatrix} \\ \\ \begin{bmatrix} 1 \\ 1 \\ 0 \\ 0 \\ 0 \\ 1 \end{bmatrix} & \begin{bmatrix} 1 \\ 0 \\ 1 \\ 0 \\ 0 \\ 1 \end{bmatrix} & \begin{bmatrix} 1 \\ 0 \\ 0 \\ 1 \\ 0 \\ 1 \end{bmatrix} & \begin{bmatrix} 0 \\ 1 \\ 1 \\ 0 \\ 0 \\ 1 \end{bmatrix} & \begin{bmatrix} 0 \\ 1 \\ 0 \\ 1 \\ 0 \\ 1 \end{bmatrix} & \begin{bmatrix} 0 \\ 0 \\ 1 \\ 1 \\ 0 \\ 1 \end{bmatrix} \end{matrix} \right\}.$$

The following proposition is often useful. It says that in a uniform randomized experiment, the assignments in different strata are independent of each other. For the elementary proof, see Problem 3.

Proposition 1 *In a uniform randomized experiment, the* $\mathbf{Z}_1, \ldots, \mathbf{Z}_S$ *are mutually independent, and* $\text{prob}(\mathbf{Z}_s = \mathbf{z}_s) = 1/\binom{n_s}{m_s}$ *for each* n_s-*tuple* $\mathbf{z}_s$ *containing* m_s *ones and* $n_s - m_s$ *zeros.*

The uniform randomized designs are by far the most common randomized experiments involving two treatments, but others are also used, particularly in clinical trials. It is useful to mention one of these methods of randomization to underscore the point that randomized experiments need not give every treatment assignment $\mathbf{z} \in \Omega_0$ the same probability. A distinguishing feature of many clinical trials is that the units are patients who arrive for treatment over a period of months or years. As a result, the number n_s of people who will fall in stratum s will not be known at the start of the experiment, so a randomized block experiment is not possible. Efron (1971) proposed the following method. Fix a probability p with $1/2 < p < 1$. When the ith patient belonging to stratum s first arrives, calculate a current measure of imbalance in stratum s, IMBAL_{si}, defined to be the number of patients so far assigned to treatment in this stratum minus the number so far assigned to control. It is easy to check that

$\text{IMBAL}_{si} = \sum_{j=1}^{i-1}(2Z_{sj} - 1)$. If $\text{IMBAL}_{si} = 0$, assign the new patient to treatment or control each with probability $1/2$. If $\text{IMBAL}_{si} < 0$, so there are too few treated patients in this stratum, then assign the new patient to treatment with probability p and to control with probability $1 - p$. If $\text{IMBAL}_{si} > 0$, so there are too many treated patients, then assign the new patient to treatment with probability $1 - p$ and to control with probability p. Efron examines various aspects of this method. In particular, he shows that it is much better than independent assignment in producing balanced treated and control groups, that is, treated and control groups with similar numbers of patients from each stratum. He also examines potential biases due to the experimenter's knowledge of IMBAL_{si}. Zelen (1974) surveys a number of related methods with similar objectives.

2.4   Testing the Hypothesis of No Treatment Effect

2.4.1   The Distribution of a Test Statistic When the Treatment Is Without Effect

In the theory of experimental design, a special place is given to the test of the hypothesis that the treatment is entirely without effect. The reason is that, in a randomized experiment, this test may be performed virtually without assumptions of any kind, that is, relying just on the random assignment of treatments. Fisher discussed the Lady and her tea with such care to demonstrate this. Other activities, such as estimating treatment effects or building confidence intervals, do require some assumptions, often innocuous assumptions, but assumptions nonetheless. The contribution of randomization to formal inference is most clear when expressed in terms of the test of no effect. Does this mean that such tests are of greater practical importance than point or interval estimates? Certainly not. It is simply that the theory of such tests is less cluttered, and so it sets randomized and nonrandomized studies in sharper contrast. The important point is that, in the absence of difficulties such as noncompliance or loss to follow-up, assumptions play a minor role in randomized experiments, and no role at all in randomization tests of the hypothesis of no effect. In contrast, inference in a nonrandomized experiment requires assumptions that are not at all innocuous. So let us follow Fisher and develop this point with care.

Each unit exhibits a response that is observed some time after treatment. To say that the treatment has no effect on this response is to say that each unit would exhibit the same value of the response whether assigned to treatment or control. If the treatment has no effect on a patient's survival, then the patient would live the same number of months under treatment or under control. This is the definition of "no effect." If changing the treatment assigned to a unit changed that unit's response, then certainly the treatment has at least some effect. If a patient would live one more month under treatment than under control, then the treatment has some effect on that patient.

In the traditional development of randomization inference, chance and probability enter only through the random assignment of treatments, that is, through the known mechanism that selects the treatment assignment $\mathbf{Z}$ from Ω. The only random quantities are $\mathbf{Z}$ and quantities that depend on $\mathbf{Z}$. When the treatment is without effect, the response of a unit is fixed, in the sense that this response would not change if a different treatment assignment $\mathbf{Z}$ were selected from Ω. Again, this is simply what it means for a treatment to be without effect. When testing the null hypothesis of no effect, the response of the ith unit in stratum s is written r_{si} and the N-tuple of responses for all N units is written $\mathbf{r}$. The lowercase notation for r_{si} emphasizes that, under the null hypothesis, r_{si} is a fixed quantity and not a random variable. Later on, when discussing treatments with effects, a different notation is needed.

A test statistic $t(\mathbf{Z}, \mathbf{r})$ is a quantity computed from the treatment assignment $\mathbf{Z}$ and the response $\mathbf{r}$. For instance, the treated-minus-control difference in sample means is the test statistic

$$t(\mathbf{Z}, \mathbf{r}) = \frac{\mathbf{Z}^T \mathbf{r}}{\mathbf{Z}^T \mathbf{1}} - \frac{(\mathbf{1} - \mathbf{Z})^T \mathbf{r}}{(\mathbf{1} - \mathbf{Z})^T \mathbf{1}}, \tag{2.2}$$

where $\mathbf{1}$ is an N-tuple of 1s. Other statistics are discussed shortly.

Given any test statistic $t(\mathbf{Z}, \mathbf{r})$, the task is to compute a significance level for a test that rejects the null hypothesis of no treatment effect when $t(\mathbf{Z}, \mathbf{r})$ is large. More precisely:

(i) The null hypotheses of no effect is tentatively assumed to hold, so $\mathbf{r}$ is fixed.

(ii) A treatment assignment $\mathbf{Z}$ has been selected from Ω using a known random mechanism.

(iii) The observed value, say T, of the test statistic $t(\mathbf{Z}, \mathbf{r})$ has been calculated.

(iv) We seek the probability of a value of the test statistic as large or larger than that observed if the null hypothesis were true.

The significance level is simply the sum of the randomization probabilities of assignments $\mathbf{z} \in \Omega$ that lead to values of $t(\mathbf{z}, \mathbf{r})$ greater than or equal to the observed value T, namely,

$$\text{prob}\{t(\mathbf{Z}, \mathbf{r}) \geq T\} = \sum_{\mathbf{z} \in \Omega} [t(\mathbf{z}, \mathbf{r}) \geq T] \cdot \text{prob}(\mathbf{Z} = \mathbf{z}), \tag{2.3}$$

$$\text{where} \quad [\text{event}] = \begin{cases} 1 & \text{if event occurs,} \\ 0 & \text{otherwise,} \end{cases} \tag{2.4}$$

and prob$(\mathbf{Z} = \mathbf{z})$ is determined by the known random mechanism that assigned treatments. This is a direct calculation, though not always a straightforward one when Ω is extremely large.

In the case of a uniform randomized experiment, there is a simpler expression for the significance level (2.3) since prob$(\mathbf{Z} = \mathbf{z}) = 1/K = 1/|\Omega|$. It is the proportion of treatment assignments $\mathbf{z} \in \Omega$ giving values of the test statistic $t(\mathbf{z}, \mathbf{r})$ greater than or equal to T, namely,

$$\text{prob}\{t(\mathbf{Z}, \mathbf{r}) \geq T\} = \frac{|\{\mathbf{z} \in \Omega : t(\mathbf{z}, \mathbf{r}) \geq T\}|}{K}. \tag{2.5}$$

2.4.2 More Tea

To illustrate, consider again Fisher's example of the Lady who tastes $N = 8$ cups of tea, all in a single stratum, so $S = 1$. A treatment assignment is an 8-tuple containing four 1s and four 0s. For instance, the assignment $\mathbf{Z} = (1, 0, 0, 1, 1, 0, 0, 1)^{\text{T}}$ would signify that cups 1, 4, 5, and 8 had milk added first and the other cups had tea added first. The set of treatment assignments Ω contains all possible 8-tuples containing four 1s and four 0s, so Ω contains $|\Omega| = K = \binom{8}{4} = 70$ such 8-tuples. The actual assignment was selected at random in the sense that prob$(\mathbf{Z} = \mathbf{z}) = 1/K = 1/70$ for all $\mathbf{z} \in \Omega$. Notice that $\mathbf{z}^{\text{T}}\mathbf{1} = 4$ for all $\mathbf{z} \in \Omega$.

The Lady's response for cup i is either $r_i = 1$ signifying that she classifies this cup as milk first or $r_i = 0$ signifying that she classifies it as tea first. Then $\mathbf{r} = (r_1, \dots, r_8)^{\text{T}}$. Recall that she must classify exactly four cups as milk first, so $\mathbf{1}^{\text{T}}\mathbf{r} = 4$. The test statistic is the number of cups correctly identified, and this is written formally as $t(\mathbf{Z}, \mathbf{r}) = \mathbf{Z}^{\text{T}}\mathbf{r} + (1 - \mathbf{Z})^{\text{T}}(1 - \mathbf{r}) = 2\mathbf{Z}^{\text{T}}\mathbf{r}$, where the second equality follows from $\mathbf{1}^{\text{T}}\mathbf{1} = 8$, $\mathbf{Z}^{\text{T}}\mathbf{1} = 4$, and $\mathbf{1}^{\text{T}}\mathbf{r} = 4$. To make this illustration concrete, suppose that $\mathbf{r} = (1, 1, 0, 0, 0, 1, 1, 0)$, so the Lady classifies the first, second, sixth, and seventh cups as milk first. To say that the treatment has no effect is to say that she would give this classification no matter how milk was added to the cups, that is, no matter how treatments were assigned to cups. If changing the cups to which milk is added first changes her responses, then she is discerning something, and the treatment has some effect, however slight or erratic.

There is only one treatment assignment $\mathbf{z} \in \Omega$ leading to perfect agreement with the Lady's responses, namely, $\mathbf{z} = (1, 1, 0, 0, 0, 1, 1, 0)$, so if $t(\mathbf{Z}, \mathbf{r}) = 8$ the significance level (2.5) is prob$\{t(\mathbf{Z}, \mathbf{r}) \geq 8\} = 1/70$. This says that the chance of perfect agreement by accident is $1/70 = 0.014$, a small chance. In other words, if the treatment is without effect, the chance that a random assignment of treatments will just happen to produce perfect agreement is $1/70$.

It is not possible to have seven agreements since to err once is to err twice. How many assignments $\mathbf{z} \in \Omega$ lead to exactly $t(\mathbf{Z}, \mathbf{r}) = $ six agreements? One such assignment with six agreements is $\mathbf{z} = (1, 0, 1, 0, 0, 1, 1, 0)$. Starting with perfect agreement, $\mathbf{z} = (1,1,0,0,0,1,1,0)$, any one of the four 1s may be made a 0 and any of the four 0s may be made a 1, so there are $16 = 4 \times 4$ assignments with exactly $t(\mathbf{Z},\mathbf{r}) = 6$ agreements. Hence, there are 17 assignments leading to six or more agreements. With six agreements the significance level (2.5) is $\text{prob}\{t(\mathbf{Z},\mathbf{r}) \geq 6\} = 17/70 = 0.24$, no longer a small probability. It would not be surprising to see six or more agreements if the treatment were without effect—it happens by chance as frequently as seeing two heads when flipping two coins.

The key point deserves repeating. Probability enters the calculation only through the random assignment of treatments. The needed probability distribution is known, not assumed. The resulting significance level does not depend upon assumptions of any kind. If the same calculation were performed in a nonrandomized study, it would require an assumption that the distribution of treatment assignments, $\text{prob}(\mathbf{Z} = \mathbf{z})$, is some particular distribution, perhaps the assumption that all assignments are equally probable, $\text{prob}(\mathbf{Z} = \mathbf{z}) = 1/K$. In a nonrandomized study, there may be little basis on which to ground or defend this assumption, it may be wrong, and it will certainly be open to responsible challenge and debate. In other words, the importance of the argument just considered is that it is one way of formally expressing the claim that randomized experiments are not open to certain challenges that can legitimately be made to nonrandomized studies.

2.4.3 Some Common Randomization Tests

Many commonly used tests are randomization tests in that their significance levels can be calculated using (2.5), though the tests are sometimes derived in other ways as well. This section briefly recalls and reviews some of these tests. The purpose of the section is to provide a reference for these methods in a common terminology so they may be discussed and used at later stages. Though invented at different times, it is natural to see the methods as members of a few classes whose properties are similar, and this is done beginning in §2.4.4. In most cases, the methods described have various optimality properties which are not discussed here; see Cox (1970) for the optimality properties of the procedures for binary outcomes and Lehmann (1975) for optimality properties of the nonparametric procedures. In all cases, the experiment is the uniform randomized experiment in §2.3.2 with $\text{prob}(\mathbf{Z} = \mathbf{z}) = 1/K$ for all $\mathbf{z} \in \Omega$.

Fisher's exact test for a 2×2 contingency table is, in fact, the test just used for the example of the Lady and her tea. Here, there is one stratum, $S = 1$; the outcome r_i is *binary*, that is, $r_i = 1$ or $r_i = 0$; and the test statistic is the number of responses equal to 1 in the treated group, that is, $t(\mathbf{Z},\mathbf{r}) = \mathbf{Z}^{\mathrm{T}}\mathbf{r}$. The 2×2 contingency table records the values of Z_i and r_i, as shown in Table 2.2 for Fisher's example. Notice that the marginal totals in this table are fixed by the structure of the experiment, because

TABLE 2.2. The 2×2 Table for Fisher's Exact Test for the Lady Tasting Tea.

		Response, r_i		
		1	0	Total
Treatment or control, Z_i	1	$\mathbf{Z}^T\mathbf{r}$	$4 - \mathbf{Z}^T\mathbf{r}$	4
	0	$4 - \mathbf{Z}^T\mathbf{r}$	$\mathbf{Z}^T\mathbf{r}$	4
	Total	4	4	8

$N = 8$ cups, $\mathbf{1}^T\mathbf{r} = 4$ and $\mathbf{1}^T\mathbf{Z} = 4$ are fixed in this experiment. Under the hypothesis of no effect, the randomization distribution of the test statistic $\mathbf{Z}^T\mathbf{r}$ is the hypergeometric distribution. The usual *chi-square* test for a 2×2 table is an approximation to the randomization significance level when N is large.

The *Mantel–Haenszel (1959) statistic* is the analogue of Fisher's exact test when there are two or more strata, $S \geq 2$, and the outcome r_{si} is binary. It is extensively used in epidemiology and certain other fields. The data may be recorded in a $2 \times 2 \times S$ contingency table giving treatment Z by outcome r by stratum s. The test statistic is again the number of 1 responses among treated units, $t(\mathbf{Z}, \mathbf{r}) = \mathbf{Z}^T\mathbf{r} = \sum_{s=1}^{S} \sum_{i=1}^{n_s} Z_{si} r_{si}$. Under the null hypothesis, the contribution from stratum s, that is, $\sum_{i=1}^{n_s} Z_{si} r_{si}$, again has a hypergeometric distribution, and (2.5) is the distribution of the sum of S independent hypergeometric variables. The Mantel–Haenszel statistic yields an approximation to the distribution of $\mathbf{Z}^T\mathbf{r}$ based on its expectation and variance, as described in more general terms in the next section. One technical attraction of this statistic is that the large sample approximation tends to work well for a $2 \times 2 \times S$ table ·with large N even if S is also large, so there may be few subjects in each of the S tables. In particular, the statistic is widely used in matching with multiple controls, in which case $m_s = 1$ for each s.

McNemar's (1947) test is for paired binary data, that is, for S pairs with $n_s = 2$, $m_s = 1$, and $r_{si} = 1$ or $r_{si} = 0$. The statistic is, yet again, the number of 1 responses among treated units; that is, $t(\mathbf{Z}, \mathbf{r}) = \mathbf{Z}^T\mathbf{r}$. McNemar's statistic is, in fact, a special case of the Mantel–Haenszel statistic, though the $2 \times 2 \times S$ table now describes S pairs and certain simplifications are possible. In particular, the distribution of $\mathbf{Z}^T\mathbf{r}$ in (2.5) is that of a constant plus a certain binomial random variable.

Developing these methods for $2 \times 2 \times S$ tables in a different way, Birch (1964) and Cox (1966, 1970) show that these three tests with binary responses possess an optimality property, so there is a sense in which Fisher's exact test, the Mantel–Haenszel test, and McNemar's test are the best tests for the problems they address. Specifically, they show that the test statistic $t(\mathbf{Z}, \mathbf{r}) = \mathbf{Z}^T\mathbf{r}$ together with the significance level (2.5) is a uniformly most powerful unbiased test against alternatives defined in terms of constant odds ratios.

Mantel's (1963) extension of the Mantel–Haenszel test is for responses r_{si} that are confined to a small number of values representing a numerical scoring of several ordered categories. As an example of such an outcome, the New York Heart Association classifies coronary patients into one of four categories based on the degree to which the patient is limited in physical activity by coronary symptoms such as chest pain. The categories are:

(1) no limitation of physical activity;

(2) slight limitation, comfortable at rest, but ordinary physical activity results in pain or other symptoms;

(3) marked limitation, minor activities result in coronary symptoms; and

(4) unable to carry on any physical activity without discomfort, which may be present even at rest.

The outcome r_{si} for a patient is then one of the integers 1, 2, 3, or 4. In this case the data might be recorded as a $2 \times 4 \times S$ contingency table for $Z \times r \times s$. Mantel's test statistic is the sum of the response scores for treated units; that is, $t(\mathbf{Z}, \mathbf{r}) = \mathbf{Z}^T \mathbf{r}$. Birch (1965) shows that the test is optimal in a certain sense.

In the case of a single stratum, $S = 1$, *Wilcoxon's (1945) rank sum test* is commonly used to compare outcomes taking many numerical values. In this test, the responses are ranked from smallest to largest. If all N responses were different numbers, the ranks would be the numbers $1, 2, \ldots, N$. If some of the responses were equal, then the average of their ranks would be used. Write q_i for the rank of r_i, and write $\mathbf{q} = (q_1, \ldots, q_N)^T$. For instance, if $N = 4$, and $r_1 = 2.3$, $r_2 = 1.1$, $r_3 = 2.3$, and $r_4 = 7.9$, then $q_1 = 2.5$, $q_2 = 1$, $q_3 = 2.5$, and $q_4 = 4$, since r_2 is smallest, r_4 is largest, and r_1 and r_3 share the ranks 2 and 3 whose average rank is $2.5 = (2 + 3)/2$. Note that the ranks $\mathbf{q}$ are a function of the responses $\mathbf{r}$ which are fixed if the treatment has no effect, so $\mathbf{q}$ is also fixed. The rank sum statistic is the sum of the ranks of the treated observations, $t(\mathbf{Z}, \mathbf{r}) = \mathbf{Z}^T \mathbf{q}$, and its significance level is determined from (2.5). The properties of the rank sum test have been extensively studied; for instance, see Lehmann (1975, §1) or Hettmansperger (1984, §3). Wilcoxon's rank sum test is equivalent to the *Mann and Whitney (1947) test*.

In the case of S matched pairs with $n_s = 2$ and $m_s = 1$ for $s = 1, \ldots, S$, *Wilcoxon's (1945) signed rank test* is commonly used for responses taking many values. Here, $(Z_{s1}, Z_{s2}) = (1, 0)$ if the first unit in pair s received the treatment or $(Z_{s1}, Z_{s2}) = (0, 1)$ if the second unit received the treatment. In this test, the absolute differences in responses within pairs $|r_{s1} - r_{s2}|$ are ranked from 1 to S, with average ranks used for ties. Let d_s be the rank of $|r_{s1} - r_{s2}|$ thus obtained. The signed rank statistic is the sum of the ranks for pairs in which the treated unit had a higher response than the control unit.

To write this formally, let $c_{s1} = 1$ if $r_{s1} > r_{s2}$ and $c_{s1} = 0$ otherwise, and similarly, let $c_{s2} = 1$ if $r_{s2} > r_{s1}$ and $c_{s2} = 0$ otherwise, so $c_{s1} = c_{s2} = 0$ if $r_{s1} = r_{s2}$. Then $Z_{s1}c_{s1} + Z_{s2}c_{s2}$ equals 1 if the treated unit in pair s had a higher response than the control unit, and equals zero otherwise. It follows that the signed rank statistic is $\sum_{s=1}^{S} d_s \sum_{i=1}^{2} c_{si} Z_{si}$. Note that d_s and c_{si} are functions of $\mathbf{r}$ and so are fixed under the null hypothesis of no treatment effect. Also, if $r_{s1} = r_{s2}$, then pair s contributes zero to the value of the statistic no matter how treatments are assigned. As with the rank sum test, the signed rank test is widely used and has been extensively studied; for instance, see Lehmann (1975, §3) or Hettmansperger (1984, §2). Section 3.2.4 below contains a numerical example using the sign-rank statistic in an observational study.

For stratified responses, a method that is sometimes used involves calculating the rank sum statistic separately in each of the S strata, and taking the sum of these S rank sums as the test statistic. This is the *stratified rank sum statistic*. It is easily checked that this statistic has the form $t(\mathbf{Z}, \mathbf{r}) = \mathbf{Z}^{\mathrm{T}}\mathbf{q}$ resembling the rank sum statistic; however, the ranks in $\mathbf{q}$ are no longer a permutation of the numbers $1, 2, \ldots, N$, but rather of the numbers $1, \ldots, n_1, 1, \ldots, n_2, \ldots, 1, \ldots, n_S$, with adjustments for ties if needed. Also Ω has changed.

Hodges and Lehmann (1962) find the stratified rank sum statistic to be inefficient when S is large compared to N. In particular, for paired data with $S = N/2$, the stratified rank test is equivalent to the sign test, which in turn is substantially less efficient than the signed rank test for data from short-tailed distributions such as the Normal. They suggest as an alternative the method of aligned ranks: the mean in each stratum is subtracted from the responses in that stratum creating aligned responses that are ranked from 1 to N, momentarily ignoring the strata. Writing $\mathbf{q}$ for these aligned ranks, the *aligned rank statistic* is the sum of the aligned ranks in the treated group, $t(\mathbf{Z}, \mathbf{r}) = \mathbf{Z}^{\mathrm{T}}\mathbf{q}$. See also Lehmann (1975, §3.3).

Another statistic is the *median test*. Let $c_{si} = 1$ if r_{si} is greater than the median of the responses in stratum s and let $c_{si} = 0$ otherwise, and let $\mathbf{c}$ be the N-tuple containing the c_{si}. Then $t(\mathbf{Z}, \mathbf{r}) = \mathbf{Z}^{\mathrm{T}}\mathbf{c}$ is the number of treated responses that exceed their stratum medians. With a single stratum, $S = 1$, the median test is quite good in large samples if the responses have a double exponential distribution, a distribution with a thicker tail than the normal; see, for instance, Hettmansperger (1984, §3.4, p. 146) and the more critical comments by Freidlin and Gastwirth (2000). In this test, the median is sometimes replaced by other quantiles or other measures of location.

Start with any statistic $t(\mathbf{Z}, \mathbf{r})$ and the randomization distribution of $t(\mathbf{Z}, \mathbf{r})$ may be determined from (2.5). This is true even of statistics that are commonly referred to a theoretical distribution instead, for instance, the *two-sample* or *paired t-tests*, among others. Welch (1937) and Wilk (1955) studied the relationship between the randomization distribution and the theoretical distribution of statistics that were initially derived from assumptions of Normally and independently distributed responses. They suggest that the theoretical distribution may be viewed as a computationally convenient approximation to the desired but computationally difficult randomization distribution. That is, they suggest that t-tests, like rank tests or Mantel–Haenszel tests, may be justified solely on the basis of the use of randomization in the design of an experiment, without reference to Normal independent errors. These findings depend on the good behavior of moments of sums of squares of responses over the randomization distribution; therefore, they depend on the absence of extreme responses. Still, the results are important as a conceptual link between Normal theory and randomization inference.

2.4.4 Classes of Test Statistics

The similarity among the commonly used test statistics in §2.4.3 is striking but not accidental. In this book, these statistics are not discussed individually, except when used in examples. The important properties of the methods are shared by large classes of statistics, so it is both simpler and less repetitive to discuss the classes.

Though invented by different people at different times for different purposes, the commonly used statistics in §2.4.3 are similar for good reason. As the sample size N increases, the number K of treatment assignments in Ω grows very rapidly, and the direct calculation in (2.5) becomes very difficult to perform, even with the fastest computers. To see why this is true, take the simplest case consisting of one stratum, $S = 1$, and an equal division of the n subjects into $m = n/2$ treated subjects and $m = n/2$ controls. Then there are $K = \binom{n}{n/2}$ treatment assignments in Ω. If one more unit is added to the experiment, increasing the sample size to $n+1$, then K is increased by a factor of $(n+1)/\{(n/2)+1\}$, that is, K nearly doubles. Roughly speaking, if the fastest computer can calculate (2.5) directly for at most a sample of size n, and if computing power doubles every year for 10 years, then 10 years hence computing power will be $2^{10} = 1024$ times greater than today and it will be possible to handle a sample of size $n+10$. Direct calculation of (2.5) is not practical for large n.

The usual solution to this problem is to approximate (2.5) using a large sample or asymptotic approximation. The most common approximations use the moments of the test statistic, its expectation and variance, and sometimes higher moments. The needed moments are easily derived for certain classes of statistics, including all those in §2.4.3.

As an alternative to asymptotic approximation, there are several proposals for computing (2.5) exactly, but they are not, as yet, commonly used. One is to compute (2.5) exactly but indirectly using clever computations that avoid working with the set Ω. For some statistics this can be done by calculating the characteristic function of the test statistic and inverting it using the fast Fourier transform; see Pagano and Tritchler (1983). A second approach is to design experiments differently so that Ω is a much smaller set, perhaps containing 10,000 or 100,000 treatment assignments. In this case, direct calculation is possible and any test statistic may be used; see Tukey (1985) for discussion.

The first class of statistics will be called *sum statistics* and they are of the form $t(\mathbf{Z}, \mathbf{r}) = \mathbf{Z}^{\mathrm{T}}\mathbf{q}$, where $\mathbf{q}$ is some function of $\mathbf{r}$. A sum statistic sums the scores q_{si} for treated units. All of the statistics in §2.4.4 are sum statistics for suitable choices of $\mathbf{q}$. In Fisher's exact test, the Mantel–Haenszel test, and McNemar's test, $\mathbf{q}$ is simply equal to $\mathbf{r}$. In the rank sum test, $\mathbf{q}$ contains the ranks of $\mathbf{r}$. In the median test, $\mathbf{q}$ is the vector of ones and zeros identifying responses r_{si} that exceed stratum medians. In the signed rank statistic, $q_{si} = d_s c_{si}$.

Simple formulas exist for the moments of sum statistics under the null hypothesis that the treatment is without effect. In this case, $\mathbf{r}$ is fixed, so $\mathbf{q}$ is also fixed. The moment formulas use the properties of simple random sampling without replacement. Recall that a simple random sample without replacement of size m from a population of size n is a random subset of m elements from a set with n elements where each of the $\binom{n}{m}$ subsets of size m has the same probability $1/\binom{n}{m}$. Cochran (1963) discusses simple random sampling. In a uniform randomized experiment, the m_s treated units in stratum s are a simple random sample without replacement from the n_s units in stratum s. The following proposition is proved in Problem 4.

Proposition 2 *In a uniform randomized experiment, if the treatment has no effect, the expectation and variance of a sum statistic* $\mathbf{Z}^{\mathrm{T}}\mathbf{q}$ *are*

$$E(\mathbf{Z}^{\mathrm{T}}\mathbf{q}) = \sum_{s=1}^{S} m_s \bar{q}_s,$$

and

$$\mathrm{var}(\mathbf{Z}^{\mathrm{T}}\mathbf{q}) = \sum_{s=1}^{S} \frac{m_s(n_s - m_s)}{n_s(n_s - 1)} \sum_{i=1}^{n_s} (q_{si} - \bar{q}_s)^2,$$

where

$$\bar{q}_s = \frac{1}{n_s} \sum_{i=1}^{n_s} q_{si}.$$

Moments are easily determined for sum statistics, but other classes of statistics have other useful properties. The first such class, the *sign-score statistics*, is a subset of the sum statistics. A statistic is a sign-score statistic if it is of the form $t(\mathbf{Z}, \mathbf{r}) = \sum_{s=1}^{S} d_s \sum_{i=1}^{n_s} c_{si} Z_{si}$, where c_{si} is binary, $c_{si} = 1$ or $c_{si} = 0$, and both d_s and c_{si} are functions of $\mathbf{r}$. Fisher's exact test, the

Mantel–Haenszel test, and McNemar's test are sign-score statistics with $d_s = 1$ and $c_{si} = r_{si}$. The signed rank and median test statistics are also sign-score statistics, but with c_{si} and d_s defined differently. A sign-score statistic is a sum statistic with $q_{si} = d_s c_{si}$, but many sum statistics, including the rank sum statistic, are not sign-score statistics. In Chapter 4, certain calculations are simpler for sign-score statistics than for certain other sum statistics, and this motivates the distinction.

Another important class of statistics is the class of *arrangement increasing functions* of $\mathbf{Z}$ and $\mathbf{r}$, which are defined in a moment. Informally, a statistic $t(\mathbf{Z}, \mathbf{r})$ is arrangement-increasing if it increases in value as the coordinates of $\mathbf{Z}$ and $\mathbf{r}$ are rearranged into an increasingly similar order within each stratum. In fact, all of the statistics in §2.4.3 are arrangement-increasing, so anything that is true of arrangement-increasing statistics is true of all the commonly used statistics in §2.3.2. Hollander, Proschan, and Sethuraman (1977) discuss many properties of arrangement-increasing functions.

A few preliminary terms are useful. The numbers S and n_s, $s = 1, \ldots, S$ with $N = \sum n_s$, are taken as given. A *stratified N-tuple* $\mathbf{a}$ is an N-tuple in which the N coordinates are divided into S strata with n_s coordinates in stratum s, where a_{si} is the ith of the n_s coordinates in stratum s. For instance, $\mathbf{Z}$ and $\mathbf{r}$ are each stratified N-tuples. If $\mathbf{a}$ is a stratified N-tuple, and if i and j are different positive integers less than or equal to n_s, then let $\mathbf{a}_{sij}$ be the stratified N-tuple formed from $\mathbf{a}$ by interchanging a_{si} and a_{sj}, that is, by placing the value a_{sj} in the ith position in stratum s and placing the value a_{si} in the jth position in stratum s. To avoid repetition, whenever the symbol $\mathbf{a}_{sij}$ appears, it is assumed without explicit mention that the subscripts are appropriate, so s is a positive integer between 1 and S and i and j are different positive integers less than or equal to n_s. A function $f(\mathbf{a}, \mathbf{b})$ of two stratified N-tuples is *invariant* if $f(\mathbf{a}, \mathbf{b}) = f(\mathbf{a}_{sij}, \mathbf{b}_{sij})$ for all s, i, j, so renumbering units in the same stratum does not change the value of $f(\mathbf{a}, \mathbf{b})$. For instance, the function $\mathbf{z}^T \mathbf{q}$ is an invariant function of $\mathbf{z}$ and $\mathbf{q}$.

Definition 3 *An invariant function $f(\mathbf{a}, \mathbf{b})$ of two stratified N-tuples is arrangement-increasing (or AI) if $f(\mathbf{a}, \mathbf{b}_{sij}) \geq f(\mathbf{a}, \mathbf{b})$ whenever*

$$(a_{si} - a_{sj}) \cdot (b_{si} - b_{sj}) \leq 0.$$

Notice what this definition says. Consider the ith and jth unit in stratum s. If $(a_{si} - a_{sj})(b_{si} - b_{sj}) < 0$, then of these two units, the one with the higher value of a has the lower value of b, so these two coordinates are out of order. However, in $\mathbf{a}$ and $\mathbf{b}_{sij}$, these two coordinates are in the same order, for b_{si} and b_{sj} have been interchanged. The definition says that an arrangement increasing function will be larger, or at least no smaller, when these two coordinates are switched into the same order.

TABLE 2.3. A Hypothetical Example Showing an Arrangement-Increasing Statistic.

i		$\mathbf{z}$	$\mathbf{q}$	$\mathbf{q}_{23}$
1	Treated	1	4	4
2	Treated	1	2	3
3	Control	0	3	2
4	Control	0	1	1
Rank sum			6	7

Notice also what the definition says when $(a_{si}-a_{sj})(b_{si}-b_{sj}) = 0$. In this case, either $a_{si} = a_{sj}$ or $b_{si} = b_{sj}$ or both. In this case, $f(\mathbf{a}, \mathbf{b}_{sij}) = f(\mathbf{a}, \mathbf{b})$.

Consider some examples. The function $\mathbf{z}^T\mathbf{q}$ is arrangement-increasing as a function of $\mathbf{z}$ and $\mathbf{q}$. To see this, note that $\mathbf{z}^T\mathbf{q}_{sij}-\mathbf{z}^T\mathbf{q} = (z_{si}q_{sj}+z_{sj}q_{si})-(z_{si}q_{si}+z_{sj}q_{sj}) = -(z_{si}-z_{sj})(q_{si}-q_{sj})$, so if $(z_{si}-z_{sj})(q_{si}-q_{sj}) \leq 0$ then $\mathbf{z}^T\mathbf{q}_{sij} - \mathbf{z}^T\mathbf{q} \geq 0$. This shows $\mathbf{z}^T\mathbf{q}$ is arrangement-increasing.

Table 2.3 is a small illustration for the rank sum statistic with a single stratum, $S = 1$, $n = 4$ units, of whom $m = 2$ received the treatment. Here, $(z_2 - z_3)(q_2 - q_3) = (1 - 0)(2 - 3) = -1 \leq 0$, and the rank sum $\mathbf{z}^T\mathbf{q} = 6$ is increased to $\mathbf{z}^T\mathbf{q}_{23} = 7$ by interchanging q_2 and q_3.

As a second example, consider the function $t(\mathbf{z}, \mathbf{r}) = \mathbf{z}^T\mathbf{q}$, where $\mathbf{q}$ is a function of $\mathbf{r}$, which may be written explicitly as $\mathbf{q}(\mathbf{r})$. Then $t(\mathbf{z}, \mathbf{r})$ may or may not be arrangement-increasing in $\mathbf{z}$ and $\mathbf{r}$ depending upon how $\mathbf{q}(\mathbf{r})$ varies with $\mathbf{r}$. The common statistics in §2.4.3 all have the following two properties:

(i) permute $\mathbf{r}$ within strata and $\mathbf{q}$ is permuted in the same way; and

(ii) within each stratum, larger r_{si} receive larger q_{si}.

One readily checks that $t(\mathbf{z}, \mathbf{r}) = \mathbf{z}^T\mathbf{q}$ is arrangement-increasing if $\mathbf{q}(\mathbf{r})$ has these two properties, because the first property ensures that $t(\mathbf{z}, \mathbf{r})$ is invariant, and the second ensures that $r_{si} - r_{sj} \geq 0$ implies $q_{si} - q_{sj} \geq 0$, so $(z_{si} - z_{sj})(r_{si} - r_{sj}) \leq 0$ implies $(z_{si} - z_{sj})(q_{si} - q_{sj}) \leq 0$, and the argument of the previous paragraph applies. The important conclusion is that all of the statistics in §2.4.3 are arrangement-increasing.

In describing the behavior of a statistic when the null hypothesis does not hold and instead the treatment has an effect, a final class of statistics is useful. Many statistics that measure the size of the difference between treated and control groups would tend to increase in value if responses in the treated group were increased and those in the control group were decreased. Statistics with this property will be called effect increasing, and the idea will now be expressed this formally. A treated unit has $2Z_{si}-1 = 1$, since $Z_{si} = 1$, and a control unit has $2Z_{si}-1 = -1$ since $Z_{si} = 0$. Let $\mathbf{z} \in \Omega$

TABLE 2.4. Hypothetical Example of an Effect Increasing Statistic.

i		z_i	$2z_i - 1$	r_i	r_i^*
1	Treated	1	1	5	6
2	Treated	1	1	2	4
3	Control	0	−1	3	2
4	Control	0	−1	1	1
Rank sum				6	7

be a possible treatment assignment and let $\mathbf{r}$ and $\mathbf{r}^*$ be two possible values of the N-tuple of responses such that $(r_{si}^* - r_{si})(2z_{si} - 1) \geq 0$ for all s, i. With treatments given by $\mathbf{z}$, this says that $r_{si}^* \geq r_{si}$ for every treated unit and $r_{si}^* \leq r_{si}$ for every control unit. In words, if higher responses indicated favorable outcomes, then every treated unit does better with $\mathbf{r}^*$ than with $\mathbf{r}$, and every control does worse with $\mathbf{r}^*$ than with $\mathbf{r}$. That is, the difference between treated and control groups looks larger with $\mathbf{r}^*$ than with $\mathbf{r}$. The test statistic is *effect increasing* if $t(\mathbf{z}, \mathbf{r}) \leq t(\mathbf{z}, \mathbf{r}^*)$ whenever $\mathbf{r}$ and $\mathbf{r}^*$ are two possible values of the response such that $(r_{si}^* - r_{si})(2z_{si} - 1) \geq 0$ for all s, i. All of the commonly used statistics in §2.4.3 are effect increasing.

Table 2.4 contains a small hypothetical example to illustrate the idea of an effect increasing statistic. Here there is a single stratum, $S = 1$, and four subjects, $n = 4$, of whom $m = 2$ received the treatment. Notice that when r_i and r_i^* are compared, treated subjects have $r_i^* \geq r_i$ while controls have $r_i^* \leq r_i$. If the responses are ranked 1, 2, 3, 4, and the ranks in the treated group are summed to give Wilcoxon's rank sum statistic, then the rank sum is larger for r_i^* than for r_i.

In summary, this section has considered four classes of statistics:

(i) the sum statistics;

(ii) the arrangement-increasing statistics;

(iii) the effect increasing statistics; and

(iv) the sign-score statistics.

All of the commonly used statistics in §2.4.3 are members of the first three classes, and most are sign-score statistics; however, the rank sum statistic, the stratified rank sum statistic, and Mantel's extension are not sign-score statistics.

2.4.5   *No Effect Means No Effect

No effect means no effect. A nonzero effect that varies from one unit to the next and that is hard to fathom or predict is, nonetheless, a nonzero effect. It may not be an immediately useful effect, but it is an effect, perhaps an effect that can someday be understood, tamed, and made useful.

Empirically, it may be difficult to discern erratic unsystematic effects, but logically they are distinct from no effect.

To emphasize this point, consider the extreme case. Suppose that we somehow discerned that the treatment erratically benefits some patients and harms others, but that we have no way of predicting who will benefit or who will be harmed, so the average effect of the treatment is essentially zero in every large group of patients defined by pretreatment variables. In point of fact, it is very difficult to discern something like this, unless we covertly introduce more information that does distinguish these supposedly indistinguishable patients. Suppose, however, we can discern this, perhaps because the treatment produces one of two easily distinguished biochemical reactions, one beneficial, the other harmful, and neither reaction is ever seen among controls; however, we are completely at a loss to identify in advance those patients who will have beneficial reactions. This is a nonzero treatment effect, perhaps not a very useful one given current knowledge, but a nonzero effect nonetheless. What would a scientist do with such an effect? Might the scientist sometimes return with the treatment to the laboratory in an effort to understand why only some patients exhibit the beneficial biochemical reaction? In contrast, no treatment effect—really no treatment effect—would send the scientist in search of another treatment.

No effect is one hypothesis among many. It is rarely, perhaps never, sufficient to know whether the null hypothesis of no treatment effect is compatible with observed data. And yet, it is typically of interest to know this along with much more. Section 2.5 and Chapter 5 discuss models for treatment effects and associated methods of inference, including confidence intervals.

Fisher (1935) and Neyman (1935), two brilliant founders of statistics, did not agree about the meaning of the null hypothesis of no treatment effect. The hypothesis of no effect as I have described it is Fisher's version. Fisher's conception is particular: randomization justifies causal inferences about particular treatment effects, on particular units, at a particular time, under particular circumstances. Change the units or the times or the circumstances and the findings may change to an extent not adequately addressed by statistical standard errors. These standard errors measure one very important source of uncertainty, namely, uncertainty about how units would have responded to a treatment they did not receive, that is, uncertainty about the effects caused by the treatment. Campbell and Stanley (1963) say that randomization ensures *internal validity* but not *external validity*; see §2.7.1 and the discussion of efficacy and effectiveness in §5.4. Neyman's (1935, p. 110) conception is general: we can "repeat the experiment indefinitely without any change of vegetative conditions or of arrangement so that ... the yields from this plot will form a population ...." For Neyman, the variations we do not understand become, by assumption, variations from sampling a population. In point of fact, we cannot repeat the experiment indefinitely, and we cannot ensure the same experimental conditions, but this conception concerns a hypothetical world in which we can. This was not a disagreement about matters of fact, but about matters of art, the art of developing statistical concepts for scientific applications.

In most cases, their disagreement is entirely without technical consequence: the same procedures are used, and the same conclusions are reached. Perhaps this is expressed most beautifully by Lehmann (1959, §5). First, Lehmann (1959, §5.7, Theorem 3) shows that inferences under a population model can be distribution-free only if they are made particular by conditioning on observed responses, yielding Fisher's randomization test. Lehmann (1959, §5.8) then uses a population model and the Neyman–Pearson lemma to obtain most powerful permutation tests; that is, he uses Neyman's conception to obtain the best tests of the type Fisher was proposing. Whatever Fisher and Neyman may have thought, in Lehmann's text they work together. The importance to mathematical statistics and to science of infinite population models and Neyman's contributions are, today, surely unquestioned.

And yet, when one is thinking about the science of an experiment, it is surely true that random assignment of treatments justifies inferences that are particular, that is, particular to certain units at certain times under certain circumstances. If the inference reaches beyond that to infinite populations extending into the indefinite future, then this has been accomplished by assuming those populations into existence, and assuming away much that is true of the world we actually inhabit. In those instances where their conceptions point in scientifically different directions—for instance, the unpredictable but distinguishable biochemical reactions above—it seems to me that Fisher's conception more closely describes how scientists think and work. Much that we cannot currently predict and do not currently fathom is not random error. The variation we do not fathom today we intend to decipher tomorrow.

2.5   Simple Models for Treatment Effects

2.5.1   Responses When the Treatment Has an Effect

If the treatment has an effect, then the observed N-tuple of responses for the N units will be different for different treatment assignments $\mathbf{z} \in \Omega$—this is what it means to say the treatment has an effect. In earlier sections, the null hypothesis of no treatment effect was assumed to hold, so the observed responses were fixed, not varying with $\mathbf{z}$, and the response was written $\mathbf{r}$. When the null hypothesis of no effect is not assumed to hold, the response changes with $\mathbf{z}$, and the response observed when the treatment assignment is $\mathbf{z} \in \Omega$ will be written $\mathbf{r_z}$. The null hypothesis of no treatment effect says that $\mathbf{r_z}$ does not vary with $\mathbf{z}$, and instead $\mathbf{r_z}$ is a constant the same for all $\mathbf{z}$; in this case, $\mathbf{r}$ was written for this constant. Notice that, for each

$\mathbf{z} \in \Omega$, the response $\mathbf{r_z}$ is some nonrandom N-tuple—probability has not yet entered the discussion. Write r_{siz} for the (s, i) coordinate of $\mathbf{r_z}$, that is, for the response of the ith unit in stratum s when the N units receive the treatment assignment $\mathbf{z}$.

To make this definite, return for a moment to Fisher's Lady tasting tea. If the Lady could not discriminate at all, then no matter how milk is added to the cup—that is, no matter what $\mathbf{z}$ is—she will classify the cups in the same way; that is, she will give the same binary 8-tuple of responses $\mathbf{r}$. On the other hand, if she discriminates perfectly, always classifying cups correctly, then her 8-tuple of responses will vary with $\mathbf{z}$; indeed, the responses will match the treatment assignments so that $\mathbf{r_z} = \mathbf{z}$.

If treatments are randomly assigned, then the treatment assignment $\mathbf{Z}$ is a random variable, so the observed responses are also random variables as they depend on $\mathbf{Z}$. Specifically, the observed response is the random variable $\mathbf{r_Z}$, that is, one of the many possible $\mathbf{r_z}$, $\mathbf{z} \in \Omega$, selected by picking a treatment assignment $\mathbf{Z}$ by the random mechanism that governs the experiment. Write $\mathbf{R} = \mathbf{r_Z}$ for the observed response, where $\mathbf{R}$ like $\mathbf{Z}$ is a random variable.

In principle, each possible treatment assignment $\mathbf{z} \in \Omega$ might yield a pattern of responses $\mathbf{r_z}$ that is unrelated to the pattern observed with another $\mathbf{z}$. For instance, in a completely randomized experiment with 50 subjects divided into two groups of 25, there might be $|\Omega| = \binom{50}{25} \doteq 1.3 \times 10^{14}$ different and unrelated 50-tuples $\mathbf{r_z}$. Since it is difficult to comprehend a treatment effect in such terms, we look for regularities, patterns, or models of the behavior of $\mathbf{r_z}$ as $\mathbf{z}$ varies over Ω. The remainder of §2.5 discusses the most basic models for $\mathbf{r_z}$ as $\mathbf{z}$ varies over Ω. Chapter 5 discusses additional models for treatment effects.

2.5.2 No Interference Between Units

A first model is that of "no interference between units" which means that "the observation on one unit should be unaffected by the particular assignment of treatments to the other units" (Cox, 1958a, §2.4). Rubin (1986) calls this SUTVA for the "stable unit treatment value assumption." Formally, no interference means that r_{siz} varies with z_{si} but not with the other coordinates of $\mathbf{z}$. In other words, the response of the ith unit in stratum s depends on the treatment assigned to this unit, but not on the treatments assigned to other units, so this unit has only two possible values of the response rather than $|\Omega|$ possible values. When this model is assumed, write r_{Tsi} and r_{Csi} for the responses of the ith unit in stratum s when assigned, respectively, to treatment or control; that is, r_{Tsi} is the common value of $\mathbf{r}_{siz}$ for all $\mathbf{z} \in \Omega$ with $z_{si} = 1$, and r_{Csi} is the common value of $\mathbf{r}_{siz}$ for all $\mathbf{z} \in \Omega$ with $z_{si} = 0$. Then the observed response from the ith unit in stratum s is $R_{si} = r_{Tsi}$ if $Z_{si} = 1$ or $R_{si} = r_{Csi}$ if $Z_{si} = 0$, which may also be written $R_{si} = Z_{si}r_{Tsi} + (1 - Z_{si})r_{Csi}$. This model, with one potential response for each unit under each treatment, has been important both to experimental design—see Neyman (1923), Welch (1937), Wilk (1955), Cox (1958b, §5), and Robinson (1973)—and to causal inference more generally—see Rubin (1974, 1977) and Holland (1986). When there is no interference between units, write $\mathbf{r}_T$ for $(r_{T11}, \ldots, r_{TS,n_S})^T$ and $\mathbf{r}_C$ for $(r_{C11}, \ldots, r_{CS,n_S})^T$.

"No interference between units" is a model and it can be false. No interference is often plausible when the units are different people and the treatment is a medical intervention with a biological response. In this case, no interference means that a medical treatment given to one patient affects only that patient, not other patients. That is often true. However, a vaccine given to many people may protect unvaccinated individuals by reducing the spread of a virus (so called herd immunity) and this is a form of interference. No interference is less plausible in some social settings, such a workplace or a classroom, where a reward given to one person may be visible to others, and may affect their behavior. No interference is often implausible when the strata are people and the units are repeated measures on a person; then a treatment given at one time may affect responses at later times; see Problem 2. In randomized single subject experiments, such as the Lady tasting tea, no interference is typically implausible.

2.5.3   The Model of an Additive Effect, and Related Models

The model of an additive treatment effect assumes units do not interfere with each other, and the administration of the treatment raises the response of a unit by a constant amount τ, so that $r_{Tsi} = r_{Csi} + \tau$ for each s, i. The principal attraction of the model is that there is a definite parameter to estimate, namely, the additive treatment effect τ. As seen in §2.7, in a uniform randomized experiment, many estimators do indeed estimate τ when this model holds.

In understanding the model of an additive treatment effect, it is important to keep in mind that the pair of responses, (r_{Tsi}, r_{Csi}), is never jointly observed for one unit (s, i). Therefore the model of an additive effect, $r_{Tsi} = r_{Csi} + \tau$, cannot be checked directly by comparing r_{Tsi} and r_{Csi} for particular units. The treatment Z_{si} and the observed response $R_{si} = Z_{si}r_{Tsi} + (1 - Z_{si})r_{Csi}$ are observed, and one can check what the model, $r_{Tsi} = r_{Csi} + \tau$, implies about these observable quantities. In a completely randomized experiment with a single stratum, $S = 1$, dropping the s, the model of an additive treatment effect, $r_{Ti} = r_{Ci} + \tau$, implies that, as sample sizes m and $n - m$ increase, the distribution of observed responses R_i for treated units $Z_i = 1$ will be shifted by τ when compared to the distribution of observed responses R_i for controls $Z_i = 0$, so the distributions will have the same shape and dispersion. That is, the histograms or boxplots would look the same, but one would be moved left or right relative to the other. This is a shift model, commonly used in nonparametrics; see

Lehmann (1975). In a uniform randomized experiment with several strata $S > 1$ and $r_{Tsi} = r_{Csi} + \tau$, the distribution of responses may have different shapes and dispersions in different strata, but within each stratum, the treated and control distributions are shifted by τ. This is a fairly weak form of no interaction between treatment group and stratum in the $2 \times S$ table of observable distributions, and it implies much less about observable distributions than the analogous nonparametric analysis of variance model, which typically assumes a common shape and dispersion in all $2S$ cells. If the only data are (Z_{si}, R_{si}), does the additive model have content beyond its implications for observable distributions? See Problem 7.

Under the additive model, the observed response from the ith unit in stratum s is $R_{si} = r_{Csi} + \tau Z_{si}$, or $\mathbf{R} = \mathbf{r}_C + \tau \mathbf{Z}$. It follows that the *adjusted responses*, $\mathbf{R} - \tau \mathbf{Z} = \mathbf{r}_C$, are fixed, not varying with the treatment assignment, $\mathbf{Z}$, so the adjusted responses satisfy the null hypothesis of no effect. This fact will be useful in drawing inferences about τ.

There are many similar models, including the model of a multiplicative effect, $r_{Tsi} = \sigma r_{Csi}$. Chapter 5 discusses quite different models for treatment effects.

2.5.4   *Positive Effects and Larger Effects

The model of an additive effect assumes a great deal about the relationship between r_{Tsi} and r_{Csi}. At times, it is desirable to describe the behavior of statistical procedures while assuming much less. When there is no interference between units, an effect is a pair $(\mathbf{r}_T, \mathbf{r}_C)$ giving the responses of each unit under each treatment. Two useful concepts are positive effects and larger effects. Unlike the model of an additive treatment effect, positive effects and larger effects are meaningful not just for continuous responses, but also for binary responses, for ordinal responses, and as seen later in §2.8, for censored responses and multivariate responses.

A treatment has a *positive effect* if $r_{Tsi} \geq r_{Csi}$ for all units (s, i) with strict inequality for at least one unit. A more compact way of writing this is that $(\mathbf{r}_T, \mathbf{r}_C)$ is a positive effect if $\mathbf{r}_T \geq \mathbf{r}_C$ with $\mathbf{r}_T \neq \mathbf{r}_C$. This says that application of the treatment never decreases a unit's response and sometimes increases it. For instance, there is a positive effect if the effect is additive and $\tau > 0$. Hamilton (1979) discusses this model in detail when the outcome is binary.

Consider two possible effects, say $(\mathbf{r}_T, \mathbf{r}_C)$ and $(\mathbf{r}_T^*, \mathbf{r}_C^*)$. Then $(\mathbf{r}_T^*, \mathbf{r}_C^*)$ is a *larger effect* than $(\mathbf{r}_T, \mathbf{r}_C)$ if $r_{Tsi}^* \geq r_{Tsi}$ and $r_{Csi}^* \leq r_{Csi}$ for all s, i. For instance, the simplest example occurs when the treatment effect is additive with the same responses under control, namely, $\mathbf{r}_C^* = \mathbf{r}_C$, $\mathbf{r}_T = \mathbf{r}_C + \tau \mathbf{1}$, and $\mathbf{r}_T^* = \mathbf{r}_C + \tau^* \mathbf{1}$, for in this case $(\mathbf{r}_T^*, \mathbf{r}_C^*)$ exhibits a *larger effect* than $(\mathbf{r}_T, \mathbf{r}_C)$ if $\tau^* \geq \tau$. In general, write $\mathbf{R}$ and $\mathbf{R}^*$ for the observed responses from, respectively, the effects $(\mathbf{r}_T, \mathbf{r}_C)$ and $(\mathbf{r}_T^*, \mathbf{r}_C^*)$, so $R_{si}^* = r_{Tsi}^*$ if $Z_{si} = 1$ and $R_{si}^* = r_{Csi}^*$ if $Z_{si} = 0$.

If a statistical test rejects the null hypothesis 5% of the time when it is true, one would hope that it would reject at least 5% of the time when it is false in the anticipated direction. Recall that a statistical test is *unbiased* against a collection of alternative hypotheses if the test is at least as likely to reject the null hypothesis when one of the alternatives is true as when the null hypothesis is true. The next proposition says that all of the common tests in §2.4.3 are unbiased tests against positive treatment effects, and the test statistic is larger when the effect is larger. The proposition is proved in somewhat more general terms in the appendix, §2.9.

Proposition 4 *In a randomized experiment, a test statistic that is effect increasing yields an unbiased test of no effect against the alternative of a positive effect, and if $(\mathbf{r}_T^*, \mathbf{r}_C^*)$ is a larger effect than $(\mathbf{r}_T, \mathbf{r}_C)$, then $t(\mathbf{Z}, \mathbf{R}^*) \geq t(\mathbf{Z}, \mathbf{R})$.*

2.6 Confidence Intervals

2.6.1 Testing General Hypotheses

So far, the test statistic $t(\mathbf{Z}, \mathbf{R})$ has been used to test the null hypothesis of no treatment effect. There is an extension to test hypotheses that specify a particular treatment effect. In §2.6.2, this extension is used to construct confidence intervals. As always, the confidence interval is the set of hypotheses not rejected by a test.

Consider testing the hypothesis $H_0 : \tau = \tau_0$ in the model of an additive effect, $\mathbf{R} = \mathbf{r}_C + \tau \mathbf{Z}$. The idea is as follows. If the null hypothesis $H_0 : \tau = \tau_0$ were true, then $\mathbf{r}_C = \mathbf{R} - \tau_0 \mathbf{Z}$, so testing $H_0 : \tau = \tau_0$ is the same as testing that $\mathbf{R} - \tau_0 \mathbf{Z}$ satisfies the null hypothesis of no treatment effect.

More precisely, if $\mathbf{r}_C$ were known, the probability, say α, that $t(\mathbf{Z}, \mathbf{r}_C)$ is greater than or equal to some fixed number T could be determined from (2.3). If the null hypothesis were true, then $\mathbf{r}_C$ would equal the *adjusted responses*, $\mathbf{R} - \tau_0 \mathbf{Z}$, so under the null hypothesis, $\mathbf{r}_C$ can be calculated from τ_0 and the observed data. If the hypothesis $H_0 : \tau = \tau_0$ is true, then the chance that $t(\mathbf{Z}, \mathbf{R} - \tau_0 \mathbf{Z}) \geq T$ is α, where α is calculated as described above with $\mathbf{r}_C = \mathbf{R} - \tau_0 \mathbf{Z}$.

Now, suppose the null hypothesis is not true, say instead $\tau > \tau_0$, and consider the behavior of the above test. In this case, $\mathbf{R} = \mathbf{r}_C + \tau \mathbf{Z}$ and the adjusted responses $\mathbf{R} - \tau_0 \mathbf{Z}$ equal $\mathbf{r}_C + (\tau - \tau_0)\mathbf{Z}$, so the adjusted responses will vary with the assigned treatment $\mathbf{Z}$. If a unit receives the treatment, it will have an adjusted response that is $\tau - \tau_0$ higher than if this unit receives the control. If the test statistic is effect increasing, as is true of all the statistics in §2.4.3, then $t(\mathbf{Z}, \mathbf{R} - \tau_0 \mathbf{Z}) = t\{\mathbf{Z}, \mathbf{r}_C + (\tau - \tau_0)\mathbf{Z}\} \geq t(\mathbf{Z}, \mathbf{r}_C) = t(\mathbf{Z}, \mathbf{R} - \tau \mathbf{Z})$, where the inequality follows from the definition of an effect increasing statistic. In words, if the null hypothesis is false and

TABLE 2.5. Example of Confidence Interval Computations.

Unit	Control Response	Group	Observed Response	Adjusted Response	Ranks of Adjusted Responses
i	r_{Ci}	Z_i	$R_i = r_{Ci} + \tau Z_i$	$R_i - \tau_0 Z_i$	q_i
1	2	1	9	8	7
2	1	0	1	1	1
3	3	0	3	3	2
4	4	0	4	4	3
5	0	1	7	6	5
6	4	1	11	10	8
7	1	1	8	7	6
8	5	0	5	5	4

$$\tau = 7, \tau_0 = 1$$

instead $\tau > \tau_0$, then an effect increasing test statistic $t(\mathbf{Z}, \mathbf{R} - \tau_0 \mathbf{Z})$ will be larger with the incorrect τ_0 than it would have been had we tested the correct value τ.

Table 2.5 illustrates these computations with a rank sum test. It is a hypothetical uniform randomized experiment with $N = 8$ units, all in a single stratum $S = 1$, with $m = 4$ units assigned to treatment, and an additive treatment effect $\tau = 7$, though the null hypothesis incorrectly says $H_0 : \tau = \tau_0 = 1$. The rank sum computed from the adjusted responses $\mathbf{R} - 1\mathbf{Z}$ is $7 + 5 + 8 + 6 = 26$, which is the largest possible rank sum for $N = 8$, $m = 4$, and the one-sided significance level is $\binom{8}{4}^{-1} = 1/70 = 0.014$. The two-sided significance level is $2 \times 0.014 = 0.028$. After removing the hypothesized $\tau_0 = 1$ from treated units, the treated units continue to have higher responses than the controls.

2.6.2 Confidence Intervals by Inverting a Test

Under the model of an additive treatment effect, $\mathbf{R} = \mathbf{r}_C + \tau\mathbf{Z}$, a $1 - \alpha$ confidence set for τ is obtaining by testing each value of τ as in §2.6.1 and collecting all values not rejected at level α into a set A. More precisely, A is the set of values of τ that, when tested, yield significance levels or P-values greater than or equal to α. For instance, in the example in Table 2.5, the value $\tau = 1$ would not be contained in a 95% confidence set. When the true value τ is tested, it is rejected with probability no greater than α, so the random set A contains the true τ with probability at least $1 - \alpha$. This is called "inverting" a test, and it is the standard way of obtaining a confidence set from a test; see, for instance, Cox and Hinkley (1974, §7.2) or Lehmann (1959, §3.5). For many test statistics, a two-sided test yields a confidence set that is an interval, whose endpoints may be determined by a line search, as illustrated in §4.3.5. Section 3.2.4 uses this confidence interval in an observational study of lead in the blood of children.

2.7   Point Estimates

2.7.1   Unbiased Estimates of the Average Effect

The most quoted fact about randomized experiments is that they lead to unbiased estimates of the average treatment effect. Take the simplest case, a uniform randomized experiment with a single stratum, with no interference between units. In this case, there are m treated units, $N - m$ control units, $E(Z_i) = m/N$, $R_i = r_{Ti}$ if $Z_i = 1$, and $R_i = r_{Ci}$ if $Z_i = 0$. The difference between the mean response in the treated group, namely, $(1/m) \sum Z_i R_i$, and the mean response in the control group, namely, $\{1/(N-m)\} \sum (1 - Z_i) R_i$, has expectation

$$E \left\{ \sum \frac{Z_i R_i}{m} - \frac{(1 - Z_i) R_i}{N - m} \right\} = E \left\{ \sum \frac{Z_i r_{Ti}}{m} - \frac{(1 - Z_i) r_{Ci}}{N - m} \right\}$$

$$= \sum \frac{(m/N) r_{Ti}}{m} - \frac{(1 - m/N) r_{Ci}}{N - m} = \frac{1}{N} \sum r_{Ti} - r_{Ci},$$

and the last term is the average of the N treatment effects $r_{Ti} - r_{Ci}$ for the N experimental units. In words, the difference in sample means is unbiased for the average effect of the treatment. Notice carefully that this is true assuming only that there is no interference between units. There is no assumption that the treatment effect $r_{Ti} - r_{Ci}$ is constant from unit to unit, no assumption about interactions.

The estimate is unbiased for the average effect on the N units in this study, namely, $(1/N) \sum r_{Ti} - r_{Ci}$, but this says nothing about the effect on other units not in the study. Campbell and Stanley (1963) say that a randomized experiment has *internal validity* in permitting inferences about effects for the N units in the study, but it need not have *external validity* in that there is no guarantee that the treatment will be equally effective for other units outside the study; see also §2.4.5. The related issue of efficacy and effectiveness is discussed in §5.4.

The difference in sample means may be biased when there are two or more strata and the experimenter assigns disproportionately more subjects to the treatment in some strata than in others. However, there is an unbiased estimate that corrects the imbalance. It consists of calculating, within stratum s, the difference between the average response in the treated group, namely, $(1/m_s) \sum_i Z_{si} R_{si}$, and the average response in the control group, namely, $\{1/(n_s - m_s)\}\sum_i(1 - Z_{si})R_{si}$, and weighting this difference by the proportion of units in stratum s, namely, n_s/N. The estimate, called *direct adjustment*, is then:

$$\sum_{s=1}^{S}\frac{n_s}{N}\sum_{i=1}^{n_s}\left\{\frac{Z_{si}R_{si}}{m_s} - \frac{(1-Z_{si})R_{si}}{n_s - m_s}\right\}. \tag{2.6}$$

To check that (2.6) is unbiased, recall that, in a uniform randomized experiment, Z_{si} has expectation m_s/n_s. It follows that (2.6) has expectation

$$E\left[\sum_{s=1}^{S}\frac{n_s}{N}\sum_{i=1}^{n_s}\left\{\frac{Z_{si}R_{si}}{m_s} - \frac{(1-Z_{si})R_{si}}{n_s - m_s}\right\}\right]$$

$$= E\left[\sum_{s=1}^{S}\frac{n_s}{N}\sum_{i=1}^{n_s}\left\{\frac{Z_{si}r_{Tsi}}{m_s} - \frac{(1-Z_{si})r_{Csi}}{n_s - m_s}\right\}\right]$$

$$= \sum_{s=1}^{S}\frac{n_s}{N}\sum_{i=1}^{n_s}\left\{\frac{(m_s/n_s)r_{Tsi}}{m_s} - \frac{(1-m_s/n_s)r_{Csi}}{n_s - m_s}\right\}$$

$$= \frac{1}{N}\sum r_{Ti} - r_{Ci},$$

so direct adjustment does indeed give an unbiased estimate of the average effect. In a very clear discussion, Rubin (1977) does calculations of this kind.

In effect, direct adjustment views the treated units and the control units as two stratified random samples from the N units in the experiment. Then (2.6) is the usual stratified estimate of mean response to treatment in the population of N units minus the usual estimate of the mean response to control in the population of N units. Notice again that direct adjustment is unbiased for the average treatment effect even if that effect varies from unit to unit or from stratum to stratum. On the other hand, the average effect is but a summary of the effects, and not a complete description, when the effect varies from one stratum to another.

2.7.2 Hodges–Lehmann Estimates of an Additive Effect

Under the model of an additive effect, $\mathbf{R} = \mathbf{r}_C + \tau\mathbf{Z}$, there are many estimates of τ. One due to Hodges and Lehmann (1963) is closely tied to the test in §2.4 and the confidence interval in §2.6. Recall that $H_0 : \tau = \tau_0$ is tested using $t(\mathbf{Z}, \mathbf{R} - \tau_0\mathbf{Z})$, that is, by subtracting the hypothesized treatment effect $\tau_0\mathbf{Z}$ from the observed responses $\mathbf{R}$, and asking whether the adjusted responses $\mathbf{R} - \tau_0\mathbf{Z}$ appear to be free of a treatment effect. The Hodges–Lehmann estimate of τ is that value $\hat{\tau}$ such that the adjusted responses $\mathbf{R} - \hat{\tau}\mathbf{Z}$ appear to be exactly free of a treatment effect. Consider this in detail. Throughout this section, the experiment is a uniform randomized experiment.

Suppose that we can determine the expectation, say $\bar{t}$, of the statistic $t(\mathbf{Z}, \mathbf{R} - \tau\mathbf{Z})$ when calculated using the correct τ, that is, when calculated from responses $\mathbf{R} - \tau\mathbf{Z}$ that have been adjusted so they are free of a treatment effect. For instance, in an experiment with a single stratum, the rank sum statistic has expectation $\bar{t} = m(N + 1)/2$ if the treatment has no effect. This is true because, in the absence of a treatment effect, the rank sum statistic is the sum of m scores randomly selected from N scores whose mean is $(N + 1)/2$. In the same way, in a stratified experiment, the stratified rank sum statistic has expectation $\bar{t} = \frac{1}{2}\sum m_s(n_s + 1)$ in the absence of a treatment effect. In an experiment comprised of S pairs, in the absence of a treatment effect, the expectation of the signed rank statistic is $\bar{t} = (S + 1)/4$, since we expect to sum half of S scores which average $(S + 1)/2$. In the absence of an effect, in an experiment with a single stratum, the difference in sample means (2.2) has expectation $\bar{t} = 0$. In each of these cases, $\bar{t}$ may be determined without knowing τ, so there is a Hodges–Lehmann estimate.

Roughly speaking, the Hodges–Lehmann estimate is the solution $\hat{\tau}$ of the equation $\bar{t} = t(\mathbf{Z}, \mathbf{R} - \hat{\tau}\mathbf{Z})$. In other words, $\hat{\tau}$ is the value such that the adjusted responses $\mathbf{R} - \hat{\tau}\mathbf{Z}$ appear to be entirely free of a treatment effect, in the sense that the test statistic $t(\mathbf{Z}, \mathbf{R} - \hat{\tau}\mathbf{Z})$ exactly equals its expectation in the absence of an effect.

If $t(\cdot, \cdot)$ is an effect increasing statistic, as is true of all of the statistics in §2.3, then $t(\mathbf{Z}, \mathbf{R} - \hat{\tau}\mathbf{Z})$ is monotone decreasing as a function of $\hat{\tau}$ with $\mathbf{Z}$ and $\mathbf{R}$ fixed. This says: The larger the treatment effect $\hat{\tau}\mathbf{Z}$ removed from the observed responses $\mathbf{R}$, the smaller the statistic becomes. This is useful in solving $\bar{t} = t(\mathbf{Z}, \mathbf{R} - \hat{\tau}\mathbf{Z})$. If a $\hat{\tau}$ has been tried such that $\bar{t} < t(\mathbf{Z}, \mathbf{R} - \hat{\tau}\mathbf{Z})$, then a larger $\hat{\tau}$ will tend to make $t(\mathbf{Z}, \mathbf{R} - \hat{\tau}\mathbf{Z})$ smaller, moving it toward $\bar{t}$. Similarly, if $\bar{t} > t(\mathbf{Z}, \mathbf{R} - \hat{\tau}\mathbf{Z})$, then a smaller $\hat{\tau}$ is needed.

Problems arise immediately. For rank statistics, such as the rank sum and the signed rank, $t(\mathbf{Z}, \mathbf{R} - \hat{\tau}\mathbf{Z})$ varies in discrete jumps as $\hat{\tau}$ is varied, so there may be no value $\hat{\tau}$ such that $\bar{t} = t(\mathbf{Z}, \mathbf{R} - \hat{\tau}\mathbf{Z})$. To see this, take a trivial case, a uniform experiment in one stratum, sample size $N = 2$, one treated unit $m = 1$. Then the rank sum statistic is either 1 or 2 depending upon which of the two units receive the treatment, but $\bar{t} = 1.5$, so it is not possible to find a $\hat{\tau}$ such that $\bar{t} = t(\mathbf{Z}, \mathbf{R} - \hat{\tau}\mathbf{Z})$.

Not only may $\bar{t} = t(\mathbf{Z}, \mathbf{R} - \hat{\tau}\mathbf{Z})$ have no solution $\hat{\tau}$, but it may have infinitely many solutions. If $t(\mathbf{Z}, \mathbf{R} - \hat{\tau}\mathbf{Z})$ varies in discrete jumps, it will be constant for intervals of values of $\hat{\tau}$.

Hodges and Lehmann resolve these problems in the following way. They define the solution of an equation $\bar{t} = t(\mathbf{Z}, \mathbf{R} - \hat{\tau}\mathbf{Z})$ as SOLVE$\{\bar{t} = t(\mathbf{Z}, \mathbf{R} -$

TABLE 2.6. Computing a Hodges–Lehmann Estimate.

τ	4.9999	5	5.0001	5.9999	6	6.0001
$t(\mathbf{Z}, \mathbf{R} - \tau\mathbf{Z})$	20	19	18	18	17	15

$\hat{\tau}\mathbf{Z})\}$ defined by

$$
\begin{aligned}
\hat{\tau} &= \text{SOLVE}\{\bar{\bar{t}} = t(\mathbf{Z}, \mathbf{R} - \hat{\tau}\mathbf{Z})\} \\
&= \frac{\inf\{\tau : \bar{\bar{t}} > t(\mathbf{Z}, \mathbf{R} - \hat{\tau}\mathbf{Z})\} + \sup\{\tau : \bar{\bar{t}} < t(\mathbf{Z}, \mathbf{R} - \tau\mathbf{Z})\}}{2}.
\end{aligned}
$$

This defines the Hodges–Lehmann estimate. Roughly speaking, if there is no exact solution, then average the smallest τ that is too large and the largest τ that is too small.

Consider the small example in Table 2.5. Under the null hypothesis of no effect, the rank sum statistic has expectation $\bar{\bar{t}} = m(N+1)/2 = 4(8+1)/2 = 18$, that is, half of the sum of all eight ranks, $36 = 1 + 2 + \cdots + 8$. Table 2.6 gives values of $t(\mathbf{Z}, \mathbf{R} - \tau\mathbf{Z})$ for several values of τ. As noted, since $t(\cdot, \cdot)$ is effect increasing, in Table 2.6, $t(\mathbf{Z}, \mathbf{R} - \tau\mathbf{Z})$ decreases in τ. We want as our estimate a value $\hat{\tau}$ such that $18 = t(\mathbf{Z}, \mathbf{R} - \hat{\tau}\mathbf{Z})$, but the table indicates that any value between 5 and 6 will do. As the table suggests, $\inf\{\tau : \bar{\bar{t}} > t(\mathbf{Z}, \mathbf{R} - \tau\mathbf{Z})\} = 6$ and $\sup\{\tau : \bar{\bar{t}} < t(\mathbf{Z}, \mathbf{R} - \tau\mathbf{Z})\} = 5$, so the Hodges–Lehmann estimate is $\hat{\tau} = (6 + 5)/2 = 5.5$.

For particular test statistics, there are other ways of computing $\hat{\tau}$. This is true, for instance, for a single stratum using the rank sum test. In this case, it may be shown that $\hat{\tau}$ is the median of the $m(N - m)$ pairwise differences formed by taking each of the m treated responses and subtracting each of the $N - m$ control responses.

The Hodges–Lehmann estimate $\hat{\tau}$ inherits properties from the test statistic $t(\cdot, \cdot)$. Consistency is one such property. Recall that a test is *consistent* if the probability of rejecting each false hypothesis tends to one as the sample size increases. Recall that an estimate is consistent if the probability that it is close to the true value tends to one as the sample size increases. As one would expect, these ideas are interconnected. A test that rejects incorrect values of τ leads to an estimate that moves away from these incorrect values. In other words, under mild conditions, consistent tests lead to consistent Hodges–Lehmann estimates; see Maritz (1981, §1.4) for some details.

2.8 *More Complex Outcomes

2.8.1 *Partially Ordered Outcomes

So far, the outcome R_{si} has been a number, possibly a continuous measurement, possibly a binary event, possibly a discrete score, but always a single number. However, for more complex responses, much of the earlier discussion continues to apply with little or no change. The purpose of §2.8 is to discuss issues that arise with certain complex outcomes, including multivariate responses and censored observations.

When the outcome R_{si} is a single number, it is clear what it means to speak of a high or low response, and it is clear what it means to ask whether responses are typically higher among treated units than among controls. For more complex responses, it may happen that some responses are higher than some others; and yet not every pair of possible responses can be ordered. For example, unit 1 may have a more favorable outcome than units 2 and 3, but units 2 and 3 may have different outcomes neither of which can be described as entirely more favorable than the other. For instance, patient 1 may live longer and have a better quality of life than patients 2 and 3, but patient 2 may outlive patient 3 though patient 3 had a better quality of life than patient 2. In this case, outcomes may be partially ordered rather than totally ordered, an idea that is formalized in a moment. Common examples are given in §2.8.2 and 2.8.3.

A *partially ordered set* or *poset* is a set A together with a relation $\precsim$ on A such that three conditions hold:

(i) $a \precsim a$ for all $a \in A$;

(ii) $a \precsim b$ and $b \precsim a$ implies $a = b$ for all $a, b \in A$; and

(iii) if $a \precsim b$ and $b \precsim c$ then $a \precsim c$ for all $a, b, c \in A$.

There is *strict inequality* between a and b if $a \precsim b$ and $a \neq b$. A poset A is *totally ordered* if $a \precsim b$ or $b \precsim a$ for every $a, b \in A$. The real numbers with conventional inequality $\leq$ are totally ordered. If A is partially ordered but not totally ordered, then for some $a, b \in A$, $a \neq b$, neither a nor b is higher than the other; that is, neither $a \precsim b$ nor $b \precsim a$. Sections 2.8.2 and 2.8.3 discuss two common examples of partially ordered outcomes, namely, censored and multivariate outcomes. Following this, in §2.8.4, general methods for partially ordered outcomes are discussed.

2.8.2 *Censored Outcomes

In some experiments, an outcome records the time to some event. In a clinical trial, the outcome may be the time between a patient's entry into the trial and the patient's death. In a psychological experiment, the outcome may be the time lapse between administration of a stimulus by the experimenter and the production of a response by an experimental subject. In a study of remedial education, the outcome may be the time until a certain level of proficiency in reading is reached.

Times may be censored in the sense that, when data analysis begins, the event may not yet have occurred. The patient may be alive at the close of the study. The stimulus may never elicit a response. The student may not develop proficiency in reading during the period under study.

If the event occurs for a unit after, say, 3 months, the unit's response is written 3. If the unit entered the study 3 months ago, if the event has not yet occurred, and if the analysis is done today, then the unit's response is written 3+ signifying that the event has not occurred in the initial 3 months.

Censored times are partially ordered. To see this, consider a simple illustration. In a clinical trial, patient 1 died at 3 months, patient 2 died at 12 months, and patient 3 entered the study 6 months ago and is alive today yielding a survival of 6+ months. Then patient 1 had a shorter survival than patients 2 and 3, but it is not possible to say whether patient 2 survived longer than patient 3 because we do not know whether patient 3 will survive for a full year.

The set A of censored survival times contains the nonnegative real numbers together with the nonnegative real numbers with a plus appended. Define the partial order $\lesssim$ on A as follows: if a and b are nonnegative real numbers, then:

(i) $a \lesssim b$ if and only if $a \leq b$;

(ii) $a \lesssim b+$ if and only if $a \leq b$; and

(iii) $a \lesssim a$ and $a+ \lesssim a+$.

Here, (i) indicates that "a" and "b" are both deaths and "a" died first. In (ii), "a" died before "b" was censored, so "b" certainly outlived "a." Of course, (iii) is just the case of equality—every censored time is equal to itself, and so is less than or equal to itself. It is easy to check that this is indeed a partial order, and that strict inequality indicates certainty about who died first.

2.8.3   *Multivariate Outcomes and Other Partially Ordered Outcomes

Quite often, a single number is not enough to describe the outcome for a unit. In an educational intervention, there may be test scores in several areas, such as reading and mathematics. In a clinical trial, the outcome may involve both survival and quality of life. A multivariate response is a p-tuple of outcomes describing an individual. If the p components are numbers, then the multivariate response inherits a partial order as follows: $(a_1, \ldots, a_p) \lesssim (b_1, \ldots, b_p)$ if and only if $a_1 \leq b_1, a_2 \leq b_2, \ldots$, and $a_p \leq b_p$. It is easy to check that this defines a partial order. As an example, if the outcome is the 2-tuple consisting of a reading score and a mathematics score, then one student has a higher multivariate response than another only if the first student did at least as well as the second student on both tests.

In fact, the components of the p-tuple need not be numbers—rather they may be any partially ordered outcomes. In the same way, the p-tuple inherits a partial order from the partial orders of individual outcomes. For instance, the outcome might be a 2-tuple consisting of a censored survival time and a number measuring quality of life. The censored survival times are partially but not totally ordered. In this case, a patient who died early with a poor quality of life would have a lower outcome than a patient who was censored late with a good quality of life.

Multivariate responses may be given other partial orders appropriate to particular contexts. Here is one that gives greatest emphasis to the first coordinate and about equal emphasis to the other two: $(a_1, a_2, a_3) \lesssim (b_1, b_2, b_3)$ if $a_1 \leq b_1$ or if $\{a_1 = b_1 \text{ and } a_2 \leq b_2 \text{ and } a_3 \leq b_3\}$. In an educational setting, this might say that a student who graduates had a better outcome than one who did not regardless of test scores, but among those who graduate, one student is better than another only if both reading and math scores are as good or better.

2.8.4   *A Test Statistic for Partially Ordered Outcomes

The task is to test the null hypothesis of no treatment effect against the alternative that treated units tend to have higher responses than controls in the sense of a partial order $\lesssim$ on the outcomes. For this purpose, define indicators L_{sij} for $s = 1, \ldots, S$, $i = 1, \ldots, n_s$, $j = 1, \ldots, n_s$, as follows:

$$L_{sij} = \begin{cases} 1 & \text{if } R_{sj} \lesssim R_{si} \text{ with } R_{si} \neq R_{sj}, \\ -1 & \text{if } R_{si} \lesssim R_{sj} \text{ with } R_{si} \neq R_{sj}, \\ 0 & \text{otherwise.} \end{cases} \qquad (2.7)$$

In words, L_{sij} compares the ith and jth units in stratum s, and L_{sij} is 1 if the ith is strictly greater than the jth, is -1 if the ith is strictly smaller than the jth, and is zero in all other cases. The statistic is

$$t(\mathbf{Z}, \mathbf{R}) = \sum_{s=1}^{S} \sum_{i=1}^{n_s} \sum_{j=1}^{n_s} Z_{si}(1 - Z_{sj}) L_{sij}. \qquad (2.8)$$

Consider the statistic in detail. The term $Z_{si}(1 - Z_{sj}) L_{sij}$ equals 1 if, in stratum s, the ith unit received the treatment, the jth unit received the control, and these two units had unequal responses with the treated unit having a higher response, $R_{sj} \lesssim R_{si}$. Similarly, $Z_{si}(1 - Z_{sj})L_{sij}$ equals -1 if, in stratum s, the ith unit is treated, the jth is a control, and the control had the higher response, $R_{si} \lesssim R_{sj}$. In all other cases, $Z_{si}(1 - Z_{sj})L_{sij}$ equals zero. So the test statistic is the number of comparisons of treated and control units in the same stratum in which the treated unit had the higher response minus the number in which the control unit had the higher response.

This statistic generalizes several familiar statistics. If the outcome is a single number and the partial order $\lesssim$ is ordinary inequality $\leq$, then (2.8) is equivalent to the Mann–Whitney (1947) statistic and the Wilcoxon (1945) rank sum statistic. If the outcome is censored and $\lesssim$ is the partial order in §2.8.2, then the statistic is Gehan's (1965) statistic.

A device due to Mantel (1967) shows that (2.8) is a sum statistic. The steps are as follows. First note that, for any subset B of $\{1, 2, \ldots, n_s\}$,

$$\sum_{i \in B} \sum_{j \in B} L_{sij} = 0 \tag{2.9}$$

since L_{sij} and L_{sji} both appear in the sum, with $L_{sij} = -L_{sji}$, and they cancel. Using this fact with $B = \{i : 1 \leq i \leq n_s \text{ with } Z_{si} = 1\}$ yields

$$0 = \sum_{i \in B} \sum_{j \in B} L_{sij} = \sum_{i=1}^{n_s} \sum_{j=1}^{n_s} Z_{si} Z_{sj} L_{sij},$$

which permits the test statistic (2.8) to be rewritten as the sum statistic

$$t(\mathbf{Z}, \mathbf{R}) = \sum_{s=1}^{S} \sum_{i=1}^{n_s} Z_{si} \sum_{j=1}^{n_s} L_{sij} = \sum_{s=1}^{S} \sum_{i=1}^{n_s} Z_{si} q_{si} \quad \text{with} \quad q_{si} = \sum_{j=1}^{n_s} L_{sij}.$$

As a result, the expectation and variance of the test statistic under the null hypothesis are given by Proposition 2. In fact, in that Proposition, $\bar{q}_s = 0$ for each s using (2.9).

The score q_{si} has an interpretation. It is the number of units in stratum s with outcomes less than unit i minus the number with outcomes greater than i. The score q_{si} is large if unit i has a response larger than that of most units in stratum s. For instance, in Gehan's statistic for censored outcomes, the score q_{si} is the number of patients in stratum s who definitely died before patient i minus the number who definitely died after patient i.

2.8.5 *Effect Increasing Statistics, Positive Effects, Larger Effects

In §2.4 and 2.5, three terms were discussed, namely, effect increasing statistics, positive effects, and larger effects. These terms apply to partially ordered outcomes with virtually no change, as shown in a moment. In each case, the definitions in §2.4 and 2.5 are the special case of the definitions in this section with the partial order $\lesssim$ given by ordinary inequality $\leq$ of real numbers.

Let $\mathbf{r}$ and $\mathbf{r}^*$ be two possible values of the N-tuple of partially ordered outcomes. If $r_{si} \lesssim r_{si}^*$ for every treated unit and $r_{si}^* \lesssim r_{si}$ for every control unit, then the treated and control groups appear farther apart for outcome $\mathbf{r}^*$ than for outcome $\mathbf{r}$. A test statistic $t(\cdot, \cdot)$ is *effect increasing* if $t(\mathbf{z}, \mathbf{r}) \leq t(\mathbf{z}, \mathbf{r}^*)$ whenever $\mathbf{r}$ and $\mathbf{r}^*$ are two possible values of the response such that $r_{si} \lesssim r_{si}^*$ if $z_{si} = 1$ and $r_{si}^* \lesssim r_{si}$ if $z_{si} = 0$ for all s, i. In words, the statistic is larger when the outcomes in treated and control groups are farther apart. The statistic in §2.8.4 is effect increasing; see Problem 6.

If there is no interference between units, then $(\mathbf{r}_T, \mathbf{r}_C)$ is a *positive effect* if $\mathbf{r}_T \neq \mathbf{r}_C$ and $r_{Csi} \lesssim r_{Tsi}$ for every s, i. In the case of censored survival times, this would mean that each patient would definitely survive at least as long under the treatment as under the control, or else would continue to be censored at the same time due to the end of the study. An effect $(\mathbf{r}_T^*, \mathbf{r}_C^*)$ is a *larger effect* than $(\mathbf{r}_T, \mathbf{r}_C)$ if $r_{Tsi} \lesssim r_{Tsi}^*$ and $r_{Csi}^* \lesssim r_{Csi}$, for all s, i, that is, if the treated responses are higher and the control responses are lower.

The following proposition is the extension of Proposition 4 to partially ordered responses. Again, the proof is given in the appendix, §2.9.

Proposition 5 *In a randomized experiment, a test statistic that is effect increasing yields an unbiased test of no effect against the alternative of a positive effect, and if $(\mathbf{r}_T^*, \mathbf{r}_C^*)$ is a larger effect than $(\mathbf{r}_T, \mathbf{r}_C)$ then $t(\mathbf{Z}, \mathbf{R}^*) \geq t(\mathbf{Z}, \mathbf{R})$.*

2.9 *Appendix: Effect Increasing Tests Under Alternatives

This appendix proves Propositions 4 and 5 which describe the behavior of effect increasing test statistics under the alternative hypotheses of positive effects or larger effects. It may be of interest to contrast these propositions with a result in Lehmann (1959, §5.8, Lemma 2) which is similar in spirit though quite different in detail. It suffices to prove Proposition 5 since Proposition 4 is the special case of the former in which the partial order is ordinary inequality. The proof depends on the following lemma.

Lemma 6 *Let $t(\cdot, \cdot)$ be effect increasing. If $(\mathbf{r}_T, \mathbf{r}_C)$ is a positive effect, then $t(\mathbf{z}, \mathbf{r}_z) \geq t(\mathbf{z}, \mathbf{r}_a)$ for all $\mathbf{z}$, $\mathbf{a} \in \Omega$. If $(\mathbf{r}_T^*, \mathbf{r}_C^*)$ is a larger effect than $(\mathbf{r}_T, \mathbf{r}_C)$, then $t(\mathbf{z}, \mathbf{r}_z^*) \geq t(\mathbf{z}, \mathbf{r}_z)$ for all $\mathbf{z} \in \Omega$.*

Proof of Lemma. Let $(\mathbf{r}_T, \mathbf{r}_C)$ be a positive effect, let $\mathbf{z}$, $\mathbf{a} \in \Omega$, and consider $\mathbf{r}_z$ and $\mathbf{r}_a$. If $z_{si} = 1$, then $r_{siz} = r_{Tsi}$ while r_{sia} may equal either

r_{Tsi} or r_{Csi} depending on a_{si}, but in either case $r_{sia} \lesssim r_{siz}$ since $(\mathbf{r}_T, \mathbf{r}_C)$ is a positive effect. Similarly, if $z_{si} = 0$, then $r_{siz} = r_{Csi} \lesssim r_{sia}$. Since $t(\cdot, \cdot)$ is effect increasing, this implies $t(\mathbf{z}, \mathbf{r_z}) \geq t(\mathbf{z}, \mathbf{r_a})$, proving the first part of the lemma.

Now let $\mathbf{z} \in \Omega$, let $(\mathbf{r}_T^*, \mathbf{r}_C^*)$ be a larger effect than $(\mathbf{r}_T, \mathbf{r}_C)$, and consider $\mathbf{r}_\mathbf{z}^*$ and $\mathbf{r_z}$. If $z_{si} = 1$, then $r_{siz} = r_{Tsi} \lesssim r_{Tsi}^* = r_{siz}^*$. If $z_{si} = 0$, then $r_{siz}^* = r_{Csi}^* \lesssim r_{Csi} = r_{siz}$. Hence $t(\mathbf{z}, \mathbf{r_z^*}) \geq t(\mathbf{z}, \mathbf{r_z})$ since $t(\cdot, \cdot)$ is effect increasing, completing the proof. ∎

Proof of Proposition 5. The lemma directly shows that if $(\mathbf{r}_T^*, \mathbf{r}_C^*)$ is a larger effect than $(\mathbf{r}_T, \mathbf{r}_C)$, then $t(\mathbf{Z}, \mathbf{R}^*) \geq t(\mathbf{Z}, \mathbf{R})$. To prove unbiasedness, let $\mathbf{Z}$ be randomly selected from Ω where $\text{prob}(\mathbf{Z} = \mathbf{z})$ is known but need not be uniform. If the random treatment assignment turns out to be $\mathbf{Z} = \mathbf{a}$, then the observed outcome is $\mathbf{R} = \mathbf{r_a}$. If the null hypothesis were true, if the treatment had no effect, the observed response would be the same $\mathbf{r_a}$ no matter how treatments were assigned, that is, the observed response would be $\mathbf{R} = \mathbf{r_a}$ no matter what value $\mathbf{Z}$ assumed. If the null hypothesis were false and the treatment had a positive effect, the observed response would vary depending upon the treatment assignment, $\mathbf{R} = \mathbf{r_z}$ if $\mathbf{Z} = \mathbf{z}$. For any fixed number T

$$\text{prob}\{t(\mathbf{Z}, \mathbf{R}) \geq T\}$$
$$= \sum_{\mathbf{z} \in \Omega} [t(\mathbf{z}, \mathbf{r_z}) \geq T] \ \text{prob}(\mathbf{Z} = \mathbf{z})$$
$$\geq \sum_{\mathbf{z} \in \Omega} [t(\mathbf{z}, \mathbf{r_a}) \geq T] \ \text{prob}(\mathbf{Z} = \mathbf{z}) \qquad \text{for } \mathbf{a} \in \Omega \text{ by the lemma.}$$

In other words, the chance that the test statistic $t(\mathbf{Z}, \mathbf{R})$ exceeds any number T is at least as great under the alternative hypothesis of a positive effect as under the null hypothesis of no effect, proving unbiasedness. ∎

2.10 *Appendix: The Set of Treatment Assignments

2.10.1 *Outline and Motivation: The Special Structure of Ω

The set Ω of treatment assignments plays an important role both in randomized experiments and in the discussion of observational studies in later chapters. This set Ω possess a special structure, first noted by Savage (1964). Using this structure, a single theorem may refer to large classes of test statistics and to all of the simple designs, including matched pairs, matching with multiple controls, two-group comparisons, and stratified comparisons. The purpose of this section is to describe the special structure of Ω. Appendices in later chapters refer back to this appendix.

Savage (1964) observed that the set Ω is a finite distributive lattice. This is useful because there are tidy theorems about probability distributions on a finite distributive lattice, including the FKG inequality and Holley's inequality. This section:

(i) offers a little motivation;

(ii) reviews the definition of a distributive lattice;

(iii) shows that Ω is indeed such a lattice; and

(iv) discusses the relevant probability inequalities.

The material in this appendix may be read without previous experience with lattices.

For motivation, consider a simple case. There is a single stratum, $S = 1$, so the s subscript is dropped in this example, and there are $n = 4$ units of which $m = 2$ receive the treatment. Then Ω contains $\binom{4}{2} = 6$ possible treatment assignments. Assume for this motivating example that the null hypothesis of no treatment effect holds, and renumber the four subjects so their observed responses are in decreasing order, $r_1 \geq r_2 \geq r_3 \geq r_4$. Since no quantity we calculate ever depends on the numbering of subjects, this renumbering changes nothing, but it is notationally convenient. The six possible treatment assignments appear in (2.10).

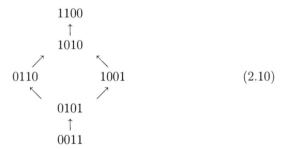

$$(2.10)$$

The treatment assignment $\mathbf{z} = (1,1,0,0)$ at the top in (2.10) is the one that would suggest the largest positive treatment effect, since this assignment places the two largest responses, r_1 and r_2, in the treated group. The assignment below this, namely, $\mathbf{z} = (1,0,1,0)$ would suggest a smaller treatment effect than $(1, 1, 0, 0)$, since r_3 has replaced r_2, but it would suggest a larger treatment effect than any other assignment. The assignments $(0, 1, 1, 0)$ and $(1, 0, 0, 1)$ are not directly comparable to each other, since the latter places the largest and smallest responses in the treated group while the former places the two middle responses in the treated group; however, both are lower than $(1, 0, 1, 0)$ and both are higher than $(0, 1, 0, 1)$.

Consider the behavior of a test statistic $t(\mathbf{z}, \mathbf{r})$ as we move through (2.10). Suppose, for instance, there are no ties among the responses, $r_1 > r_2 >$

$r_3 > r_4$, and $t(\mathbf{z}, \mathbf{r})$ is the rank sum statistic. Then $t(\mathbf{z}, \mathbf{r}) = 7$ for $\mathbf{z} = 1100$, $t(\mathbf{z}, \mathbf{r}) = 6$ for 1010, $t(\mathbf{z}, \mathbf{r}) = 5$ for both 1001 and 0110, $t(\mathbf{z}, \mathbf{r}) = 4$ for 0101, and $t(\mathbf{z}, \mathbf{r}) = 3$ for 0011, so $t(\mathbf{z}, \mathbf{r})$ increases steadily along upward paths in (2.10). If, instead, $t(\mathbf{z}, \mathbf{r})$ were the difference between the mean response in treated and control groups, it would again be increasing along upward paths.

Suppose, instead, that r_2 and r_3 were tied, so $r_1 > r_2 = r_3 > r_4$. In this case, the rank sum statistic would give average rank 2.5 to both r_2 and r_3, so moving from 1100 to 1010 would not change $t(\mathbf{z}, \mathbf{r})$. Notice, however, that even with ties, $t(\mathbf{z}, \mathbf{r})$ is monotone increasing (i.e., nondecreasing) along upward paths.

Actually, the order in (2.10) applies to many statistics whether ties are present or not. If $t(\mathbf{z}, \mathbf{r})$ is any arrangement-increasing statistic, then $t(\mathbf{z}, \mathbf{r})$ is monotone-increasing on upward paths in (2.10). Most reasonable statistics will assign a higher value to 1100 than to 1010, but reasonable statistics can differ in how they order assignments that are not comparable like 1001 and 0110.

Take a look at a second example, the case of $S = 2$ matched pairs, so $n_s = 2$ and $m_s = 1$ for $s = 1, 2$. Then Ω contains $2^2 = 4$ treatment assignments $\mathbf{z} = (z_{11}, z_{12}, z_{21}, z_{22})$. Again, assume the null hypothesis of no treatment effect and renumber the units in each pair so that in the first pair $r_{11} \geq r_{12}$, and in the second pair $r_{21} \geq r_{22}$. The set Ω appears in (2.11).

$$(2.11)$$

The assignment $\mathbf{z}$ in (2.11) suggesting the largest positive treatment effect is $\mathbf{z} = (1, 0, 1, 0)$ since in both pairs the treated unit had a higher response than the control. For $\mathbf{z} = 1001$ and $\mathbf{z} = 0110$, the treated unit had the higher response in one pair and the lower response in the other. In the assignment $\mathbf{z} = 0101$ the treated unit had a lower response than the control in both pairs.

Once again, common statistics are monotone-increasing along upward paths in (2.11). For instance, this is true of the signed rank statistic, which equals zero at the bottom of (2.11), equals one or two in the middle, and equals three at the top. Indeed, all arrangement-increasing functions are monotone-increasing along upward paths in (2.11).

What does all this suggest? There are certain treatment assignments $\mathbf{z} \in \Omega$ that are higher than others, and this is true without reference to the nature of the response $\mathbf{r}$ or the specific test statistic $t(\mathbf{z}, \mathbf{r})$. The responses might be continuous or they might be discrete scores or they might be binary. The test statistic might be the signed rank statistic or the McNemar statistic. In all these cases, $\mathbf{z} = 1010$ is higher than $\mathbf{z} = 1001$ in (2.11). Certain statements about treatment assignments $\mathbf{z} \in \Omega$ should be true generally, without reference to the specific nature of the outcome or the test statistic.

2.10.2   *A Brief Review of Lattices

Briefly, a lattice is a partially ordered set in which each pair of elements has a greatest lower bound and a least upper bound. This terminology is discussed formally in a moment, but first consider what this means in (2.10). A point $\mathbf{z}$ in (2.10) is below another $\mathbf{z}^*$ if there is a path up from $\mathbf{z}$ to $\mathbf{z}^*$; for instance, 0110 is below 1100. The points 1001 and 0110 are not comparable—there is not a path up from one to the other—so Ω is partially but not totally ordered. The least upper bound of 0110 and 1001 is 1010, for it is the smallest element above both of them. The least upper bound of 1010 and 1100 is 1100. A nice introduction to lattices is given by MacLane and Birkoff (1988).

A set Ω is *partially ordered* by a relation $\precsim$ if for all $\mathbf{z}, \mathbf{z}^*, \mathbf{z}^{**} \in \Omega$:

(i) $\mathbf{z} \precsim \mathbf{z}$;

(ii) $\mathbf{z} \precsim \mathbf{z}^*$ and $\mathbf{z}^* \precsim \mathbf{z}$ implies $\mathbf{z} = \mathbf{z}^*$; and

(iii) $\mathbf{z} \precsim \mathbf{z}^*$ and $\mathbf{z}^* \precsim \mathbf{z}^{**}$ implies $\mathbf{z} \precsim \mathbf{z}^{**}$.

An *upper bound* for $\mathbf{z}, \mathbf{z}^* \in \Omega$ is an element $\mathbf{z}^{**}$ such that $\mathbf{z} \precsim \mathbf{z}^{**}$ and $\mathbf{z} \precsim \mathbf{z}^{**}$. A *least upper bound* $\mathbf{z}^{**}$ for $\mathbf{z}, \mathbf{z}^*$ is an upper bound that is below all other upper bounds for $\mathbf{z}, \mathbf{z}^*$; that is, if $\mathbf{z}^{***}$ is any upper bound for $\mathbf{z}, \mathbf{z}^*$, then $\mathbf{z}^{**} \precsim \mathbf{z}^{***}$. If a least upper bound for $\mathbf{z}, \mathbf{z}^*$ exists, then it is unique by (ii). Lower bound and *greatest lower bound* are defined similarly. A *lattice* is a partially ordered set Ω in which every pair $\mathbf{z}, \mathbf{z}^*$ of elements has a least upper bound, written $\mathbf{z} \vee \mathbf{z}^*$, and a greatest lower bound, written $\mathbf{z} \wedge \mathbf{z}^*$. A lattice Ω is *finite* if the set Ω contains only finitely many elements. In (2.10), both 1010 and 1100 are upper bounds for the pair 1001 and 0110, but the least upper bound is $1001 \vee 0110 = 1010$.

The partial order $\precsim$ and the operations $\vee$ and $\wedge$ are tied together by the following relationship: $\mathbf{z} \precsim \mathbf{z}^*$ if and only if $\mathbf{z} \vee \mathbf{z}^* = \mathbf{z}^*$ and $\mathbf{z} \wedge \mathbf{z}^* = \mathbf{z}$. In fact, using this relationship, a lattice may be defined beginning with the operations $\vee$ and $\wedge$ rather than beginning with the partial order $\precsim$, that is, defining the partial order in terms of the operations. The following theorem is well known; see MacLane and Birkoff (1988, §XIV, 2) for proof.

Theorem 7 *A set Ω with operations $\vee$ and $\wedge$ is a lattice if and only if for all $\mathbf{z}, \mathbf{z}^*, \mathbf{z}^{**} \in \Omega$:*

L1. $\mathbf{z} \vee \mathbf{z} = \mathbf{z}$ and $\mathbf{z} \wedge \mathbf{z} = \mathbf{z}$;

L2. $\mathbf{z} \vee \mathbf{z}^* = \mathbf{z}^* \vee \mathbf{z}$ and $\mathbf{z} \wedge \mathbf{z}^* = \mathbf{z}^* \wedge \mathbf{z}$;

L3. $\mathbf{z} \vee (\mathbf{z}^* \vee \mathbf{z}^{**}) = (\mathbf{z} \vee \mathbf{z}^*) \vee \mathbf{z}^{**}$ and $\mathbf{z} \wedge (\mathbf{z}^* \wedge \mathbf{z}^{**}) = (\mathbf{z} \wedge \mathbf{z}^*) \wedge \mathbf{z}^{**}$; and

L4. $\mathbf{z} \wedge (\mathbf{z} \vee \mathbf{z}^*) = \mathbf{z} \vee (\mathbf{z} \wedge \mathbf{z}^*) = \mathbf{z}$.

Here, L2 and L3 are the commutative and associate laws, L1 is called idempotence, and L4 is called absorption. A lattice is *distributive* if the distributive law also holds,

$$\mathbf{z} \vee (\mathbf{z}^* \wedge \mathbf{z}^{**}) = (\mathbf{z} \vee \mathbf{z}^*) \wedge (\mathbf{z} \vee \mathbf{z}^{**}) \qquad \text{for all} \quad \mathbf{z}, \mathbf{z}^*, \mathbf{z}^{**} \in \Omega.$$

2.10.3 *The Set of Treatment Assignments Is a Distributive Lattice

This section gives Savage's (1964) demonstration that Ω is a distributive lattice. With each N-dimensional $\mathbf{z} \in \Omega$, associate a vector $\mathbf{c}$ of dimension $\sum m_s$, as follows. The vector $\mathbf{c}$ is made up of S pieces, where piece s has m_s coordinates. It is suggestive and almost accurate to say that $\mathbf{c}$ contains the ranks of the responses of treated units, each stratum being ranked separately, the ranks being arranged in decreasing order in each stratum. This would be exactly true if there were no ties, but it is not exactly true in the case of ties. Here is the exact definition, with or without ties. If $z_{s1} = 0$, $z_{s2} = 0, \ldots, z_{s,i-1} = 0$, $z_{si} = 1$, then $c_{s1} = n_s - i + 1$. Continuing, if $z_{s,i+1} = 0, \ldots, z_{s,j-1} = 0$, $z_{sj} = 1$, then $c_{s2} = n_s - j + 1$, and so on. In terms of the $\mathbf{c}$, (2.10) becomes (2.12), and (2.11) becomes (2.13). For instance, in (2.10), $\mathbf{z} = 1100$ becomes $\mathbf{c} = 43$, since the first 1 in $\mathbf{z}$ appears in position $i = 1$, so $n - i + 1 = 4 - 1 + 1 = 4$ and the second 1 in $\mathbf{z}$ appears in position $j = 2$, so $n - j + 1 = 4 - 2 + 1 = 3$.

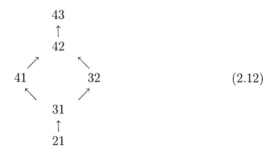

$$(2.12)$$

If there are ties among the responses in a stratum, then $\mathbf{c}$ is no longer a collection of ranks, because $\mathbf{c}$ distinguishes units with the same tied response. In the end, this is not a problem. The lattice order makes a few distinctions among treatment assignments that statistical procedures will ignore.

$$\begin{array}{ccc} & 22 & \\ \nearrow & & \searrow \\ 21 & & 12 \\ \nwarrow & & \nearrow \\ & 11 & \end{array} \qquad (2.13)$$

It is readily checked that each $\mathbf{z}$ has one and only one corresponding $\mathbf{c}$. Given $\mathbf{z}$, $\mathbf{z}^* \in \Omega$, with corresponding $\mathbf{c}$ and $\mathbf{c}^*$, the operations $\vee$ and $\wedge$ are defined as follows. Define $\mathbf{c} \vee \mathbf{c}^*$ and $\mathbf{c} \wedge \mathbf{c}^*$ as the vectors containing, respectively, $\max(c_{si}, c_{si}^*)$ and $\min(c_{si}, c_{si}^*)$. Define $\mathbf{z} \vee \mathbf{z}^*$ and $\mathbf{z} \wedge \mathbf{z}^*$ as the elements of Ω corresponding to $\mathbf{c} \vee \mathbf{c}^*$ and $\mathbf{c} \wedge \mathbf{c}^*$. It is readily checked that this definition makes sense, that is, that $\mathbf{c} \vee \mathbf{c}^*$ and $\mathbf{c} \wedge \mathbf{c}^*$ always correspond to elements of Ω. For instance, in (2.10), $\mathbf{z} = 0110$ and $\mathbf{z}^* = 1001$ correspond to $\mathbf{c} = 32$ and $\mathbf{c}^* = 41$, so $\mathbf{c} \vee \mathbf{c}^* = 42$ and $\mathbf{c} \wedge \mathbf{c}^* = 31$, so $\mathbf{z} \vee \mathbf{z}^* = 1010$ and $\mathbf{z} \wedge \mathbf{z}^* = 0101$, as is consistent with (2.10). Notice carefully that the coordinate (s, i) of $\mathbf{z} \vee \mathbf{z}^*$ is not generally equal to $\max(z_{si}, z_{si}^*)$.

To show that Ω is a lattice with these operations, one needs to check L1 to L4 in Theorem 7, but L1 to L3 hold trivially for $\max(c_{si}, c_{si}^*)$ and $\min(c_{si}, c_{si}^*)$. To show $\mathbf{z} \wedge (\mathbf{z} \vee \mathbf{z}^*) = \mathbf{z}$ in L4, it suffices to show $\mathbf{c} \wedge (\mathbf{c} \vee \mathbf{c}^*) = \mathbf{c}$. If $c_{si} \geq c_{si}^*$, then $\min\{c_{si}, \max(c_{si}, c_{si}^*)\} = \min(c_{si}, c_{si}) = c_{si}$, while if $c_{si} < c_{si}^*$, then $\min\{c_{si}, \max(c_{si}, c_{si}^*)\} = \min(c_{si}, c_{si}^*) = c_{si}$, so $\mathbf{c} \wedge (\mathbf{c} \vee \mathbf{c}^*) = \mathbf{c}$ as required. The second part of L4 is proved in the same way. So Ω is a lattice.

More than this, Ω is a distributive lattice. As proof, it suffices to show $\mathbf{c} \vee (\mathbf{c}^* \wedge \mathbf{c}^{**}) = (\mathbf{c} \vee \mathbf{c}^*) \wedge (\mathbf{c} \vee \mathbf{c}^{**})$, that is, to show

$$\max\{c_{si}, \min(c_{si}^*, c_{si}^{**})\} = \min\{\max(c_{si}, c_{si}^*), \max(c_{si}, c_{si}^{**})\}.$$

There are two cases. If $c_{si} \geq \min(c_{si}^*, c_{si}^{**})$, then $\max\{c_{si}, \min(c_{si}^*, c_{si}^{**})\} = c_{si}$, but also c_{si} is less than or equal to both $\max(c_{si}, c_{si}^*)$ and $\max(c_{si}, c_{si}^{**})$ yet it equals one of them, so

$$\min\{\max(c_{si}, c_{si}^*), \max(c_{si}, c_{si}^{**})\} = c_{si}.$$

On the other hand, if $c_{si} < \min(c_{si}^*, c_{si}^{**})$, then

$$\max\{c_{si}, \min(c_{si}^*, c_{si}^{**})\} = \min(c_{si}^*, c_{si}^{**}),$$

but $\max(c_{si}, c_{si}^*) = c_{si}^*$, and $\max(c_{si}, c_{si}^{**}) = c_{si}^{**}$, so

$$\min\{\max(c_{si}, c_{si}^*), \max(c_{si}, c_{si}^{**})\} = \min(c_{si}^*, c_{si}^{**}),$$

as required to complete the proof.

2.10.4 *Inequalities for Probability Distributions on a Lattice

This section discusses two inequalities for probability distributions on a finite distributive lattice, namely, the FKG inequality and Holley's inequality. These inequalities are the principal tool that makes use of the lattice properties of Ω. The original proofs of these inequalities are somewhat involved, but Ahlswede and Daykin (1978) developed a simpler proof involving nothing more than elementary probability. Their proof is nicely presented in several recent texts (Anderson 1987, §6, Bollobas, 1986, §19), to which the reader may refer.

A real-valued function on Ω, $f : \Omega \to \mathbb{R}$ is isotonic if $\mathbf{z} \precsim \mathbf{z}^*$ implies $f(\mathbf{z}) \leq f(\mathbf{z}^*)$. Throughout this appendix, $\mathbf{r}$ has been sorted into order within each stratum, $r_{si} \geq r_{s,i+1}$ for each s, i. With this order, the arrangement-increasing statistics $t(\mathbf{z}, \mathbf{r})$ are some of the isotonic functions on Ω. Actually, the arrangement-increasing statistics are the interesting isotonic functions, for they are the isotonic functions that are unchanged by interchanging tied responses in the same stratum. If there are ties, that is, if $r_{si} = r_{s,i+1}$ for some s and i, then there are isotonic functions that are not arrangement-increasing, specifically functions that increase when $z_{si} = 0, z_{s,i+1} = 1$ is replaced by $z_{si} = 1, z_{s,i+1} = 0$; however, these functions are not interesting as test statistics $t(\mathbf{z}, \mathbf{r})$ because they distinguish between people who gave identical responses. From a practical point of view, the important point is that a property of all isotonic functions on Ω is automatically a property of all arrangement-increasing functions, and all of the statistics in §2.4.3 are arrangement-increasing.

The first inequality is due to Fortuin, Kasteleyn, and Ginibre (1971).

Theorem 8 (The FKG Inequality) *Let $f(\cdot)$ and $g(\cdot)$ be isotonic functions on a finite distributive lattice Ω. If a random element $\mathbf{Z}$ of Ω is selected by a probability distribution satisfying*

$$\mathrm{prob}(\mathbf{Z} = \mathbf{z} \vee \mathbf{z}^*) \cdot \mathrm{prob}(\mathbf{Z} = \mathbf{z} \wedge \mathbf{z}^*) \geq \mathrm{prob}(\mathbf{Z} = \mathbf{z}) \cdot \mathrm{prob}(\mathbf{Z} = \mathbf{z}^*)$$
$$for\ all \quad \mathbf{z}, \mathbf{z}^* \in \Omega,$$

then

$$\mathrm{cov}\{f(\mathbf{Z}), g(\mathbf{Z})\} \geq 0.$$

For example, randomization gives equal probabilities to all elements of Ω, so the randomization distribution satisfies the condition for the FKG inequality. Hence, under the null hypothesis of no effect in a randomized experiment, any two arrangement-increasing statistics have a nonnegative correlation.

The next theorem is due to Holley (1974).

Theorem 9 (Holley's Inequality) *Let $f(\cdot)$ be an isotonic function on a finite distributive lattice Ω. If $\mathbf{Z}$ and $\tilde{\mathbf{Z}}$ are random elements of Ω selected* *by two probability distributions satisfying*

$$\text{prob}(\mathbf{Z} = \mathbf{z} \vee \mathbf{z}^*) \cdot \text{prob}(\tilde{\mathbf{Z}} = \mathbf{z} \wedge \mathbf{z}^*) \geq \text{prob}(\mathbf{Z} = \mathbf{z}) \cdot \text{prob}(\tilde{\mathbf{Z}} = \mathbf{z}^*)$$
$$for\ all\quad \mathbf{z}, \mathbf{z}^* \in \Omega,$$

then

$$E\{f(\mathbf{Z})\} \geq E\{f(\tilde{\mathbf{Z}})\}.$$

In other words, the premise of Holley's inequality is a sufficient condition for $\mathbf{Z}$ to be stochastically larger than $\tilde{\mathbf{Z}}$, in the sense that for every arrangement-increasing function $f(\cdot)$, the random variable $f(\mathbf{Z})$ has higher expectation than $f(\tilde{\mathbf{Z}})$. Holley's inequality helps later in comparing a nonrandom assignment of treatments to a random assignment. A related result is given by Krieger and Rosenbaum (1994). Literature related to Holley's inequality is reviewed in Rosenbaum (1999).

2.10.5   *An Identity in Ω

There is a useful identity in the set Ω of treatment assignments. The identity links $\vee$ and $\wedge$ to the addition of vectors, and therefore it is useful in verifying the conditions of the FKG inequality and Holley's inequality. It is true for this lattice, but not true generally for all lattices.

Lemma 10 *For all* $\mathbf{z}, \mathbf{z}^* \in \Omega$,

$$\mathbf{z} \vee \mathbf{z}^* + \mathbf{z} \wedge \mathbf{z}^* = \mathbf{z} + \mathbf{z}^*.$$

Proof. Fix a coordinate (s, i), so the task is to show $z_{si} + z_{si}^* = z_{\wedge si} + z_{\vee si}$, where $z_{\wedge si}$ and $z_{\vee si}$ are the (s, i) coordinates of $\mathbf{z} \wedge \mathbf{z}^*$ and $\mathbf{z} \vee \mathbf{z}^*$, respectively. Let $\mathbf{c}$ and $\mathbf{c}^*$ correspond with $\mathbf{z}$ and $\mathbf{z}^*$, respectively. There are three cases, depending upon the value of $z_{si} + z_{si}^*$.

1. If $z_{si} + z_{si}^* = 0$, then $c_{sj} \neq n_s - i + 1$ and $c_{sj}^* \neq n_s - i + 1$ for $j = 1, \ldots, m_s$, so $\max\left(c_{sj}, c_{sj}^*\right) \neq n_s - i + 1$ and $\min\left(c_{sj}, c_{sj}^*\right) \neq n_s - i + 1$ for $j = 1, \ldots, m_s$, so $z_{\wedge si} + z_{\vee si} = 0$, as required.

2. If $z_{si} + z_{si}^* = 2$, then there is a j and a k such that $c_{sj} = n_s - i + 1$ and $c_{sk}^* = n_s - i + 1$. If $j = k$, then $\max\left(c_{sj}, c_{sj}^*\right) = n_s - i + 1$ and $\min\left(c_{sj}, c_{sj}^*\right) = n_s - i + 1$, so $z_{\wedge si} = 1$ and $z_{\vee si} = 1$, so that $z_{\wedge si} + z_{\vee si} = 2$, as required. If $j < k$, then $n_s - i + 1 = c_{sj} > c_{sk}$ and $c_{sj}^* > c_{sk}^* = n_s - i + 1$, so $\min\left(c_{sj}, c_{sj}^*\right) = c_{sj} = n_s - i + 1$ and $\max\left(c_{sk}, c_{sk}^*\right) = c_{sk}^* = n_s - i + 1$, so $z_{\wedge si} = 1$ and $z_{\vee si} = 1$, so that $z_{\wedge si} + z_{\vee si} = 2$, as required. The case $j > k$ is similar.

3. If $z_{si} = 1$ and $z_{si}^* = 0$, so $z_{si} + z_{si}^* = 1$, then there is a j such that $c_{sj} = n_s - i + 1$ but $c_{sk}^* \neq n_s - i + 1$ for $k = 1, \ldots, m_s$. In this case, either $n_s - i + 1 = \max\left(c_{sj}, c_{sj}^*\right)$ or $n_s - i + 1 = \min\left(c_{sj}, c_{sj}^*\right)$ but not both, and moreover, $n_s - i + 1 \neq \max\left(c_{sk}, c_{sk}^*\right)$ and $n_s - i + 1 \neq \min\left(c_{sk}, c_{sk}^*\right)$ for all

$k \neq j$, so $z_{\wedge si} + z_{\vee si} = 1$, as required. The case $z_{si} = 0$ and $z_{si}^* = 1$ is similar. ∎

If there were no ties, so $\mathbf{c}$ and $\mathbf{c}^*$ are ranks, then Lemma 10 has the following interpretation. Within each stratum, the operations $\vee$ and $\wedge$ take the ranks in $\mathbf{c}$ and $\mathbf{c}^*$ and apportion them in forming $\mathbf{c} \vee \mathbf{c}^*$ and $\mathbf{c} \wedge \mathbf{c}^*$, but in this process they do not create or delete ranks that appear in $\mathbf{c}$ and $\mathbf{c}^*$.

2.11 Bibliographic Notes

Fisher is usually credited with the invention of randomized experiments. See, in particular, his important and influential book, *The Design of Experiments*, first published in 1935. Randomization is discussed in many articles and textbooks. In particular, see Kempthorne (1952), Cox (1958a, §5) and Cox and Reid (2000) for discussions of randomization in experimental design, and see Lehmann (1975) and Maritz (1981) for discussions of its role in nonparametrics. Mantel's (1963) paper was significant not just for the method he proposed, but also for its strengthening of the link between nonparametric methods and contingency table methods. The model for a treatment effect in §2.5.2 in which each unit has two potential responses, one under treatment and the other under control, has a long history. In an article first published in Polish and recently translated into English, Neyman (1923) used it to study the behavior of statistical tests under random assignment of treatments. Related work was done by Welch (1937), Wilk (1955), Cox (1958b, §5), and Robinson (1973), among others. Rubin (1974, 1977) first used the model in observational studies. In particular, he discussed the conditions under which matching, stratification, and covariance adjustment all estimate the same treatment effect. See also Hamilton (1979) and Holland (1986). Arrangement-increasing functions have been studied under various names by Eaton (1967), Hollander, Proshan, and Sethuraman (1977), and Marshall and Olkin (1979, §6F); see also Savage (1957). Although the Hodges-Lehmann (1963) estimates are often derived from rank tests, these *R-estimates* are very closely related to other families of estimates based on order statistics, *L-estimates*, or based on solving equations, *M-estimates*; see Gastwirth (1966) and Jureckova (1984). An attraction of R-estimates over L-estimates or M-estimates is that R-estimates have associated tests and confidence intervals that are exact, nonparametric, and explicitly linked to randomization in experiments. Sign-score statistics are discussed in Rosenbaum (1988) in connection with sensitivity analysis where these statistics permit certain simplifications. The discussion of complex outcomes in §2.8 draws from Mann and Whitney (1947), Gehan (1965), Mantel (1967), and Rosenbaum (1991, 1994). The material in §2.10 uses ideas from Savage (1964) and Rosenbaum (1989, 1995). The results in §2.10 concern permutations of vectors with binary coordinates, but some of these results extend to permutations of vectors with real coordinates; see Krieger and Rosenbaum (1994).

2.12   Problems

1. **The surprising power of the Lady tasting tea.** In §2.2, what is the power of the test? Specifically, suppose the Lady can distinguish milk first from tea first, and is always accurate. What is the power of a one-sided, 0.05 level test? Which 2×2 tables of the form Table 2.2 lead to rejection at the 0.05 level? If the Lady can distinguish, what is the chance of a table that leads to rejection?

2. **Interference between units with longitudinal data.** Suppose that there are S people, $s = 1, \ldots, S$, and person s is measured once a week for n_s consecutive weeks, $i = 1, \ldots, n_s$. Here, one unit (s, i) is one person in one week. For person s, a fixed number, m_s, of weeks are picked at random, independently for different people, and person s is treated in those weeks. Write $Z_{si} = 1$ if person s is treated in week i, $Z_{si} = 0$ otherwise, so $m_s = \sum_{i=1}^{n_s} Z_{si}$. The observed response of person s in week i is R_{si}, which may be affected by the current treatment Z_{si} and previous treatments, Z_{sj}, $j = 1, \ldots, i$. In addition, person s has a pretreatment baseline response, R_{s0}, which is unaffected by treatment, and so is fixed. Consider the model $R_{si} - R_{s,i-1} = \eta_{si} + \Delta Z_{si}$ for $i = 1, \ldots, n_s$, so the treatment produces additive gains, where Δ and the η_{si} are unknown fixed parameters. Show that this model violates the condition of "no interference between units" in §2.5.2. Let $T = t(\mathbf{Z}, \mathbf{R})$ be the stratified rank sum statistic, applied to the changes, $R_{si} - R_{s,i-1}$, so the n_s changes for person s are ranked from 1 to n_s and T is the sum of the ranks for the $\sum m_s$ treated weeks. Under the null hypothesis, $H_0 : \Delta = 0$, what is the randomization distribution of T? How does it compare to the usual randomization distribution of T of the stratified rank test? How could you use the randomization distribution of T when $\Delta = 0$ to test the general hypothesis $H_0 : \Delta = \Delta_0$? (Hint: Think about adjusted responses, $R_{si} - R_{s,i-1} - \Delta_0 Z_{si}$.) How could you use the randomization distribution of T when $\Delta = 0$ to build a confidence interval for Δ? Does interference between units preclude randomization inference?

3. **Proof of Proposition 1.** Let A and B be two finite, nonempty, disjoint sets, and let $A \times B$ be the set of all ordered pairs (a, b) with $a \in A$ and $b \in B$. If (a, b) is picked at random from $A \times B$, with each element of $A \times B$ having the same probability, show that a and b are independent. Use this to prove Proposition 1 for $S = 2$. Then use it

again to show that if Proposition 1 is true for S, then it is also true for $S + 1$.

4. **Proof of Proposition 2.** Prove Proposition 2. (Hint: Why does

$$var\left(\sum_{s=1}^{S}\sum_{i=1}^{n_s} Z_{si}q_{si}\right) = \sum_{s=1}^{S} var\left(\sum_{i=1}^{n_s} Z_{si}q_{si}\right) ?$$

Why does

$$var\left(\sum_{i=1}^{n_s} Z_{si}q_{si}\right) = var\left\{\sum_{i=1}^{n_s} Z_{si}\left(q_{si} - \overline{q}_s\right)\right\} ?$$

Remember $q_{si} - \overline{q}_s$ is fixed. What is $E\left(Z_{si}\right)$? What is

$$E\left\{\sum_{i=1}^{n_s} Z_{si}(q_{si} - \overline{q}_s)\right\} ?$$

What is $E\left(Z_{si}Z_{sj}\right)$? Be careful about $i = j$ and $i \neq j$.)

5. **Different statistics that yield the same randomization test.** Let $f\left(\cdot\right)$ be a strictly increasing function, so $x < y$ implies $f\left(x\right) < f\left(y\right)$. Show that a test that rejects at level α when $t\left(\mathbf{Z}, \mathbf{R}\right) \geq k$ is exactly the same test as the test that rejects when $f\left\{t\left(\mathbf{Z}, \mathbf{R}\right)\right\} \geq f\left(k\right)$. In a uniform randomized experiment with a single stratum, $S = 1$, dropping the s subscript, show that a randomization test of no treatment effect based on the total in the treated group, $\sum Z_i R_i$, is exactly the same test as a randomization test based on the difference between the treated and control group means,

$$t\left(\mathbf{Z}, \mathbf{R}\right) = \frac{\sum Z_i R_i}{m} - \frac{\sum\left(1 - Z_i\right) R_i}{n - m}.$$

In a uniform randomized experiment with a single stratum, $S = 1$, what is the Hodges—Lehmann estimate of an additive treatment effect, $r_{Ti} = r_{Ci} + \tau$ obtained from taking $t\left(\mathbf{Z}, \mathbf{R}\right)$ to be the difference between the treated and control group means?

6. **An effect increasing statistic with partially ordered responses.** Show that the statistic (2.8) is effect increasing. (Hint: Consider two response vectors, $\mathbf{r}$ and $\mathbf{r}^*$, and the corresponding indicators, L_{sij} and L_{sij}^*.)

7. **Metaphysics.** Section 2.5.3 discussed the distribution of observable quantities $\left(Z_{si}, R_{si}\right)$ in a uniform randomized experiment under the model of an additive treatment effect, $r_{Tsi} = r_{Csi} + \tau$. Because $\left(r_{Tsi}, r_{Csi}\right)$ is not jointly observed, one sees only $R_{si} = r_{Tsi}$ if $Z_{si} = 1$ for a treated subject, or else one sees $R_{si} = r_{Csi}$ if $Z_{si} = 0$ for a

control subject. Consider the case of a single stratum, $S = 1$, dropping the subscript s, and recall that, in a completely randomized experiment, the observable consequence of the additive effect model, $r_{Ti} = r_{Ci} + \tau$, is that the distribution of treated and control responses have the same shape and dispersion, but different locations, so the treated distribution is shifted by τ. Does the additive model $r_{Tsi} = r_{Csi} + \tau$ have content beyond its implications for observable distributions? Keep in mind that this is a problem in metaphysics, not statistics, so perhaps there is an answer, perhaps not. Hint: It is reasonable to ask of a question whether it is a reasonable question to ask. What does the phrase "content beyond" mean in this question? If "content beyond" were replaced by "observable consequences," what becomes of the question? If "content beyond" were replaced by "a mathematical form different from," what becomes of the question? In parallel, Wittgenstein (1958, #47, p22-23) writes:

> To the *philosophical* question: "Is the visual image of this tree composite, and what are its component parts?" the correct answer is "That depends upon what you understand by 'composite'." (And that is of course not an answer but a rejection of the question.)

2.13   References

Ahlswede, R. and Daykin, D. (1978) An inequality for the weights of two families of sets, their unions, and intersections. *Z. Wahrsch. Verus Gebiete*, **43**, 183–185.

Anderson, I. (1987) *Combinatorics of Finite Sets*. New York: Oxford University Press.

Birch, M. W. (1964) The detection of partial association, I: The 2×2 case. *Journal of the Royal Statistical Society*, Series **B**, **26**, 313–324.

Birch, M. W. (1965) The detection of partial association, II: The general case. *Journal of the Royal Statistical Society*, Series **B**, **27**, 111–124.

Bollobas, B. (1986) *Combinatorics*. New York: Cambridge University Press.

Campbell, D. and Stanley, J. (1963) *Experimental and Quasi-Experimental Designs for Research*. Chicago: Rand McNally.

Cochran, W. G. (1963) *Sampling Techniques*. New York: Wiley.

Cox, D. R. (1958a) *Planning of Experiments*. New York: Wiley.

Cox, D. R. (1958b) The interpretation of the effects of non-additivity in the Latin square. *Biometrika*, **45**, 69–73.

Cox, D. R. (1966) A simple example of a comparison involving quantal data. *Biometrika*, **53**, 215–220.

Cox, D. R. (1970) *The Analysis of Binary Data*. London: Methuen.

Cox, D. R. and Hinkley, D.V. (1974) *Theoretical Statistics*. London: Chapman & Hall.

Cox, D. R. and Reid, N. (2000) *The Theory of the Design of Experiments*. New York: CRC Press.

Eaton, M. (1967) Some optimum properties of ranking procedures. *Annals of Mathematical Statistics*, **38**, 124–137.

Eaton, M. (1982) A review of selected topics in probability inequalities. *Annals of Statistics*, **10**, 11–43.

Eaton, M. (1987) *Lectures on Topics in Probability Inequalities*. Amsterdam: Centrum. voor Wiskunde en Informatica.

Efron, B. (1971) Forcing a sequential experiment to be balanced. *Biometrika*, **58**, 403–417.

Fisher, R. A. (1935, 1949) *The Design of Experiments*. Edinburgh: Oliver & Boyd.

Fortuin, C., Kasteleyn, P., and Ginibre, J. (1971) Correlation inequalities on some partially ordered sets. *Communications in Mathematical Physics*, **22**, 89–103.

Freidlin, B. and Gastwirth, J. L. (2000) Should the median test be retired from general use? *American Statistician*, **54**, 161–164.

Friedman, L. M., DeMets, D. L., and Furberg, C. D. (1998) *Fundamentals of Clinical Trials*. New York: Springer-Verlag.

Gastwirth, J. L. (1966) On robust procedures. *Journal of the American Statistical Association*, **61**, 929-948.

Gehan, E. (1965) A generalized Wilcoxon test for comparing arbitrarily singly censored samples. *Biometrika*, **52**, 203–223.

Gibbons, J. D. (1982) Brown-Mood median test. In: *Encyclopedia of Statistical Sciences*, Volume 1, S. Kotz and N. Johnson, eds., New York: Wiley, pp. 322–324.

Hamilton, M. (1979) Choosing a parameter for 2×2 table or $2 \times 2 \times 2$ table analysis. *American Journal of Epidemiology*, **109**, 362–375.

Hettmansperger, T. (1984) *Statistical Inference Based on Ranks.* New York: Wiley.

Hodges, J. and Lehmann, E. (1962) Rank methods for combination of independent experiments in the analysis of variance. *Annals of Mathematical Statistics*, **33**, 482–497.

Hodges, J. and Lehmann, E. (1963) Estimates of location based on rank tests. *Annals of Mathematical Statistics*, **34**, 598–611.

Holland, P. (1986) Statistics and causal inference (with discussion). *Journal of the American Statistical Association*, **81**, 945–970.

Hollander, M., Proschan, F., and Sethuraman, J. (1977) Functions decreasing in transposition and their applications in ranking problems. *Annals of Statistics*, **5**, 722–733.

Hollander, M. and Wolfe, D. (1973) *Nonparametric Statistical Methods.* New York: Wiley.

Holley, R. (1974) Remarks on the FKG inequalities. *Communications in Mathematical Physics*, **36**, 227–231.

Jureckova, J. (1984) M-, L- and R-estimators. In: *Handbook of Statistics*, Volume IV, P. R. Krishnaiah and P. K. Sen, eds., New York: Elsevier, pp. 463–485.

Kempthorne, O. (1952) *The Design and Analysis of Experiments.* New York: Wiley.

Krieger, A. M. and Rosenbaum, P. R. (1994) A stochastic comparison for arrangement increasing functions. *Combinatorics, Probability and Computing*, **3**, 345–348.

Lehmann, E. L. (1959) *Testing Statistical Hypotheses.* New York: Wiley.

Lehmann, E. L. (1975) *Nonparametrics: Statistical Methods Based on Ranks.* San Francisco: Holden-Day.

MacLane, S. and Birkoff, G. (1988) *Algebra.* New York: Chelsea.

Mann, H. and Whitney, D. (1947) On a test of whether one of two random variables is stochastically larger than the other. *Annals of Mathematical Statistics*, **18**, 50–60.

Mantel, N. (1963) Chi-square tests with one degree of freedom: Extensions of the Mantel–Haenszel procedure. *Journal of the American Statistical Association*, **58**, 690–700.

Mantel, N. (1967) Ranking procedures for arbitrarily restricted observations. *Biometrics*, **23**, 65–78.

Mantel, N. and Haenszel, W. (1959) Statistical aspects of retrospective studies of disease. *Journal of the National Cancer Institute*, **22**, 719–748.

Maritz, J. (1981) *Distribution-Free Statistical Methods*. London: Chapman & Hall.

Marshall, A. and Olkin, I. (1979) *Inequalities: Theory of Majorization and Its Applications*. New York: Academic.

McNemar, Q. (1947) Note on the sampling error of the differences between correlated proportions or percentage. *Psychometrika*, **12**, 153–157.

Murphy, M., Hultgren, H., Detre, K., Thomsen, J., and Takaro, T. (1977) Treatment of chronic stable angina: A preliminary report of survival data of the randomized Veterans Administration Cooperative study. *New England Journal of Medicine*, **297**, 621–627.

Neyman, J. (1923) On the application of probability theory to agricultural experiments. Essay on principles. Section 9. (In Polish) *Roczniki Nauk Roiniczych, Tom X*, pp. 1–51. Reprinted in *Statistical Science 1990*, **5**, 463–480, with discussion by T. Speed and D. Rubin.

Neyman, J. (1935) Statistical problems in agricultural experimentation. *Supplement to the Journal of the Royal Statistical Society*, **2**, 107–180.

Pagano, M. and Tritchler, D. (1983) Obtaining permutation distributions in polynomial time. *Journal of the American Statistical Association*, **78**, 435–440.

Robinson, J. (1973) The large sample power of permutation tests for randomization models. *Annals of Statistics*, **1**, 291–296.

Rosenbaum, P. R. (1988) Sensitivity analysis for matching with multiple controls. *Biometrika*, **75**, 577–581.

Rosenbaum, P. R. (1989) On permutation tests for hidden biases in observational studies: An application of Holley's inequality to the Savage lattice. *Annals of Statistics*, **17**, 643–653.

Rosenbaum, P. R. (1991) Some poset statistics. *Annals of Statistics*, **19**, 1091–1097.

Rosenbaum, P. R. (1994) Coherence in observational studies. *Biometrics*, **50**, 368–374.

Rosenbaum, P. R. (1995) Quantiles in nonrandom samples and observational studies. *Journal of the American Statistical Association*, **90**, 1424–1431.

Rosenbaum, P. R. (1999) Holley's inequality. *Encyclopedia of Statistical Sciences*, Update Volume **3**, S. Kotz, C. B. Read, D. L. Banks, eds., New York: Wiley, pp. 328–331.

Rubin, D. B. (1974) Estimating the causal effects of treatments in randomized and nonrandomized studies. *Journal of Educational Psychology*, **66**, 688–701.

Rubin, D. B. (1977) Assignment to treatment group on the basis of a covariate. *Journal of Educational Statistics*, **2**, 1–26.

Rubin, D. B. (1986) Which ifs have causal answers? *Journal of the American Statistical Association*, **81**, 961–962.

Savage, I. R. (1957) Contributions to the theory of rank order statistics: The trend case. *Annals of Mathematical Statistics*, **28**, 968–977.

Savage, I. R. (1964) Contributions to the theory of rank order statistics: Applications of lattice theory. *Review of the International Statistical Institute*, **32**, 52–63.

Tukey, J. W. (1985) Improving crucial randomized experiments—especially in weather modification—by double randomization and rank combination. In: *Proceedings of the Berkeley Conference in Honor of Jerzy Neyman and Jack Kiefer*, L. Le Cam and R. Olshen, eds., Volume 1, Belmont, CA: Wadsworth, pp. 79–108.

Welch, B. L. (1937) On the z-test in randomized blocks and Latin squares. *Biometrika*, **29**, 21–52.

Wilcoxon, F. (1945) Individual comparisons by ranking methods. *Biometrics*, **1**, 8083.

Wilk, M. B. (1955) The randomization analysis of a generalized randomized block design. *Biometrika*, **42**, 70–79.

Wittgenstein, L. (1958) *Philosophical Investigations* (Third Edition). Englewood Cliffs, NJ: Prentice-Hall.

Zelen, M. (1974) The randomization and stratification of patients to clinical trials. *Journal of Chronic Diseases*, **27**, 365–375.

3
Overt Bias in Observational Studies

3.1 Introduction: An Example and Planning Adjustments

3.1.1 Outline: When Can Methods for Randomized Experiments Be Used?

An observational study is biased if the treated and control groups differ prior to treatment in ways that matter for the outcomes under study. An overt bias is one that can be seen in the data at hand—for instance, prior to treatment, treated subjects are observed to have lower incomes than controls. A hidden bias is similar but cannot be seen because the required information was not observed or recorded. Overt biases are controlled using adjustments, such as matching or stratification. In other words, treated and control subjects may be seen to differ in terms of certain observed covariates, but these visible differences may be removed by comparing treated and control subjects with the same values of the observed covariates, that is, subjects in the same matched set or stratum defined by the observed covariates. It is natural to ask when the standard methods for randomized experiments may be applied to matched or stratified data from an observational study. This chapter discusses a model for an observational study in which there is overt bias but no hidden bias. The model is, at best, one of many plausible models, but it does clarify when methods for randomized experiments may be used in observational studies, and so it becomes the starting point for thinking about hidden biases. Dealing with hidden bias is the focus of most of the later chapters. To permit discussion of conceptual issues in this chapter, Chapter 10 discusses the algorithmic issues that arise in constructing matched sets or strata with many covariates. The remainder of §3.1 considers an example and then discusses some of the planning steps that precede adjustments for covariates.

3.1.2   An Example with a Single Covariate

Cochran (1968) presents three stark examples of overt biases and their removal through adjustments. We will look at one of these. The data are from a study by Best and Walker of mortality in three groups of men: nonsmokers, cigarette smokers, and cigar and pipe smokers. Nonsmokers had a mortality rate of 20.2 deaths per 1000 people per year, cigarette smokers had 20.5 deaths, and cigar and pipe smokers had 35.5 deaths. The naive interpretation would be that cigarettes are harmless, but either cigars or pipes or both are dangerous. Cochran then gives the mean age in each group: 54.9 years for nonsmokers, 50.5 years for smokers, and 65.9 for cigar and pipe smokers. Clearly, the cigar and pipe smokers are older, so their higher death rate is not surprising, and may not reflect an effect of cigars or pipes. On the other hand, the cigarette smokers are the youngest group, and yet their mortality rate is slightly higher than the somewhat older nonsmokers. Perhaps cigarettes are not harmless.

Cochran then adjusts mortality for age, that is, removes an overt bias in the outcome by adjusting for an imbalance in a covariate. He uses age to divide the men into three strata or subclasses so that men in the same stratum have similar ages. Nonsmokers, cigarette smokers, and cigar and pipe smokers of roughly the same age are then compared to each other within each stratum, and the results are combined into a single rate using direct adjustment, essentially as described in §2.7.1. The adjusted mortality rate is 20.3 deaths per 1000 per year for nonsmokers, 28.3 for cigarette smokers, and 21.2 for cigar and pipe smokers. Now it is cigarettes that appear dangerous.

Which rates should be trusted, unadjusted or adjusted? Neither. The unadjusted rates are clearly wrong as a basis for estimating the effects of smoking, for they compare men who are not comparable in terms of one of the most important features of human mortality, namely, age. The adjusted rates are not clearly wrong. They might estimate the effects of smoking. However, it is possible that there is another covariate that was not recorded that has an impact similar to age; in this case, there would be a hidden bias. The current chapter discusses the conditions under which the methods in Chapter 2 for randomized experiments successfully estimate treatment effects in observational studies. These conditions become the basis in later chapters for thinking about hidden biases.

Cochran used three age strata. One might reasonably ask whether three strata are sufficient, whether such broad age groups suffice to remove the

overt bias due to age, and indeed this is the main question in Cochran's
paper. If instead of three strata, twelve strata are used, then the adjusted
rates are 20.2 for nonsmokers, 29.5 for cigarette smokers, and 19.8 for cigar
and pipe smokers. Three-age strata and twelve-age strata produce similar
adjusted rates, both of which are very different from the rates prior to
adjustment. Cochran presents a theoretical argument concluding that five
strata, each containing 20% of the subjects, will remove about 90% of the
bias in a single continuous covariate such as age.

3.1.3   Planning Adjustments for Overt Biases

Options narrow as an investigation proceeds. What is easy early on may become difficult or impossible later. This section discusses the earliest stages
of planning and data collection, as they relate to adjustments for bias.
The points raised are elementary, but at times ignored. When ignored, the
problems created can be far from elementary, at times insurmountable.

The control of overt biases begins before the study is designed. A first
step in planning an observational study is to determine what treatments
will be studied, and in the process to distinguish outcomes from covariates. Outcomes measure quantities that may be affected by the treatment,
while covariates are not affected; see §2.5. Cox (1958, §4.2) uses the term
concomitant observations in place of covariate and writes:

> The essential point in our assumptions about these observa-
> tions is that the value for any unit must be unaffected by the
> particular assignment of treatments to units actually used. In
> practice this means that either: (a) the concomitant observa-
> tions are taken before the assignment of treatments to units is
> made; or (b) the concomitant observations are made after the
> assignment of treatments, but before the effect of treatments
> has had time to develop ... ; or (c) we can assume from our
> knowledge of the nature of the concomitant observations con-
> cerned, that they are unaffected by the treatment.

As an example of type (c), Cox mentions the covariate that records the
relative humidity in a textile factory, where it is known that the treatments
under study could not possibly affect the relative humidity.

If adjustments are not confined to covariates, then adjustments may
remove part or all of the effect of the treatment. To illustrate, consider an
extreme, hypothetical example. Imagine a study comparing a placebo and
a drug intended to reduce blood pressure, the outcome being the incidence
of stroke. If the groups were compared after adjustment for blood pressure
levels six months after the start of treatment, then the adjusted incidence
of stroke might be similar in drug and placebo groups, not because the drug
has failed to work, but rather because the drug reduces the risk of stroke

by reducing blood pressure. If the effect of the drug on blood pressure is removed, the effect on stroke is removed with it.

While adjustments for an outcome can remove part of the treatment effect, adjustments of this sort are occasionally performed. It may be suspected that the treatment has only slight effects on a particular outcome, but this outcome may be strongly related to an important covariate that was not measured. An example occurred in the studies by Coleman, Hoffer, and Kilgore (1982) and Goldberger and Cain (1982) of the effects of Catholic versus public high schools. These studies compared cognitive test scores in the senior year of high school adjusting for various covariates, but the studies also adjusted for an outcome, namely, cognitive test scores in the sophomore year. The sophomore year test scores may already be affected by the difference between Catholic and public high schools, so they are, in principle, outcomes, not covariates. Still, it is natural to suspect that any effect of Catholic versus public high schools is produced gradually and cumulatively, and that only a part of the effect is present in the sophomore year. These studies used this outcome as a surrogate for an important covariate that was unavailable, namely, cognitive test scores prior to the start of high school. There are, then, two hazards: adjusting for sophomore test scores can remove part of the difference between Catholic and public schools; and failing to adjust for an early test score may yield a comparison of students who were not comparable in terms of their cognitive abilities prior to the start of high school. Notice that the second hazard is not present in a randomized experiment, so in an experiment, it is possible to give unequivocal advice that adjustments for outcomes should be avoided when estimating treatment effects. In an observational study, both hazards are present, and must be weighed; see Rosenbaum (1984a, §4) for discussion of alternative methods of analysis. The important point for the initial planning of observational studies is the distinction between outcomes and covariates, and their different status in adjustments.

The next step in planning is to list the covariates that will be measured. It is at this stage that biases become either overt or hidden. Since there is no way to completely address a hidden bias, a small change in this list may determine whether the study is convincing. A small oversight, easily corrected in the planning stage, may be an insurmountable problem at a later stage. In the design of randomized clinical trials, the standard practice is to begin with a written protocol that describes the data that will be collected and the main analyses that will be performed. Before the trial starts, the protocol is circulated for critical comment. Observational studies would, I believe, benefit from a written protocol and critical commentary.

Adjustments for overt biases may begin with data collection rather than with data analysis. Often treated subjects are matched to controls to form pairs or matched sets of subjects who are comparable in terms of observed covariates, and matching may take place before outcomes are measured. Chapter 10 discusses matching methods. Here, three points should be mentioned. First, unlike analytical adjustments, adjustments that are built into the study design are irrevocable. In the hypothetical example above concerning drug versus placebo to prevent stroke, it would be a mistake to adjust for blood pressure after treatment. If this mistake were made using an analytical method such as in §3.1.2, then it could be corrected by performing a different analysis, but if the mistake were made by matched sampling then it would be difficult to correct.

Second, certain covariates are more easily controlled through matching in the design than through analytical adjustments. Typically, these are covariates that classify subjects into many small categories. Matching can ensure that treated and control subjects belong to the same categories, but if matching is not used in the design of the study, some categories may have treated subjects and no controls or controls and no treated subjects. For instance, consider a study (Rosenbaum, 1986) that compared cognitive skills in what would be the senior year of high school for sophomores who dropped out of school and similar sophomores who remained in school. This was done with a national sample of students, and the high school was an important covariate with many values. The study used matched pairs of students *from the same school* having similar test scores, academic performance, and disciplinary records in the sophomore year, before the dropout left school.

Cost is an important consideration in deciding whether to match. If some covariate information is readily available, but other data are difficult or expensive to obtain, then matching becomes more attractive, but if data come with negligible costs, then matching during the design becomes somewhat less attractive. The reason is that, in many studies, some controls will be so different from treated subjects that they are of little use for comparisons (Dehejia and Wahba 1999). In the example above, many high-school students look very different from most dropouts in terms of test scores, academic performance, and disciplinary problems, so these students are of limited use in trying to determine how students who drop out would have performed had they remained in school. Matching may avoid collecting data on controls who will later be of little use.

A compromise between selecting matched pairs and using all potential controls is to match each treated subject to several controls. Ury (1975) examines the efficiency of studies that match several controls to each treated subject, finding that there is little to be gained from having more than four controls per treated subject with continuous responses. Smith (1997) presents an interesting case study of multivariate matching with multiple controls. In a single application, he compared pair matching with 1 control, matching with 8 controls and matching with 15 controls, concluding that 8 controls was best in this particular study. See also Ming and Rosenbaum (2000). Chapter 10 discusses the construction of matched sets with equal and variable numbers of controls per treated subject.

Matched studies can often be improved by a pilot study that forms a small number of matched pairs and scrutinizes those pairs using ethnographic or qualitative techniques. For instance, one might interview a few paired subjects or read the text of their hospital charts. This process may begin to reveal the hidden biases not visible in data on observed covariates, or it may suggest more accurate ways of using the data. An example is discussed in detail by Rosenbaum and Silber (2001). Emerson (1981) and Katz (2001) survey ethnographic techniques with reference to a large literature; see also Blumer (1969) and Becker (1996).

Having collected the data on covariates, the question arises: Should adjustments be made for all observed covariates? If not, how should covariates be selected for adjustment? These questions are somewhat controversial, not so much because the issues involved are unclear, but rather because there is no fully satisfactory answer. In principle, there is little or no reason to avoid adjustment for a true covariate, a variable describing subjects before treatment. There is little harm in comparing subjects who were comparable before treatment in ways that are not relevant for the outcomes of interest. In experiments, randomization tends to make treated and control groups comparable in terms of all covariates, relevant and irrelevant. In practice, the situation is often more involved, and increasing the number of covariates used in adjustments increases costs and complexities, and may make it more difficult to adjust for the most important covariates. In part, there are issues of data quality and completeness. As more covariates are collected and analyzed, it becomes increasingly difficult to ensure that all covariates meet high standards of accuracy and completeness, and increasingly difficult to ensure that each covariate receives the needed attention when used in modeling or matching. If there are many covariates, each with some missing data, there may be few subjects with complete data on all covariates, and this may make the analysis more difficult than it would otherwise be. These considerations weigh most heavily on covariates having doubtful relevance to outcomes of interest.

Perhaps the most common method for selecting covariates is also the most widely criticized. It entails comparing treated and control groups with respect to a long list of covariates, say using a t-test, and adjusting only for those covariates for which significant differences are found. There are three problems with this. First, the process does not consider the relationship between covariate and outcome. Second, there is no reason to believe that the absence of statistical significance implies the imbalance in the covariate is small enough to be ignored. Third, the process considers covariates one at a time, while the adjustments will control the covariates simultaneously. Addressing the first two problems, Cochran (1965, §3.1) studied this technique under a simple linear regression model in which all quantities are Normally distributed and a single covariate is the only source of bias. He looked at the coverage probability of the 95% confidence interval for the effect of the treatment on an outcome when no adjustment had been made for the covariate. This coverage probability was 90% or more providing the t-statistic for the covariate was less than 1.5 in absolute value and providing the squared correlation between the covariate and the outcome was 0.5 or less. This limitation on the square correlation is often reasonable for a covariate whose relevance is in doubt. He concluded: "If a single [covariate] shows a value of t above 1.5, these results suggest that we have another look at this [covariate] when the values of the [outcome] become known." Canner (1984, 1991) discusses closely related issues.

The following approach is often reasonable and practical. Begin by selecting a tentative list of covariates for adjustments using scientific knowledge of the relevant covariates together with exploratory comparisons of covariates in the treated and control groups, perhaps including some version of the technique evaluated by Cochran (1965, §3.1). With this tentative list, determine the tentative method of adjustment; that is, select the matched pairs or sets, define the strata, and determine whatever modeling technique will be used. Apply this method of adjustment to the covariates excluded from the tentative list, identifying any covariates exhibiting a large imbalance after adjustment. Reconsider the tentative list of covariates in light of this analysis. This approach addresses, at least in part, each of the three problems in the previous paragraph. The focus is on covariates known to be relevant since they are included in the initial list. At the same time, the data are given several opportunities to call attention to imbalances that might not be anticipated. Examples along these lines are discussed by Rosenbaum and Rubin (1984) and Silber et al. (2001).

3.2   Adjustments by Exact Stratification and Matching

3.2.1   Treatment Assignment with Unknown Probabilities

When is an observational study free of hidden bias? When do adjustments such as matching and stratification remove all of the bias? This section describes a model for an observational study with overt but no hidden bias. In most observational studies, this model is, at best, one of many plausible models—hidden biases are possible. The model is a start, indicating the inferences that would be appropriate were hidden biases absent. Later chapters try to determine whether hidden biases are present and ask how inferences might change if they are.

Initially, there are M units available for study, and each has a value of an observed covariate $\mathbf{x}$, which may contain several variables. Often the covariates $\mathbf{x}$ are used to reorganize the data prior to analysis, for instance, by matching or stratifying on $\mathbf{x}$. Number the M units $j = 1, \ldots, M$, so $\mathbf{x}_{[j]}$ is the covariate for the jth unit and the treatment assignment for this unit is $Z_{[j]}$. The bracketed subscript $[j]$ signifies the numbering of units before they are reorganized. After reorganization, a unit will have a different subscript without a bracket.

As a model for an observational study, imagine that unit j is assigned to treatment with probability $\pi_{[j]} = \text{prob}(Z_{[j]} = 1)$ and to control with probability $1 - \pi_{[j]} = \text{prob}(Z_{[j]} = 0)$, with assignments for distinct units being independent, and with $0 < \pi_{[j]} < 1$. The model says that treatments were assigned by flipping biased coins, possibly a different coin with a different bias for each unit, where the biases of the coins or the π's are unknown. The model says:

$$\text{prob}(Z_{[1]} = z_1, \ldots, Z_{[M]} = z_M) = \prod_{j=1}^{M} \pi_{[j]}^{z_j} \{1 - \pi_{[j]}\}^{1-z_j}. \qquad (3.1)$$

In an observational study, $\pi_{[j]}$ is unknown, so the distribution of treatment assignments $Z_{[1]}, \ldots, Z_{[M]}$ is unknown, and it is not possible to draw inferences as in Chapter 2 where randomization created a known distribution of treatment assignments.

Consider now the model for an observational study with overt biases but no hidden biases. An observational study is *free of hidden bias* if the π's, though unknown, are known to depend only on the observed covariates $\mathbf{x}_{[j]}$, so two units with the same value of $\mathbf{x}$ have the same chance π of receiving the treatment. Formally, the study is free of hidden bias if there exists a function $\lambda(\cdot)$, whose form will typically be unknown, such that $\pi_{[j]} = \lambda(\mathbf{x}_{[j]})$ for $j = 1, \ldots, M$. If the study is free of hidden bias, then (3.1) becomes

$$\text{prob}(Z_{[1]} = z_1, \ldots, Z_{[M]} = z_M) = \prod_{j=1}^{M} \lambda(\mathbf{x}_{[j]})^{z_j} \{1 - \lambda(\mathbf{x}_{[j]})\}^{1-z_j}. \qquad (3.2)$$

In short, an observational study is *free of hidden bias* when (3.2) holds. Rubin (1977) calls (3.2) "randomization on the basis of a covariate."

When the study is free of hidden bias, the function $\lambda(\mathbf{x})$ is called the propensity score. In §10.2, the propensity score $\lambda(\mathbf{x})$ is redefined so that it is still meaningful when hidden biases are present; however, in that case, $\pi_{[j]} \neq \lambda(\mathbf{x}_{[j]})$, and (3.2) does not follow from (3.1). A study is free of hidden bias when the treatment assignment probabilities $\pi_{[j]}$ are given by the propensity score $\lambda(\mathbf{x}_{[j]})$ which is always a function of the observed covariates $\mathbf{x}_{[j]}$. Chapter 4 discusses a model in which there is hidden bias and $\pi_{[j]}$ is not a function of $\mathbf{x}_{[j]}$.

A significance level, such as (2.3), cannot be calculated using (3.2) because $\lambda(\mathbf{x})$ is unknown. To adjust for overt bias in a study that is free of hidden bias is to address the fact that $\lambda(\mathbf{x})$ is unknown. The simplest approach is to stratify on $\mathbf{x}$.

3.2.2 Stratifying on $\mathbf{x}$

Often, units are grouped into strata on the basis of the covariate $\mathbf{x}$. From the M units, select $N \leq M$ units and group them into S nonoverlapping strata with n_s units in stratum s. In selecting the N units and assigning them to strata, use only the $\mathbf{x}$'s and possibly a table of random numbers. A stratification formed in this way is called a *stratification on* $\mathbf{x}$. Renumber the units so the ith unit in stratum s has treatment assignment Z_{si} and covariate $\mathbf{x}_{si}$. Using the same notation as Chapter 2, write $\mathbf{Z}$ for the N-tuple $(Z_{11}, \ldots , Z_{S,n_S})^T$. Write m_s for the number of treated units in stratum s; that is, $m_s = \sum_i Z_{si}$, and $\mathbf{m} = (m_1 \ldots , m_S)^T$.

An *exact stratification* on $\mathbf{x}$ has strata that are homogeneous in $\mathbf{x}$, so two units are in the same stratum only if they have the same value of $\mathbf{x}$, that is $\mathbf{x}_{si} = \mathbf{x}_{sj}$ for all s, i, and j. Exact stratification on $\mathbf{x}$ is practical only when $\mathbf{x}$ is of low dimension and its coordinates are discrete; otherwise, it will be difficult to locate many units with the same $\mathbf{x}$.

In an exact stratification on $\mathbf{x}$, if the study is free of hidden bias, that is, if (3.2) holds, then all units in the same stratum have the same chance of receiving the treatment. In this case, write λ_s in place of $\lambda(\mathbf{x}_{si})$, so (3.2) implies:

$$\text{prob}(\mathbf{Z} = \mathbf{z}) = \prod_{s=1}^{S} \prod_{i=1}^{n_s} \lambda_s^{z_{si}} (1 - \lambda_s)^{1 - z_{si}}. \tag{3.3}$$

In (3.3), the distribution of treatment assignments, $\text{prob}(\mathbf{Z} = \mathbf{z})$, is unknown because λ_s is unknown, and $m_s = \sum_i Z_{si}$ is a random variable. Consider the conditional distribution of $\mathbf{Z}$ given $\mathbf{m}$. It is a distribution on a set Ω whose elements are N-tuples of 0s and 1s such that $\mathbf{z} \in \Omega$ if and only if $m_s = \sum_i z_{si}$ for $s = 1, \ldots , S$, so Ω has $K = \prod_{s=1}^{S} \binom{n_s}{m_s}$ elements. Every treatment assignment $\mathbf{z} \in \Omega$ has the same unconditional probability in (3.3), namely,

$$\text{prob}(\mathbf{Z} = \mathbf{z}) = \prod_{s=1}^{S} \lambda_s^{m_s} (1 - \lambda_s)^{n_s - m_s}. \tag{3.4}$$

It follows that the conditional probability given $\mathbf{m}$ is constant, $\text{prob}(\mathbf{Z} = \mathbf{z}|\mathbf{m}) = 1/K$. Of course, this is the distribution of $\mathbf{Z}$ in a uniform randomized experiment; see §2.3.2.

In short, if an observational study is free of hidden bias, and if one stratifies exactly on $\mathbf{x}$, then the conditional distribution of the treatment assignment $\mathbf{Z}$ given the numbers $\mathbf{m}$ of treated units in each stratum, namely $\text{prob}(\mathbf{Z} = \mathbf{z}|\mathbf{m})$, is the same as the distribution of treatment assignments in a uniform randomized experiment. This is true even though the treatment assignment probabilities $\lambda(\mathbf{x})$ are unknown. In this case, given $\mathbf{m}$, the statistical procedures discussed in Chapter 2 have the properties described there. In other words, if the study is free of hidden bias and one stratifies exactly on $\mathbf{x}$, then the study may be analyzed using methods for a uniform randomized experiment.

Be clear on a key point. This result does not say that there is no difference between an experiment and an observational study. The difference is that in a uniform randomized experiment, the assignment probabilities $\text{prob}(\mathbf{Z} = \mathbf{z})$ are known to equal $1/K$ because we forced this to be true by randomizing. In an observational study, the conclusion $\text{prob}(\mathbf{Z} = \mathbf{z}|\mathbf{m}) = 1/K$ is deduced from the premise that the study is free of hidden bias, a premise we have little reason to believe. In an observational study, this premise is subjected to strict scrutiny, asking whether evidence can support or refute it, asking how findings would change if the premise were in error. This scrutiny is the focus of most later chapters.

3.2.3 Matching on $\mathbf{x}$

In §3.2.2, strata were formed using $\mathbf{x}$ alone. One way that matching differs from stratification is that there are constraints on the number m_s of treated units and the number $n_s - m_s$ of control units in a stratum. For instance, pair matching requires $n_s = 2$ and $m_s = 1$ for each s, while matching with multiple controls requires $n_s \geq 2$ and $m_s = 1$. A *matching on* $\mathbf{x}$ is a matched sample formed by:

(i) placing some restriction on S, $\mathbf{m}$ and $\mathbf{n} = (n_1, \ldots, n_S)^{\mathrm{T}}$, and

(ii) picking a stratification that meets these restrictions based exclusively on the pattern of $\mathbf{x}$s in the strata and possibly a table of random numbers.

For instance, a pair matched sample with $S = 100$ pairs would be formed by considering all possible stratifications with $n_s = 2$ and $m_s = 1$ for $m_s = 1$, $s = 1, \ldots, 100$, and selecting one of these possible stratifications based on the $\mathbf{x}$'s in the strata and possibly random numbers. An *exact matching on* $\mathbf{x}$ is a matching on $\mathbf{x}$ in which $\mathbf{x}$ is the same for all n_s units in each matched set, that is, $\mathbf{x}_{si} = \mathbf{x}_{sj}$ for $i, j = 1, \ldots, n_s$ for each s. As with exact stratification, exact matching is possible only when $\mathbf{x}$ is of low dimension and discrete.

The same argument as in §3.2.2 shows that, in an observational study that is free of hidden bias, if one matches exactly on $\mathbf{x}$, then the conditional distribution of the treatment assignment $\mathbf{Z}$ given $\mathbf{m}$ is the same as in a uniform randomized experiment, $\text{prob}(\mathbf{Z} = \mathbf{z}|\mathbf{m}) = 1/K$. If the study were free of hidden bias, it could be analyzed as if it were a matched randomized experiment, but of course, the comment at the end of §3.2.2 applies here as well.

3.2.4 An Example: Lead in the Blood of Children

Morton et al. (1982) studied lead in the blood of children whose parents worked in a factory where lead was used in making batteries. They were concerned that children were exposed to lead inadvertently brought home by their parents. Their study included 33 such children from different families—they are the exposed or treated children. The outcome R was the level of lead found in a child's blood in $\mu g/dl$ of whole blood. The covariate $\mathbf{x}$ was two-dimensional, recording age and neighborhood of residence. They matched each exposed child to one control child of the same age and neighborhood whose parents were employed in other industries not using lead. Table 3.1 shows the levels of lead found in the children's blood in $\mu g/dl$ of whole blood.

If this study were free of hidden bias, which may or may not be the case, we would be justified in analyzing Table 3.1 using methods for a uniform randomized experiment with 33 matched pairs. If the null hypothesis of no treatment effect is tested using Wilcoxon's signed rank test, the one-sided significance level is less than 0.0001. The Hodges–Lehmann estimate of the size of an additive effect is 15 $\mu g/dl$ with 95% confidence interval (9.5, 20.5). If the study were free of hidden bias, this would strongly suggest that the parents who worked with lead did raise the level of lead in their children's blood by about 15 $\mu g/dl$, a large increase compared to the level of lead found in controls. In later chapters, these data are examined again without the premise that the study is free of hidden bias.

3.2.5 Stratifying and Matching on the Propensity Score

Often, exact stratification or matching on $\mathbf{x}$ is difficult or impossible. If $\mathbf{x}$ is of high dimension or contains continuous measurements, each of the N units may have a different value of $\mathbf{x}$, so no stratum can contain a treated and control unit with the same $\mathbf{x}$. There are several questions. Do a large number of covariates—that is, a high-dimensional $\mathbf{x}$—make stratification and matching infeasible? Does close but inexact matching on $\mathbf{x}$ remove most of the bias due to $\mathbf{x}$? What algorithms produce good stratifications or matchings? The second and third questions are discussed in Chapter 10. The current section begins to answer the first question. As it turns out, there is a sense in which all matching problems are one-dimensional, so the dimensionality of $\mathbf{x}$ is not critical by itself.

Suppose an observational study is free of hidden bias, so (3.2) holds. Instead of stratifying or matching exactly on $\mathbf{x}$, imagine forming strata or matched sets in which units in the same stratum have the same chance of receiving the treatment $\lambda(\mathbf{x})$. Then within a stratum or matched set, units may have different values of $\mathbf{x}$, but they have the same propensity score $\lambda(\mathbf{x})$. Formally, it may happen that $\mathbf{x}_{si} \neq \mathbf{x}_{sj}$ but always $\lambda(\mathbf{x}_{si}) = \lambda(\mathbf{x}_{sj})$. Call this *exact matching or stratification on the propensity score*. In this

TABLE 3.1. Lead in Children's Blood ($\mu g/dl$).

Pair	Exposed	Control	Difference	Rank
1	38	16	22	22
2	23	18	5	8
3	41	18	23	23.5
4	18	24	−6	9.5
5	37	19	18	21
6	36	11	25	26
7	23	10	13	14
8	62	15	47	32
9	31	16	15	17
10	34	18	16	18.5
11	24	18	6	9.5
12	14	13	1	2.5
13	21	19	2	4
14	17	10	7	11
15	16	16	0	1
16	20	16	4	7
17	15	24	−9	12.5
18	10	13	−3	5.5
19	45	9	36	30
20	39	14	25	26
21	22	21	1	2.5
22	35	19	16	18.5
23	49	7	42	31
24	48	18	30	28
25	44	19	25	26
26	35	12	23	23.5
27	43	11	32	29
28	39	22	17	20
29	34	25	9	12.5
30	13	16	−3	5.5
31	73	13	60	33
32	25	11	14	15.5
33	27	13	14	15.5

case, the arguments in §3.2.2 and §3.2.3 go through without changes. In those arguments, equal $\mathbf{x}$'s within strata were used only to ensure equal $\lambda(\mathbf{x})$'s. In short, in an observational study free of hidden bias, exact matching or stratification on the propensity score yields a conditional distribution of treatment assignments $\mathbf{Z}$ given $\mathbf{m}$ that is the same as a uniform randomized experiment, namely $\mathrm{prob}(\mathbf{Z} = \mathbf{z}|\mathbf{m}) = 1/K$. In this case, the methods in Chapter 2 for uniform randomized experiments may be applied. The same conclusion is reached if strata or matched sets are formed based on $\lambda(\mathbf{x})$ and parts of $\mathbf{x}$, providing the strata or matched sets are homogeneous in $\lambda(\mathbf{x})$.

In practice, $\lambda(\mathbf{x})$ is unknown, so matching or stratification on $\lambda(\mathbf{x})$ is not possible. The use of estimated propensity scores in matching and stratification is discussed in Chapter 10. Also, §10.2 discusses balancing properties of the propensity score that are true whether or not the study is free of hidden bias.

3.3   Case-Referent Studies

3.3.1   Selecting Subjects Based on Their Outcomes

In Chapter 1, the study of DES and vaginal cancer is a case-referent study, and such a study has two features that distinguish it from the other observational studies in Chapter 1. First, the binary outcome, namely, vaginal cancer, is extremely rare, so a study that simply followed women until they developed the disease would need an enormous number of women to produce even a handful of cases of this rare cancer. This first feature is the motivation for conducting a case-referent study, but it is the second feature that characterizes such a study. In a case-referent study, cases are deliberately over-represented and referents are under-represented. The DES study identified cases of vaginal cancer and compared them to a small number of matched referents in terms of the frequency of maternal exposure to DES. In other words, subjects are included or excluded from the study, in part, on the basis of their outcomes. Instead of comparing the outcomes of treated and untreated groups, the case-referent study compares the frequency of exposure to the treatment among cases and referents.

At first, it is not clear that this makes sense. If the outcome $\mathbf{R} = \mathbf{r_Z}$ is affected by the treatment $\mathbf{Z}$, then selecting subjects using their outcomes may distort the frequency of exposure to the treatment. Indeed, this seems to have happened in the DES study. Of the 40 women in the study, 7 had mothers who had used DES, which exceeds the frequency of exposure to DES in the general population. When the treatment has an effect, it is related to the outcome, so selecting subjects using their outcomes changes the frequency of exposure to the treatment. How can a case-referent study be interpreted?

TABLE 3.2. Data Before Selecting Cases and Referents.

		Z	
		1	0
R	1	$\sum_{\mathbf{x}} Z_{[j]} r_{T[j]}$	$\sum_{\mathbf{x}} (1 - Z_{[j]}) r_{C[j]}$
	0	$\sum_{\mathbf{x}} Z_{[j]} (1 - r_{T[j]})$	$\sum_{\mathbf{x}} (1 - Z_{[j]})(1 - r_{C[j]})$

TABLE 3.3. Expectations in the Absence of Hidden Bias.

		Z	
		1	0
R	1	$\lambda(\mathbf{x}) \sum_{\mathbf{x}} r_{T[j]}$	$\{1 - \lambda(\mathbf{x})\} \sum_{\mathbf{x}} r_{C[j]}$
	0	$\lambda(\mathbf{x}) \sum_{\mathbf{x}} (1 - r_{T[j]})$	$\{1 - \lambda(\mathbf{x})\} \sum_{\mathbf{x}} (1 - r_{C[j]})$

3.3.2 Synthetic Case-Referent Studies

A synthetic case-referent study starts with the population of M subjects in §3.2.1, and draws a random sample of cases and a separate random sample of referents, possibly after stratification using the observed covariates $\mathbf{x}$. Synthetic case-referent studies are typically conducted when there is a computerized database describing the entire population of M subjects, but the study requires the costly collection of additional data not in the database. See Silber et al. (2001) for an example in which the population is comprised of all Medicare patients in Pennsylvania. This sort of study does occur, but far more common are case-referent studies that do not use random sampling. Synthetic case-referent studies are easier to consider theoretically because the mechanism that selects subjects has known properties. The term "synthetic" was introduced by Mantel (1973), while the odds ratio property discussed in this section is due to Cornfield (1951).

Consider the data before cases and referents are sampled, as in §3.2.1, so the jth of the M subjects has observed covariate $\mathbf{x}_{[j]}$, treatment assignment $Z_{[j]}$, and observed binary response $R_{[j]}$, which equals $r_{T[j]}$ if j is given the treatment and $r_{C[j]}$ if j is given the control. Divide the M subjects in the population into strata based on $\mathbf{x}$, and abbreviate by $\sum_{\mathbf{x}}$ a sum over all subjects j with $\mathbf{x}_{[j]} = \mathbf{x}$; that is, write $\sum_{\mathbf{x}}$ for $\sum_{j:\mathbf{x}_{[j]}=\mathbf{x}}$. The subjects in the stratum with covariate value $\mathbf{x}$ are recorded in the contingency Table 3.2.

If there is no hidden bias, then $E(Z_{[j]}) = \lambda(\mathbf{x}_{[j]})$, so the entries in Table 3.2 have as expectations the values in Table 3.3.

The odds ratio or cross-product ratio in Table 3.3 is

$$\frac{\left(\lambda(\mathbf{x}) \sum_{\mathbf{x}} r_{T[j]}\right) \left(\{1 - \lambda(\mathbf{x})\} \sum_{\mathbf{x}} (1 - r_{C[j]})\right)}{\left(\lambda(\mathbf{x}) \sum_{\mathbf{x}} (1 - r_{T[j]})\right) \left(\{1 - \lambda(\mathbf{x})\} \sum_{\mathbf{x}} r_{C[j]}\right)} = \frac{\left(\sum_{\mathbf{x}} r_{T[j]}\right) \left(\sum_{\mathbf{x}} (1 - r_{C[j]})\right)}{\left(\sum_{\mathbf{x}} (1 - r_{T[j]})\right) \left(\sum_{\mathbf{x}} r_{C[j]}\right)},$$

$$(3.5)$$

TABLE 3.4. Expected Counts in a Synthetic Case-Referent Study Absent Hidden Bias.

		Z	
		1	0
R	1	$k_{1\mathbf{x}}\lambda(\mathbf{x})\sum_{\mathbf{x}}r_{T[j]}$	$k_{1\mathbf{x}}\{1-\lambda(\mathbf{x})\}\sum_{\mathbf{x}}r_{C[j]}$
	0	$k_{0\mathbf{x}}\lambda(\mathbf{x})\sum_{\mathbf{x}}(1-r_{T[j]})$	$k_{0\mathbf{x}}\{1-\lambda(\mathbf{x})\}\sum_{\mathbf{x}}(1-r_{C[j]})$

and this odds ratio is a measure of the magnitude of a treatment effect. Notice that the odds ratio is one if there is no treatment effect in the stratum defined by $\mathbf{x}$, that is if $r_{T[j]} = r_{C[j]}$ for all $[j]$ with $\mathbf{x}_{[j]} = \mathbf{x}$, and the odds ratio is greater than one if the treatment has a positive effect in the stratum defined by $\mathbf{x}$, that is if $r_{T[j]} \geq r_{C[j]}$ for all j with $\mathbf{x}_{[j]} = \mathbf{x}$ and $r_{T[j]} \neq r_{C[j]}$ for some j with $\mathbf{x}_{[j]} = \mathbf{x}$. Hamilton (1979) discusses a wide variety of related measures under the model of a positive effect.

Tables 3.2 and 3.3 describe the initial population of M subjects. Consider a synthetic case-referent study formed from Table 3.2. Draw a random sample without replacement consisting of a fraction $k_{1\mathbf{x}}$ of the cases, the first row of the table, and a random sample consisting of a fraction $k_{0\mathbf{x}}$ from the referents, the second row, with $0 < k_{1\mathbf{x}} \leq 1$ and $0 < k_{0\mathbf{x}} \leq 1$. The resulting table of counts for the synthetic case-referent study has expectations shown in Table 3.4.

The key observation is that the odds ratio computed from Table 3.4 equals the odds ratio (3.5) before case-referent sampling. In other words, in the absence of hidden bias, the data from a synthetic case-referent study provide a direct estimate of the population odds ratio (3.5) at each $\mathbf{x}$.

As in §3.2, when attention shifts from the population of M subjects to the $N < M$ subjects included in the case-referent study, the notation in §3.2.3 is used; that is, the ith subject in the sth stratum defined by $\mathbf{x}$ has unbracketed subscript s, i. As in §3.2, let $m_s = \sum_i Z_{si}$ be the number of treated or exposed subjects in stratum s of the case-referent study. If the study were free of hidden bias, so (3.2) holds, and if the treatment had no effect, so $r_{Tsi} = r_{Csi}$ for each s, i, then after synthetic case-referent sampling, the conditional distribution of the treatment assignments $\mathbf{Z}$ given $\mathbf{m}$ is uniform on Ω. This means, for instance, that the Mantel–Haenszel statistic may be used to test the null hypothesis of no effect in a synthetic case-referent study that is free of hidden bias.

3.3.3 Selection Bias in Case-Referent Studies

Unlike the synthetic study in §3.3.2, most case-referent studies do not use random sampling. It is common to use all cases that are made available by some process, for instance, all new cases admitted to one or more hospitals in a given time interval, together with referents selected by an ostensibly similar process, for instance, patients with some other illness admitted to the same hospitals at the same time, or neighbors or coworkers of the cases.

Nonrandom selections of cases and referents may distort the odds ratio in (3.5). For instance, if cases of lung cancer at a hospital were compared to referents who were selected as patients with cardiac disease in the same hospital, the odds ratio linking lung cancer with cigarette smoking would be too small, because smoking causes both lung cancer and cardiac disease. In this case, there is a *selection bias*, that is, a bias that was not the result of the manner in which subjects were assigned to treatment in the population, but rather a bias introduced by the nonrandom process of selecting cases and referents. Selection bias is discussed further in Chapter 8.

3.4   *Small Sample Inference with an Unknown Propensity Score

3.4.1   *Conditional Inference Under a Logit Model for the Propensity Score

In §3.2, in the absence of hidden bias, the distribution of treatment assignments $\text{prob}(\mathbf{Z} = \mathbf{z})$ in (3.2) was unknown because the propensity score $\boldsymbol{\lambda} = (\lambda_1, \dots, \lambda_S)^\mathrm{T}$ was unknown. However, by conditioning on the number of treated subjects in each stratum $\mathbf{m}$, the conditional distribution of treatment assignments, $\text{prob}(\mathbf{Z} = \mathbf{z} | \mathbf{m})$, was known, and, in fact, was the distribution of treatment assignments in a uniform randomized experiment. Notice that the unknown parameter $\boldsymbol{\lambda}$ was eliminated by conditioning on a sufficient statistic $\mathbf{m}$. This line of reasoning generalizes.

Suppose that $\boldsymbol{\lambda}$ satisfies a logit model,

$$\log\left(\frac{\lambda_s}{1 - \lambda_s}\right) = \boldsymbol{\beta}^\mathrm{T}\mathbf{x}_s, \tag{3.6}$$

where $\boldsymbol{\beta}$ is an unknown parameter. Write

$$\bar{\mathbf{m}} = \sum_{s=1}^{S} m_s \mathbf{x}_s, \tag{3.7}$$

so $\bar{\mathbf{m}}$ is the sum of the $\mathbf{x}_s$ weighted by the number of treated subjects m_s in stratum s. Under the model (3.6), $\bar{\mathbf{m}}$ is sufficient for $\boldsymbol{\beta}$, so $\text{prob}(\mathbf{Z} = \mathbf{z} | \bar{\mathbf{m}})$ is a known distribution, free of the unknown parameter $\boldsymbol{\beta}$. See Cox (1970) for a detailed discussion of logit models, including a discussion of the sufficiency of $\bar{\mathbf{m}}$.

Let $\bar{\Omega}$ be the set containing all treatment assignments $\mathbf{z}$ that give rise to the same value of $\bar{\mathbf{m}}$, that is,

$$\bar{\Omega} = \left\{ \mathbf{z} : z_{si} \in \{0, 1\}, \ i = 1, \dots, n_s, s = 1, \dots, S, \ \bar{\mathbf{m}} = \sum_{s=1}^{S} \mathbf{x}_s \sum_{i=1}^{n_s} z_{si} \right\}.$$

Notice that $\bar{\Omega}$ is a larger set than Ω, in the sense that $\Omega \subseteq \bar{\Omega}$, so $\bar{\Omega}$ contains at least as many treatment assignments $\mathbf{z}$ as Ω. This is true because every $\mathbf{z} \in \Omega$ gives rise to the same $\mathbf{m}$, and hence also to the same $\bar{\mathbf{m}}$.

Under the logit model (3.6), the conditional distribution of treatment assignments given $\bar{\mathbf{m}}$ is constant on $\bar{\Omega}$,

$$\mathrm{prob}(\mathbf{Z} = \mathbf{z} | \bar{\mathbf{m}}) = \frac{1}{|\bar{\Omega}|} \qquad \text{for each} \quad \mathbf{z} \in \bar{\Omega}. \tag{3.8}$$

As a result, in the absence of hidden bias, under the model (3.6), the known distribution (3.8) forms the basis for a permutation test; in particular, a test statistic $T = t(\mathbf{Z}, \mathbf{r})$ has significance level

$$\mathrm{prob}\{t(\mathbf{Z}, \mathbf{r}) \geq T | \bar{\mathbf{m}}\} = \frac{|\{\mathbf{z} \in \bar{\Omega} : t(\mathbf{z}, \mathbf{r}) \geq T\}|}{|\bar{\Omega}|}, \tag{3.9}$$

for testing the null hypothesis of no treatment effect. The significance level (3.9) is the proportion of treatment assignments $\mathbf{z}$ in $\bar{\Omega}$ giving a value of the test statistic $t(\mathbf{z}, \mathbf{r})$ at least equal to the observed value of T.

The procedure just described is useful in small samples when the strata are so thin that the set Ω is too small to be useful. For instance, if Ω contains fewer than 20 treatment assignments, then no pattern of responses can be significant at the 0.05 level using (2.5). Such a small Ω arises when the sample size N is small and this small sample size is thinly spread over many strata; for instance, strata with $m_s = 0$ or $m_s = n_s$ contribute nothing to (2.5). If Ω is too small to permit reasonable permutation inferences, the larger set $\bar{\Omega}$ may be used instead. The size $|\bar{\Omega}|$ of $\bar{\Omega}$ depends on how the covariates $\mathbf{x}$ are coded, and finding an $\bar{\Omega}$ of appropriate size may sometimes be accomplished by adjusting the coding of the covariates. An example of the method is given in the next section.

Exact stratification and matching in §3.2 are useful only if there are many strata or matched sets that contain at least $n_s \geq 2$ subjects with at least one treated subject and one control. In contrast, the method in the current section may be used when $n_s = 1$ for every s, so each unit may have a distinct $\mathbf{x}$.

The method of this section is a generalization of the method in §3.2. Suppose that $\mathbf{x}$ simply contains indicators coding the S strata. For instance, in one such coding, $\mathbf{x}$ has $S - 1$ binary coordinates, with coordinate $x_{sij} = 1$ if $j = s$ and $x_{sij} = 0$ if $j \neq s$, for $j = 1, \dots, S - 1$. In this case, $\bar{\Omega} = \Omega$, and (3.9) equals (2.5).

TABLE 3.5. Fourteen Lung Cancer Patients from a Phase II Trial of Pacco.

ID	Tumor Response r_{si}	Treatment Z_{si}	Cell Type	Previous Treatment	Performance Status	s
1	0	0	Squamous	None	0	1
2	0	0	Large cell	None	1	2
3	0	0	Squamous	Radiation	1	3
4	0	0	Squamous	Radiation	1	3
5	0	0	Squamous	Radiation	2	4
6	1	1	Squamous	Radiation	1	3
7	0	1	Squamous	Radiation	1	3
8	0	1	Adeno.	Radiation	1	5
9	1	1	Squamous	None	1	6
10	0	1	Large cell	None	2	7
11	0	1	Squamous	Radiation and chemotherapy	2	8
12	0	1	Squamous	Chemotherapy	1	9
13	0	1	Squamous	None	0	1
14	2	1	Squamous	None	1	6

3.4.2   *An Example: A Small Observational Study Embedded in a Clinical Trial

Table 3.5 describes 14 patients taken from a clinical trial of the drug treatment combination PACCO in the treatment of nonsmall cell bronchogenic carcinoma (Whitehead, Rosenbaum, and Carbone 1984, Rosenbaum 1984b, §3). This phase II trial contained two minor variations of what was intended to be the same treatment; however, when the responses were tabulated, all of the patients who responded to therapy had received the same variation of the treatment. The question is whether this is evidence that the treatments differ, given the characteristics of the patients involved. As presented here, the example is adapted to illustrate the method.

The outcome is tumor response, where 0 signifies no response, 1 signifies a partial response, and 2 signifies a complete response. The covariate information describes the cell type, previous treatment, and performance status. In $\mathbf{x}_s$ previous treatment was coded as three binary variables, cell type as two binary variables, and performance status was taken as a single variable with three scored categories, so $\mathbf{x}_s$ has dimension six.

Table 3.6 describes the strata based on the covariates. The 14 patients are divided into nine strata, with most strata containing a single patient. The set Ω has $2 \times 1 \times 6 \times 1 \times 1 \times 1 \times 1 \times 1 = 12$ treatment assignments, so no matter what responses are observed, the smallest possible significance

TABLE 3.6. Strata for the PACCO Trial.

s	Cell Type	Previous Treatment	Performance Status	n_s	m_s	$\binom{n_s}{m_s}$
1	Squamous	None	0	2	1	2
2	Large cell	None	1	1	0	1
3	Squamous	Radiation	1	4	2	6
4	Squamous	Radiation	2	1	0	1
5	Adeno.	Radiation	1	1	1	1
6	Squamous	None	1	2	2	1
7	Large cell	None	2	1	1	1
8	Squamous	Radiation and chemotherapy	2	1	1	1
9	Squamous	Chemotherapy	1	1	1	1

level is $1/12 = 0.083$. The set Ω is too small to be useful for a permutation test.

The set $\bar{\Omega}$ is somewhat larger, containing 28 treatment assignments. These are all the treatment assignments that give rise to the observed value of $\bar{\mathbf{m}}$. This means that a treatment assignment $\mathbf{z}$ must have the correct number of treated patients with each cell type, each previous treatment, and the correct average performance status. It contains all treatment assignments $\mathbf{z}$ such that:

(i) nine patients receive treatment one, and of those nine:

(ii) one has adenocarcinoma;

(iii) one has large cell carcinoma;

(iv) seven have squamous cell carcinoma;

(v) three have had only previous radiation therapy;

(vi) one has had only previous chemotherapy;

(vii) one has had both previous chemotherapy and previous radiation therapy; and

(viii) the average performance status is $10/9 = 1.1$.

In other words, these 28 treatment assignments resemble the observed treatment assignment in the sense that similar patients received the treatment. The test statistic is the total of the response scores for treated patients, $T = t(\mathbf{Z}, \mathbf{r}) = \mathbf{Z}^T\mathbf{r} = 2 + 1 + 1 = 4$. Of the 28 treatment assignments $\mathbf{z}$ in $\bar{\Omega}$, 8 have $\mathbf{z}^T\mathbf{r} = 4$, 13 have $\mathbf{z}^T\mathbf{r} = 3$, 4 have $\mathbf{z}^T\mathbf{r} = 2$, and 3 have $\mathbf{z}^T\mathbf{r} = 1$. Therefore, under the null hypothesis, the distribution (3.9) of

$\mathbf{Z}^T\mathbf{r}$ assigns probability $8/28 = 0.29$ to 4, $13/28 = 0.46$ to 3, $4/28 = 0.14$ to 2, and $3/28 = 0.11$ to 1. In other words, it is not surprising to find that all the responses occurred in treatment group one—this would happen by chance 29% of the time if the treatments did not differ in their effects.

Algorithms for computations involving $\bar{\Omega}$ are discussed in Rosenbaum (1984b).

3.5   *Large Sample Inference with an Unknown Propensity Score

3.5.1   *Covariance Adjustment of Randomization Tests

This section describes a method of covariate adjustment for randomization tests. The method is simple to describe and to apply, and motivation follows the description. It turns out that the method is a large sample approximation to the exact test in §3.4.

As in §3.4, suppose a logit model accurately describes the propensity score,

$$\log\left(\frac{\lambda_s}{1-\lambda_s}\right) = \boldsymbol{\beta}^T\mathbf{x}_s, \tag{3.10}$$

and suppose the study is free of hidden bias, so that $\mathrm{prob}(Z_{si} = 1) = \pi_{si} = \lambda(\mathbf{x}_s) = \lambda_s$. As in §3.4.1, each stratum may consist of a single unit, $n_s = 1$, so that each unit may have a distinct value of $\mathbf{x}$.

Under the null hypothesis of no treatment effect, $H_0 : r_{Tsi} = r_{Csi} = r_{si}$ for all s, i, the observed responses $\mathbf{R}$ equal a fixed vector $\mathbf{r}$ not varying with $\mathbf{Z}$. As in Chapter 2 and §3.2, when the null hypothesis of no effect is true, the fixed $\mathbf{R} = \mathbf{r}$ is observed no matter what $\mathbf{Z}$ is, so r_{si} does not predict Z_{si} at each $\mathbf{x}_s$. On the other hand, if the null hypothesis were false and instead $r_{Tsi} > r_{Csi}$ for each s, i, then the observed response, $R_{si} = Z_{si}r_{Tsi} + (1 - Z_{si})r_{Csi}$ would be positively related to Z_{si} at each $\mathbf{x}_s$. So the task is to check for a relationship between treatment Z_{si} and observed response R_{si} given $\mathbf{x}_s$, exploiting the assumed form (3.10).

Let $\mathbf{q}(\mathbf{R})$ be some way of scoring the observed responses, such as their ranks, so that under the null hypothesis of no treatment effect, $\mathbf{q}(\mathbf{R}) = \mathbf{q}(\mathbf{r}) = \mathbf{q}$, say, is fixed. Consider the model:

$$\log\left\{\frac{\mathrm{prob}(Z_{si} = 1)}{\mathrm{prob}(Z_{si} = 0)}\right\} = \boldsymbol{\beta}^T\mathbf{x}_s + \theta q_{si}. \tag{3.11}$$

By what has been said, under the null hypothesis of no treatment effect, in the absence of hidden bias, assuming model (3.10), it follows that $\theta = 0$ in (3.11).

There are several ways to test $H_0 : \theta = 0$ in (3.11). There is an exact, uniformly most powerful unbiased test of $H_0 : \theta = 0$ versus $H_A : \theta > 0$; see Cox (1970, §4.2). As it turns out, this test is precisely the test in §3.4 with significance level (3.9) providing $t(\mathbf{Z}, \mathbf{r}) = \mathbf{Z}^T \mathbf{q}$; see Rosenbaum (1984b, §4.2). In other words, so far, this section has not yielded a new procedure, but rather a new motivation for the procedure in §3.4. The most powerful unbiased test is not always practical, however. For instance, $\bar{\Omega}$ may be too large or too small to permit practical use of (3.9).

When N is large compared to the dimension of $\mathbf{x}$, it is more common to test a hypothesis about a coefficient in a logit model, such as $H_0 : \theta = 0$ in (3.11), using one of several large sample tests associated with maximum likelihood estimation of the model. These tests may also be applied here.

Suppose, again, that the study is free of hidden bias and (3.10) describes the propensity score, but instead of the null hypothesis of no effect, consider the model of an additive treatment effect, $H_0 : r_{Tsi} = r_{Csi} + \tau$. In this case, the observed response $R_{si} = r_{Csi} + \tau Z_{si}$ will be related to Z_{si} at each $\mathbf{x}_s$, positively related for $\tau > 0$, and negatively related for $\tau < 0$. Consider testing the hypothesis $H_0 : \tau = \tau_0$. The procedure is analogous to that in §2.6.1. Specifically, calculate the adjusted responses, $\mathbf{r} = \mathbf{R} - \tau_0 \mathbf{Z}$, which are fixed when the null hypothesis is true, and apply the method above with $\mathbf{q} = \mathbf{q}(\mathbf{r}) = \mathbf{q}(\mathbf{R} - \tau_0 \mathbf{Z})$. A confidence interval for τ is found by inverting the test.

3.5.2   *Example: Benzene Exposure Among Shoe Workers

Tunca and Egeli (1996) studied the effects of benzene exposure among shoe workers. The $58 = \sum m_s$ shoe workers, $Z = 1$, in Bursa, Turkey were exposed to benzene from glues. They were compared to $20 = \sum n_s - m_s$ controls, $Z = 0$, from the same region who were not exposed to benzene. The outcome, R, is one measure of chromosome damage, namely, the percentage of cells with breaks. Adjustment is made for a three-dimensional covariate, $\mathbf{x}$, giving age, alcohol intake ($+$ or $-$), and smoking in packs per day. Table 3.7 gives the data for 6 of the $78 = 58 + 20$ subjects, with data for all subjects given in Tunca and Egeli (1996). Casual examination of means shows the shoe workers have many more breaks than the controls, but they are also older (37.2 years versus 28.5 years), smoke more (0.8 packs per day versus .2 packs per day), and are more likely to drink alcohol (40% versus 5%).

To test the null hypothesis of no treatment effect, the model (3.11) was fit, predicting treatment Z from the three covariates and the ranks of the percentage of breaks R. When a logit model is fitted to independent binary trials by maximum likelihood, a common test of the hypothesis that a coefficient is zero uses the ratio of the coefficient to its approximate standard error, and compares that ratio to the standard Normal distribution. For age, alcohol, smoking, and the ranks of the percentages of breaks, these

TABLE 3.7. Chromosome Breaks Among Shoe Workers and Controls.

Treatment Z	Breaks R	Age	Alcohol	Smoking
0	0.00	24	−	0
0	0.00	31	−	0
⋮	⋮	⋮	⋮	⋮
0	5.00	23	−	0.1
⋮	⋮	⋮	⋮	⋮
1	11.11	26	−	1.0
⋮	⋮	⋮	⋮	⋮
1	0.00	50	+	1.5
1	9.09	40	−	0

ratios are, respectively, 2.96, 1.72, 1.59, and 2.18. Because $2.18 \geq 1.65$, and $1 - \Phi(1.65) = 0.05$, where $\Phi(\cdot)$ is the standard Normal cumulative distribution, the hypothesis $H_0 : \theta = 0$ may be rejected at the 0.05 level in a one-sided test. This is an appropriate large sample test of the null hypothesis of no treatment effect assuming the study is free of hidden bias and (3.10) accurately describes the propensity score.

The model of an additive treatment effect, $H_0 : r_{Tsi} = r_{Csi} + \tau$, is not useful—indeed, it is misleading—in this example. Many of the shoe workers and nearly all of the controls had $R = 0.00$ breaks, even though many other shoe workers had substantial numbers of breaks. As just noted, the hypothesis $H_0 : \tau = 0$ was rejected at the 0.05 level. However, the hypothesis $H_0 : \tau = \tau_0$ is not rejected in a one-sided 0.05 level test for *every* $\tau_0 > 0$, no matter how small τ_0 is. The reason is that if a positive quantity is subtracted from the many $R = 0.00$ values among shoe workers, these values become strictly smaller than all control responses. A better model for the treatment effect in data of this sort is discussed in §5.6.

3.6   *Inexact Matching Followed by Stratification

3.6.1   *An Example: Vasectomy and Myocardial Infarction

Walker et al. (1981) studied the possible effect of vasectomy on increased risk of myocardial infarction (MI), a possibility suggested by animal studies where increased risks were observed. The study contained 4830 pairs of men, one vasectomized and one control, matched for year of birth and calendar time of follow-up. The data were not matched for two other variables, obesity and smoking history recorded as binary traits, both of which are believed to be related to the risk of MI. This section describes a method for controlling covariates that were not controlled by the matching.

The outcome is a binary variable indicating whether an MI occurred during the follow-up period; however, the method discussed here may be used with outcomes of any kind. McNemar's test statistic is used; see §2.4.3 for discussion of this test. In most of the 4830 pairs, no MI occurred. Pairs containing no MI or two MIs are said to be concordant, and it is not difficult to verify that these pairs do not contribute to McNemar's test. There were 36 discordant pairs, that is, pairs in which one person had an MI and the other did not—only these affect the test. Walker's (1982) data for these 36 pairs of men are given in Table 3.8. The score q_{si} is 1 if the ith man in pair s had an MI and is 0 otherwise.

In each pair, there are two possibilities. A pair may be exactly matched for obesity and smoking, so the two matched men are the same on these variables, or else the men may differ. If they differ, then there are six ways they may differ; that is, there are six patterns of imbalance in the covariates. For instance, one possible imbalance occurs if one man in a pair is a nonsmoker who is not obese and the other is a smoker who is not obese; this is (os, oS) in Table 3.8. Notice that, in counting the patterns of imbalance in covariates, we consider only the covariates and not vasectomy or MI. The 36 pairs are grouped into seven classes, one class that is perfectly matched and six classes for the six patterns of imbalance.

3.6.2 *Adjusting Inexactly Matched Pairs by Stratifying Pairs

Suppose that an observational study is free of hidden bias, so (3.2) holds, and pairs of treated and control units are matched inexactly for $\mathbf{x}$, so $\mathbf{x}_{s1}$ may not equal $\mathbf{x}_{s2}$. How does one control imbalances in $\mathbf{x}$ that remain after matching? The method described in this section is useful when matching has failed to control a few coordinates of $\mathbf{x}$ containing discrete variables, as in §3.6.1. It involves grouping the matched pairs into classes so that the pairs in the same class have the same pattern of covariate imbalance.

Using (3.2), for any two distinct units j and k, $\text{prob}(Z_{[j]} = 1, Z_{[k]} = 0) = \lambda\left(\mathbf{x}_{[j]}\right)\left\{1 - \lambda\left(\mathbf{x}_{[k]}\right)\right\}$ and $\text{prob}(Z_{[j]} = 0, Z_{[k]} = 1) = \lambda\left(\mathbf{x}_{[k]}\right)\left\{1 - \lambda\left(\mathbf{x}_{[j]}\right)\right\}$. Pair matching selects units so that $m_s = Z_{s1} + Z_{s2} = 1$ for each s. Therefore, conditionally given that a pair contains exactly one treated unit, the chance that the first unit is the treated unit is

$$\text{prob}(Z_{s1} = 1 | Z_{s1} + Z_{s2} = 1) = \frac{\lambda(\mathbf{x}_{s1})\{1 - \lambda(\mathbf{x}_{s2})\}}{\lambda(\mathbf{x}_{s1})\{1 - \lambda(\mathbf{x}_{s2})\} + \lambda(\mathbf{x}_{s2})\{1 - \lambda(\mathbf{x}_{s1})\}}.$$
(3.12)

In many applications, some coordinates of $\mathbf{x}$ are matched and others are not. In the vasectomy and MI example, men were matched for year of birth and calendar time of follow-up, but not for smoking and obesity. In this situation, it seems natural to let the matching control the matched coordinates of $\mathbf{x}$ and to make additional adjustments only for the unmatched

TABLE 3.8. Vasectomy and Myocardial Infarction

Class c	Covariate imbalance $(\mathbf{x}_{s1}, \mathbf{x}_{s2})$	q_{s1}	q_{s2}	Z_{s1}	Walker's id#, s
0	No imbalance	1	0	0	4
		1	0	1	8
		1	0	1	10
		0	1	1	11
		1	0	0	12
		1	0	1	16
		0	1	0	17
		0	1	0	23
		1	0	0	26
		0	1	0	27
		1	0	1	30
		0	1	0	35
		1	0	1	36
1	(os, oS)	0	1	0	3
		1	0	1	6
		0	1	1	9
		0	1	1	14
		0	1	0	15
		0	1	1	20
		0	1	1	21
		1	0	0	22
		0	1	0	24
		0	1	1	29
		0	1	0	32
2	(os, Os)	0	1	1	1
		0	1	0	28
		1	0	0	33
3	(oS, OS)	0	1	0	19
		0	1	0	25
4	(oS, Os)	0	1	0	2
		0	1	0	7
		1	0	0	34
5	(oS, OS)	0	1	0	13
		1	0	0	18
		0	1	1	31
6	(Os, OS)	0	1	1	5

Key: O = obese, o = not obese, S = smoker, s = nonsmoker.

coordinates of $\mathbf{x}$. Write $\mathbf{x} = (\overline{\mathbf{x}}, \widetilde{\mathbf{x}})$ where subjects are matched for $\overline{\mathbf{x}}$ but not for $\widetilde{\mathbf{x}}$. Then $\overline{\mathbf{x}}_{s1} = \overline{\mathbf{x}}_{s2}$ for every s, but $\widetilde{\mathbf{x}}_{s1} \neq \widetilde{\mathbf{x}}_{s2}$ for at least some s. Ideally, matched pairs could be grouped into classes based on the imbalances in the unmatched coordinates, $\widetilde{\mathbf{x}}_{s1}$ and $\widetilde{\mathbf{x}}_{s2}$, ignoring the matched coordinates, $\overline{\mathbf{x}}_{s1} = \overline{\mathbf{x}}_{s2}$, as was done in Table 3.12 where pairs are grouped based on obesity and smoking, ignoring year of birth and calendar time of follow-up. As intuition might suggest, we can control imbalances in $\overline{\mathbf{x}}$ by matching on $\overline{\mathbf{x}}$, then separately control the remaining imbalances in $\widetilde{\mathbf{x}}$ by classifying the pairs, only if $\overline{\mathbf{x}}$ and $\widetilde{\mathbf{x}}$ do not interact with each other in determining $\lambda(\mathbf{x})$. Specifically, consider the following additive logit model for the treatment assignment probabilities, $\lambda(\mathbf{x})$:

$$\lambda(\mathbf{x}) = \frac{\exp\{\xi(\overline{\mathbf{x}}) + \zeta(\widetilde{\mathbf{x}})\}}{1 + \exp\{\xi(\overline{\mathbf{x}}) + \zeta(\widetilde{\mathbf{x}})\}} \tag{3.13}$$

for some unknown functions $\xi(\cdot)$ and $\zeta(\cdot)$. Because $\overline{\mathbf{x}}_{s1} = \overline{\mathbf{x}}_{s2}$, it follows that $\xi(\overline{\mathbf{x}}_{s1}) = \xi(\overline{\mathbf{x}}_{s2})$, so that, substituting (3.13) into (3.12) and simplifying yields:

$$\mathrm{prob}(Z_{s1} = 1 | Z_{s1} + Z_{s2} = 1) = \frac{\exp\{\zeta(\widetilde{\mathbf{x}}_{s1})\}}{\exp\{\zeta(\widetilde{\mathbf{x}}_{s1})\} + \exp\{\zeta(\widetilde{\mathbf{x}}_{s2})\}}, \tag{3.14}$$

which depends only on the unmatched coordinates, $\widetilde{\mathbf{x}}$, not on the matched coordinates, $\overline{\mathbf{x}}$. This sort of simplification is often possible with logit models; see Cox (1970). In other words, under model (3.13), matching on $\overline{\mathbf{x}}$ has removed all of the bias due to $\overline{\mathbf{x}}$, and only the bias due to $\widetilde{\mathbf{x}}$ remains. When model (3.13) holds, covariate imbalance refers to $\widetilde{\mathbf{x}}$ only, so the 11 pairs in Table 3.8 with a nonobese nonsmoker matched to a nonobese smoker, (os, oS), all have the same pattern of covariate imbalance, and all 11 pairs may be placed in the same class for adjustments. When model (3.13) does not hold, the method of this section may still be applied, but many more classes are needed. This is because pairs with different values of the matched covariates, $\overline{\mathbf{x}}$, must be placed in different classes, so the 11 pairs in Table 3.8 with a nonobese nonsmoker matched to a nonobese smoker, (os, oS), would have to be divided up into different classes based on year of birth and calendar time of follow-up. The discussion that follows applies to both cases, that is, whether or not model (3.13) holds; however, the definition of covariate imbalance and the classes that control it do depend on whether model (3.13) holds. Specifically, when model (3.13) holds, a pattern of covariate imbalance refers to unmatched coordinates $(\widetilde{\mathbf{x}}_{s1}, \widetilde{\mathbf{x}}_{s2})$ only, but when this model does not hold, a pattern of covariate imbalance refers to all coordinates $(\mathbf{x}_{s1}, \mathbf{x}_{s2})$ whether matched or not.

Divide the S pairs into $C + 1$ classes, where class 0 contains the exactly matched pairs with $\mathbf{x}_{s1} = \mathbf{x}_{s2}$, and the other C classes contain the C patterns of imbalance in $\mathbf{x}$. Let l_c be the set of pairs exhibiting imbalance c, so that $l_0 \cup l_1 \cup \cdots \cup l_C = \{1, \ldots, S\}$. In class $c = 0$, there is an exact match; that is, $\mathbf{x}_{s1} = \mathbf{x}_{s2}$ for $s \in l_0$. In the other classes, there is an imbalance, $\mathbf{x}_{s1} \neq \mathbf{x}_{s2}$ for $s \in l_c$ for $1 \leq c \leq C$. Write $\tilde{n}_c$ for the number of pairs in l_c.

Renumber the two units in each pair so that every pair s in class c has the same value of $\tilde{\mathbf{x}}_{s1}$, and every pair in class c has the same value of $\tilde{\mathbf{x}}_{s2}$, as in Table 3.8. This notational change simplifies the appearance of various quantities but it does not change their values.

With this notation, (3.12) takes the same value for all pairs in the same class. Write ρ_c for the common value of (3.12) or (3.14) in class c; that is, $\mathrm{prob}(Z_{s1} = 1 | Z_{s1} + Z_{s2} = 1) = \rho_c$ for all $s \in l_c$. Notice that $\rho_0 = \frac{1}{2}$, but the other ρ_c are unknown since $\lambda(\mathbf{x})$ is unknown.

For $c = 1, \ldots, C$, write $\tilde{m}_c$ for the number of pairs in class s in which the treated unit is the first unit,

$$\tilde{m}_c = \sum_{s \in l_c} Z_{s1} \quad \text{and write} \quad \tilde{\mathbf{m}} = \begin{bmatrix} \tilde{m}_1 \\ \vdots \\ \tilde{m}_C \end{bmatrix}.$$

Then $\tilde{m}_c$, like m_s, is a random variable, since it depends on the Z's. Note the distinction between $\mathbf{m}$ and $\tilde{\mathbf{m}}$, both of which are used below. With this notation, the distribution of treatment assignments within pairs is

$$\mathrm{pr}(\mathbf{Z} = \mathbf{z} | \mathbf{m}) = \prod_{c=0}^{C} \prod_{s \in l_c} \rho_c^{z_{s1}} \{1 - \rho_c\}^{z_{s2}},$$

$$= \left(\frac{1}{2}\right)^{\tilde{n}_0} \prod_{c=1}^{C} \rho_c^{\tilde{m}_c} \{1 - \rho_c\}^{\tilde{n}_c - \tilde{m}_c}. \tag{3.15}$$

Unlike the distribution of treatment assignments in §3.2.3 for exact matching, the distribution (3.15) for inexact matching involves unknown parameters, the ρ_c, reflecting the remaining imbalances in $\mathbf{x}$.

Consider the conditional distribution of the treatment assignments $\mathbf{Z}$ given both $\mathbf{m}$ and $\tilde{\mathbf{m}}$. As will now be seen, conditioning on both $\mathbf{m}$ and $\tilde{\mathbf{m}}$ yields a known distribution free of the unknown ρ_c. It is a distribution on a set $\tilde{\Omega} \subseteq \Omega$, where $\mathbf{z} \in \tilde{\Omega}$ if and only if:

(i) $\mathbf{z}$ is a treatment assignment for S matched pairs; that is, $z_{s1} + z_{s2} = m_s = 1$ or, equivalently, $\mathbf{z} \in \Omega$; and

(ii) $\mathbf{z}$ exhibits the same degree of imbalance as the observed data; that is, $\tilde{m}_c = \sum_{s \in l_c} z_{s1}$ for $c = 1, \ldots, C$.

The set $\tilde{\Omega}$ has

$$\tilde{K} = 2^{\tilde{n}_0} \prod_{c=1}^{C} \binom{\tilde{n}_c}{\tilde{m}_c}$$

elements, and $\text{pr}(\mathbf{Z} = \mathbf{z}|\mathbf{m}, \tilde{\mathbf{m}}) = 1/\tilde{K}$ for each $\mathbf{z} \in \tilde{\Omega}$. In class $c = 0$ containing $\tilde{n}_0$ pairs, all $2^{\tilde{n}_0}$ possible assignments are equally likely. In class $c \geq 1$ containing $\tilde{n}_c$ pairs, the assignments give the treatment to the first unit in exactly $\tilde{m}_c$ pairs, and there are $\binom{\tilde{n}_c}{\tilde{m}_c}$ such assignments.

Using the known distribution $\text{pr}(\mathbf{Z} = \mathbf{z}|\mathbf{m}, \tilde{\mathbf{m}})$, significance levels are obtained in a manner similar to that in §2.4.1. Under the null hypothesis of no treatment effect, the statistic $T = t(\mathbf{Z}, \mathbf{r})$ has significance level

$$\text{prob}\{t(\mathbf{Z}, \mathbf{r}) \geq T|\mathbf{m}, \tilde{\mathbf{m}}\} = \frac{|\{\mathbf{z} \in \tilde{\Omega} : t(\mathbf{z}, \mathbf{r}) \geq T\}|}{\tilde{K}}, \quad (3.16)$$

which parallels (2.5) and is simply the proportion of treatment assignments in $\tilde{\Omega}$ giving values of the test statistic at least as large as the observed value.

If the test statistic is a sum statistic in the sense of §2.4.4, that is, if $t(\mathbf{Z}, \mathbf{r}) = \mathbf{Z}^{\mathrm{T}}\mathbf{q}$, where $\mathbf{q}$ is a function of $\mathbf{r}$, then the null expectation and variance of the test statistic are given in the following proposition. The proposition assumes the study is free of hidden bias in the sense that (3.2) holds, and it concerns the conditional expectation and variance of a sum statistic given $\mathbf{m}$, $\tilde{\mathbf{m}}$, that is, the expectation and variance of $t(\mathbf{Z}, \mathbf{r}) = \mathbf{Z}^{\mathrm{T}}\mathbf{q}$ over the distribution (3.16).

Proposition 11 *If the study is free of hidden bias, then the null expectation and variance of* $\mathbf{Z}^{\mathrm{T}}\mathbf{q}$ *are*

$$E(\mathbf{Z}^{\mathrm{T}}\mathbf{q}|\mathbf{m}, \tilde{\mathbf{m}}) = \sum_{s \in l_0} \frac{q_{s1} + q_{s2}}{2} + \sum_{c=1}^{C} \tilde{m}_c \tilde{\mu}_{c1} + (\tilde{n}_c - \tilde{m}_c)\tilde{\mu}_{c2}$$

$$\text{var}(\mathbf{Z}^{\mathrm{T}}\mathbf{q}|\mathbf{m}\tilde{\mathbf{m}}) = \sum_{s \in l_0} \frac{(q_{s1} - q_{s2})^2}{4} + \sum_{c=1}^{C} \frac{\tilde{m}_c(\tilde{n}_c - \tilde{m}_c)\tilde{\sigma}_c^2}{\tilde{n}_c}, \quad (3.17)$$

where

$$\tilde{\mu}_{ci} = \frac{1}{\tilde{n}_c} \sum_{s \in l_c} q_{si} \quad for \quad i = 1, 2$$

and

$$\tilde{\sigma}_c^2 = \frac{1}{\tilde{n}_c - 1} \sum_{s \in l_c} \{(q_{s1} - q_{s2}) - (\tilde{\mu}_{c1} - \tilde{\mu}_{c2})\}^2.$$

Proof. The proof makes use of several elementary observations. First, the conditional distribution of $\mathbf{Z}$ given $\mathbf{m}$, $\tilde{\mathbf{m}}$ is uniform on $\tilde{\Omega}$, and as a result, the Z_{si}'s in different classes are independent of each other. Second, write $\mathbf{Z}^{\mathrm{T}}\mathbf{q}$ as

$$\mathbf{Z}^{\mathrm{T}}\mathbf{q} = \sum_{c=0}^{C} \sum_{s \in l_c} Z_{s1}(q_{s1} - q_{s2}) + \sum_{c=0}^{C} \sum_{s \in l_c} q_{s2}, \quad (3.18)$$

where the second sum on the right is a constant since it does not involve $\mathbf{Z}$. In class $c = 0$, the Z_{s1}'s are independent of each other given $\mathbf{m}$, $\tilde{\mathbf{m}}$, and each Z_{s1} equals 1 or 0 with probability $1/2$. For $c \geq 1$, the sum $\sum_{s \in l_c} Z_{s1}(q_{s1} - q_{s2})$ in (3.18) is the sum of $\tilde{m}_c$ of the $(q_{s1} - q_{s2})$'s randomly selected without replacement from among the $\tilde{n}_c$ pairs $s \in l_c$. The proposition then follows directly from standard facts about simple random sampling without replacement. ∎

3.6.3 *Return to the Example of Vasectomy and Myocardial Infarction

The 36 pairs in Table 3.8 are divided into seven classes, numbered $0, 1, \ldots, 6$, based on the pattern of imbalance in two unmatched covariates, obesity and smoking. In class 0, the two men are exactly matched for obesity and smoking. There are $\tilde{n}_0 = 13$ exactly matched pairs in class 0. In class 1, labeled (os, oS), each pair contains two men who are not obese, of whom exactly one is a smoker, and there are $\tilde{n}_1 = 11$ pairs with this imbalance, and so on. Pair $s = 3$ is the first pair in class $c = 1$, labeled (os, oS), and in this pair, the second man is a nonobese smoker, oS, who had an MI, $q_{32} = 1$, and a vasectomy, $Z_{32} = 1 = 1 - Z_{31}$.

McNemar's statistic is the number of vasectomized men who had an MI, namely, $\sum_s \sum_i Z_{si}q_{si} = 20$ of a possible 36. Proposition 11 gives the moments of the statistic adjusting for obesity and smoking in addition to the variables used to form matched pairs. Table 3.9 gives a few intermediate calculations. Using these in (3.17) gives an expectation of 19.015 and a variance of 6.813 for the McNemar statistic under the null hypothesis of no treatment effect. In words, the number of vasectomized men who had an MI, namely, 20, is quite close to the expectation 19.015 in the absence of a treatment effect. The standardized deviate with continuity correction is $(20 - 19.015 - \frac{1}{2})/\sqrt{6.813} = 0.186$, and this is small when compared with the standard normal distribution, so there is no indication of an effect of vasectomy on the risk of MI. This would be a correct test if the study were free of hidden bias once adjustments had been made for the two matched and the two unmatched covariates.

Had smoking and obesity been ignored, the usual expectation for McNemar's statistic would have been $36/2 = 18$ vasectomized MIs rather than 19.015, so the adjustment for smoking and obesity moved the expected count closer to the observed 20. In other words, if vasectomy had no effect on the risk of MI, we would nonetheless have expected more than half of the vasectomized men to exhibit MIs, because both vasectomy and MI are related to smoking and obesity in the data shown in Table 3.9.

TABLE 3.9. Intermediate Calculations for Vasectomy and Myocardial Infarction.

c	$\tilde{n}_c$	$\tilde{m}_c$	$\tilde{\mu}_{c1}$	$\tilde{\mu}_{c2}$	$\tilde{\sigma}_c^2$
0	13	6			
1	11	6	0.182	0.818	0.655
2	3	1	0.333	0.667	1.333
3	2	0	0.000	1.000	0.000
4	3	0	0.333	0.667	1.333
5	3	1	0.333	0.667	1.333
6	1	1	0.000	1.000	0.000

3.7 Bibliographic Notes

Direct adjustment is surveyed by Bishop, Fienberg, and Holland (1975, §4.3), Cochran and Rubin (1973), Fleiss (1981, §14), Kitagawa (1964), and Mosteller and Tukey (1977, §11). The analysis of matched pairs and matched sets is surveyed by Breslow and Day (1980, §5), Cochran and Rubin (1973), Fleiss (1981, §8), Gastwirth (1988, §11), and Kleinbaum, Kupper, and Morgenstern (1982, §18). Statistical procedures that may be derived from an assumption of random assignment of treatments within strata have long been applied to observational studies; see, for instance, the influential paper by Mantel and Haenszel (1959). Cochran (1965, §3.2) viewed matching, subclassification, and model-based adjustments as different ways of doing the same thing. In an important paper, Rubin (1977) demonstrated that if treatments are randomly assigned on the basis of a covariate, then adjustments for the covariate produce appropriate estimates of a treatment effect. Rubin (1978) develops related ideas from a Bayesian view. Rosenbaum (1984b, 1987a) obtains known permutation distributions by conditioning on a sufficient statistic for unknown assignment probabilities, as in §3.2.

The propensity score is proposed in Rosenbaum and Rubin (1983), and its link in §3.2.5 to permutation inference is discussed by Rosenbaum (1984b). Joffe and Rosenbaum (1999) survey and extend propensity score methods, discussing in particular propensity scores for doses of treatment and propensity scores in case-cohort studies; see also Imbens (2000).

Cole (1979, p. 16) says the first true case-referent study was conducted in 1926. Cornfield (1951) showed that case-referent sampling did not alter the odds ratio. He also argued that if the disease is rare, as it is in most case-referent studies, then the odds ratio approximates another measure, the relative risk. See also Greenhouse (1982). Mantel's (1973) paper is a careful discussion of case-referent studies; see also Hamilton (1979), Rosenbaum (1987b), and Holland and Rubin (1988). Further references concerning case-referent studies are given in Chapter 7.

Conditional tests given a sufficient statistic for the propensity score in §3.4 are discussed in Rosenbaum (1984b), along with the large sample approximation in §3.5; see also Robins, Mark, and Newey (1992) and Robins and Ritov (1997). Inexact matching followed by stratification in §3.6 is discussed in Rosenbaum (1988); however, that paper mistakenly does not mention the need for assumption (3.13) in the presented analysis of Walker's data, where the classes were based only on the unmatched covariates.

3.8 Problems

1. Problems 1 and 2 consider a test statistic motivated by linear regression, but instead of referring the statistic to a theoretical distribution, these problems consider its permutation distribution. Write $\mathbf{X}$ for the matrix with N rows, numbered $(s, i), s = 1, \ldots, S, i = 1, \ldots, n_s$ where row (s, i) is $\mathbf{x}_s^T$. Assume that $\mathbf{X}$ has more rows than columns and has rank equal to the number of columns. Consider the linear regression model $\mathbf{R} = \mathbf{X}\theta + \mathbf{Z}\tau + \mathbf{e}$, where $\mathbf{e}$ is a vector of unobserved errors, but do not assume the model is correct. The least squares estimate of τ under this model is

$$t(\mathbf{Z}, \mathbf{R}) = \frac{\mathbf{R}^T(\mathbf{I} - \mathbf{H})\mathbf{Z}}{\mathbf{Z}^T(\mathbf{I} - \mathbf{H})\mathbf{Z}}, \qquad \text{where} \quad \mathbf{H} = \mathbf{X}(\mathbf{X}^T\mathbf{X})^{-1}\mathbf{X}^T. \quad (3.19)$$

In a uniform randomized experiment (or in an observational study free of hidden bias), consider the permutation distribution (2.5) of this $t(\mathbf{Z}, \mathbf{R})$ under the null hypothesis of no treatment effect. Show that the significance level (2.5) for the covariance adjusted estimate $t(\mathbf{Z}, \mathbf{R})$ equals the significance level (2.5) for the total response in the treated group, $\mathbf{Z}^T\mathbf{r}$, which in turn equals the significance level for the difference in sample means (2.5). (Hint: How does $\mathbf{X}^T\mathbf{z}$ vary as $\mathbf{z}$ ranges over Ω?)

2. Continuing Problem 1, show that the same conclusion holds if the significance level is based on (3.9) rather than (2.5). That is, show that the significance level (3.9) with $t(\mathbf{Z}, \mathbf{R})$ given by the covariate adjusted estimate (3.19) equals the significance level (3.9) with $t(\mathbf{Z}, \mathbf{R})$ given by $\mathbf{Z}^T\mathbf{r}$. (Hint: How does $\mathbf{X}^T\mathbf{z}$ vary as $\mathbf{z}$ ranges over $\bar{\Omega}$?)   (Rosenbaum 1984b, §2.5)

3.9 References

Becker, H. S. (1996) The epistemology of qualitative research. In: *Ethnography and Human Development*, R. Jessor, A. Colby, and R. Shweder, eds., Chicago: University of Chicago Press, pp. 53–72.

Bishop, Y., Fienberg, S., and Holland, P. (1975) *Discrete Multivariate Analysis*. Cambridge, MA: MIT Press.

Blumer, H. (1969) The methodological position of symbolic interactionism. In: H. Blumer, *Symbolic Interactionism: Perspective and Method*. Berkeley: University of California Press.

Breslow, N. and Day, N. (1980) *The Analysis of Case-Control Studies*. Volume I of *Statistical Methods in Cancer Research*. Lyon, France: International Agency for Research on Cancer of the World Health Organization.

Canner, P. (1984) How much data should be collected in a clinical trial? *Statistics in Medicine*, **3**, 423–432.

Canner, P. (1991) Covariate adjustment of treatment effects in clinical trials. *Controlled Clinical Trials*, **12**, 359–366.

Cochran, W. G. (1957) The analysis of covariance. *Biometrics*, **13**, 261–281.

Cochran, W. G. (1965) The planning of observational studies of human populations (with Discussion). *Journal of the Royal Statistical Society*, Series **A**, **128**, 134–155.

Cochran, W. G. (1968) The effectiveness of adjustment by subclassification in removing bias in observational studies. *Biometrics*, **24**, 205–213.

Cochran, W. G. and Rubin, D. B. (1973) Controlling bias in observational studies: A review. *Sankya*, Series **A**, **35**, 417–446.

Cole, P. (1979) The evolving case-control study. *Journal of Chronic Diseases*, **32**, 15–27.

Coleman, J., Hoffer, T., and Kilgore, S. (1982) Cognitive outcomes in public and private schools. *Sociology of Education*, **55**, 65–76.

Cornfield, J. (1951) A method of estimating comparative rates from clinical data: Applications to cancer of the lung, breast and cervix. *Journal of the National Cancer Institute*, **11**, 1269–1275.

Cox, D. R. (1970) *The Analysis of Binary Data*. London: Methuen.

Cox, D. R. (1958) *The Planning of Experiments*. New York: Wiley.

Dehejia, R. H. and Wahba, S. (1999) Causal effects in nonexperimental studies: Reevaluating the evaluation of training programs. *Journal of the American Statistical Association*, **94**, 1053–1062.

Emerson, R. M. (1981) Observational field work. *Annual Review of Sociology*, **7**, 351–378.

Fleiss, J. (1981) *Statistical Methods for Rates and Proportions.* New York: Wiley.

Gastwirth, J. (1988) *Statistical Reasoning in Law and Public Policy.* New York: Academic.

Goldberger, A. and Cain, G. (1982) The causal analysis of cognitive outcomes in the Coleman, Hoffer, and Kilgore report. *Sociology of Education*, **55**, 103–122.

Greenhouse, S. (1982) Cornfield's contributions to epidemiology. *Biometrics*, **38S**, 33–46.

Hamilton, M. (1979) Choosing a parameter for 2×2 table or $2 \times 2 \times 2$ table analysis. *American Journal of Epidemiology*, **109**, 362–375.

Heckman, J. J., Ichimura, H., and Todd, P. (1998) Matching as an econometric evaluation estimator. *Review of Economic Studies*, **65**, 261–294.

Hirano, K., Imbens, G. W., and Ridder, G. (2000) Efficient estimation of average treatment effects using the estimated propensity score. National Bureau of Economic Research, Working Paper T0251.

Holland, P. and Rubin, D. (1988) Causal inference in retrospective studies. *Evaluation Review*, **12**, 203–231.

Imbens, G. W. (2000) The role of the propensity score in estimating dose-response functions. *Biometrika*, **87**, 706–710.

Joffe, M. M. and Rosenbaum, P. R. (1999) Propensity scores. *American Journal of Epidemiology*, **150**, 327–333.

Katz, J. (2001) Analytic induction. In: *International Encyclopedia of Social and Behavioral Sciences,* Oxford: Elsevier Science Limited, to appear.

Kitagawa, E. (1964) Standardized comparisons in population research. *Demography*, **1**, 296–315.

Kleinbaum, D., Kupper, L., and Morgenstern, H. (1982) *Epidemiologic Research.* Belmont, CA: Wadsworth.

Mantel, N. (1973) Synthetic retrospective studies. *Biometrics*, **29**, 479–486.

Mantel, N. and Haenszel, W. (1959) Statistical aspects of retrospective studies of disease. *Journal of the National Cancer Institute*, **22**, 719–748.

Ming, K. and Rosenbaum, P. R. (2000) Substantial gains in bias reduction from matching with a variable number of controls. *Biometrics*, **56**, 118–124.

Morton, D., Saah, A., Silberg, S., Owens, W., Roberts, M., and Saah, M. (1982) Lead absorption in children of employees in a lead related industry. *American Journal of Epidemiology*, **115**, 549–555.

Mosteller, F. and Tukey, J. (1977) *Data Analysis and Regression.* Reading, MA: Addison–Wesley.

Petersen, L. A., Normand, S., Daley, J., and McNeil, B. (2000) Outcome of myocardial infarction in Veterans Health Administration patients as compared with Medicare patients. *New England Journal of Medicine*, **343**, December 28, 1934–1941.

Robins, J. M. and Ritov, Y. (1997) Toward a curse of dimensionality appropriate asymptotic theory for semi-parametric models. *Statistics in Medicine*, **16**, 285–319.

Robins, J. M., Mark, S. D. and Newey, W. K. (1992) Estimating exposure effects by modeling the expectation of exposure conditional on confounders. *Biometrics*, **48**, 479–495.

Rosenbaum, P. R. (1984a) The consequences of adjustment for a concomitant variable that has been affected by the treatment. *Journal of the Royal Statistical Society*, Series A, **147**, 656–666.

Rosenbaum, P. R. (1984b) Conditional permutation tests and the propensity score in observational studies. *Journal of the American Statistical Association*, **79**, 565–574.

Rosenbaum, P. R. (1986) Dropping out of high school in the United States: An observational study. *Journal of Educational Statistics*, **11**, 207–224.

Rosenbaum, P. R. (1987a) Model-based direct adjustment. *Journal of the American Statistical Association*, **82**, 387–394.

Rosenbaum, P. R. (1987b) The role of a second control group in an observational study (with Discussion). *Statistical Science*, **2**, 292–316.

Rosenbaum, P. R. (1988) Permutation tests for matched pairs with adjustments for covariates. *Applied Statistics*, **37**, 401–411.

Rosenbaum, P. R. and Rubin, D. B. (1983) The central role of the propensity score in observational studies for causal effects. *Biometrika*, **70**, 41–55.

Rosenbaum, P. and Rubin, D. (1984) Reducing bias in observational studies using subclassification on the propensity score. *Journal of the American Statistical Association*, **79**, 516–524.

Rosenbaum, P. R. and Silber, J. H. (2001) Matching and thick description in an observational study of mortality after surgery. *Biostatistics*, **2**, 217-232.

Rubin, D.B. (1977) Assignment to treatment group on the basis of a covariate. *Journal of Educational Statistics*, **2**, 1–26.

Rubin, D.B. (1978) Bayesian inference for causal effects: The role of randomization. *Annals of Statistics*, **6**, 34–58.

Silber, J. H., Rosenbaum, P. R., Trudeau, M. E., Even-Shoshan, O., Chen, W., Zhang, X., and Mosher, R. E. (2001) Multivariate matching and bias reduction in the surgical outcomes study. *Medical Care*, to appear.

Smith, H. L. (1997) Matching with multiple controls to estimate treatment effects in observational studies. *Sociological Methodology*, 27, 325–353.

Tunca, B. T. and Egeli, U. (1996) Cytogenetic findings on shoe workers exposed long-term to benzene. *Environmental Health Perspectives*, **104**, supplement 6, 1313–1317.

Ury, H. (1975) Efficiency of case-control studies with multiple controls per case: Continuous or dichotomous data. *Biometrics*, **31**, 643–649.

Walker, A. (1982) Efficient assessment of confounder effects in matched follow-up studies. *Applied Statistics*, **31**, 293–297.

Walker, A., Jick, H., Hunter, J., Danford, A., Watkins, R., Alhadeff, L., and Rothman, K. (1981) Vasectomy and non-fatal myocardial infarction. *Lancet*, 13–15.

Whitehead, R., Rosenbaum, P., and Carbone, P. (1984) Cisplatin, doxorubicin, cyclophosphamide, lomustine, and vincristine (PACCO) in the treatment of nonsmall cell bronchogenic carcinoma. *Cancer Treatment Reports*, **68**, 771–773.

4

Sensitivity to Hidden Bias

4.1 What Is a Sensitivity Analysis?

Cornfield, Haenszel, Hammond, Lilienfeld, Shimkin, and Wynder (1959) first conducted a formal sensitivity analysis in an observational study. Their paper is a survey of the evidence available in 1959 linking smoking with lung cancer. The paper asks whether the association between smoking and lung cancer is an effect caused by smoking or whether it is instead due to a hidden bias. Can lung cancer be prevented by not smoking? Or are the higher rates of lung cancer among smokers due to some other difference between smokers and nonsmokers?

In their effort to sort through conflicting claims, Cornfield et al. (1959) derived an inequality for a risk ratio defined as the ratio of the probability of death from lung cancer for smokers divided by the probability of death from lung cancer for nonsmokers. Specifically, they write:

> If an agent, A, with no causal effect upon the risk of a disease, nevertheless, because of a positive correlation with some other causal agent, B, shows an apparent risk, r, for those exposed to A, relative to those not so exposed, then the prevalence of B, among those exposed to A, relative to the prevalence among those not so exposed, must be greater than r.

> Thus, if cigarette smokers have 9 times the risk of nonsmokers for developing lung cancer, and this is not because cigarette smoke is a causal agent, but only because cigarette smokers produce hormone X, then the proportion of hormone X-producers

among cigarette smokers must be at least 9 times greater than that of nonsmokers. If the relative prevalence of hormone X-producers is considerably less than ninefold, then hormone X cannot account for the magnitude of the apparent effect.

Their statement is an important conceptual advance. The advance consists in replacing a general qualitative statement that applies in all observational studies by a quantitative statement that is specific to what is observed in a particular study. Instead of saying that an association between treatment and outcome does not imply causation, that hidden biases can explain observed associations, they say that to explain the association seen in a particular study, one would need a hidden bias of a particular magnitude. If the association is strong, the hidden bias needed to explain it is large.

Though an important conceptual advance, the above inequality of Cornfield et al. (1959) is limited to binary outcomes, and it ignores sampling variability which is hazardous except in extremely large studies. This chapter discusses a general method of sensitivity analysis that is similar to their method in spirit and purpose.

4.2   A Model for Sensitivity Analysis

4.2.1   The Model Expressed in Terms of Assignment Probabilities

If a study is free of hidden bias, as discussed in Chapter 3, then the chance $\pi_{[j]}$ that unit j receives the treatment is a function $\lambda(\mathbf{x}_{[j]})$ of the observed covariates $\mathbf{x}_{[j]}$ describing unit j. There is *hidden bias* if two units with the same observed covariates $\mathbf{x}$ have differing chances of receiving the treatments, that is, if $\mathbf{x}_{[j]} = \mathbf{x}_{[k]}$ but $\pi_{[j]} \neq \pi_{[k]}$ for some j and k. In this case, units that appear similar have differing chances of assignment to treatment.

A sensitivity analysis asks: How would inferences about treatment effects be altered by hidden biases of various magnitudes? Suppose the π's differ at a given $\mathbf{x}$. How large would these differences have to be to alter the qualitative conclusions of a study?

Suppose that we have two units, say j and k, with the same $\mathbf{x}$ but possibly different π's, so $\mathbf{x}_{[j]} = \mathbf{x}_{[k]}$ but possibly $\pi_{[j]} \neq \pi_{[k]}$. Then units j and k might be matched to form a matched pair or placed together in the same subclass in our attempts to control overt bias due to $\mathbf{x}$. The odds that units j and k receive the treatment are, respectively, $\pi_{[j]}/(1 - \pi_{[j]})$ and $\pi_{[k]}/(1 - \pi_{[k]})$, and the odds ratio is the ratio of these odds. Imagine that we knew that this odds ratio for units with the same $\mathbf{x}$ was at most some number $\Gamma \geq 1$; that is,

$$\frac{1}{\Gamma} \leq \frac{\pi_{[j]}(1 - \pi_{[k]})}{\pi_{[k]}(1 - \pi_{[j]})} \leq \Gamma \qquad \text{for all } j, k \text{ with} \quad \mathbf{x}_{[j]} = \mathbf{x}_{[k]}. \qquad (4.1)$$

If Γ were 1, then $\pi_{[j]} = \pi_{[k]}$ whenever $\mathbf{x}_{[j]} = \mathbf{x}_{[k]}$, so the study would be free of hidden bias, and the methods of Chapter 3 would apply. If $\Gamma = 2$, then two units who appear similar, who have the same $\mathbf{x}$, could differ in their odds of receiving the treatment by as much as a factor of 2, so one could be twice as likely as the other to receive the treatment. In other words, Γ is a measure of the degree of departure from a study that is free of hidden bias. A sensitivity analysis will consider several possible values of Γ and show how the inferences might change. A study is sensitive if values of Γ close to 1 could lead to inferences that are very different from those obtained assuming the study is free of hidden bias. A study is insensitive if extreme values of Γ are required to alter the inference.

4.2.2 The Model Expressed in Terms of an Unobserved Covariate

When speaking of hidden biases, we commonly refer to characteristics that were not observed, that are not in $\mathbf{x}$, and therefore that were not controlled by adjustments for $\mathbf{x}$. This section expresses the model of §4.2.1 in terms of an unobserved covariate, say u, that should have been controlled along with $\mathbf{x}$ but was not controlled because u was not observed. This reexpression of the model takes place without loss of generality, that is, it entails writing the same model in a different way. The models of §4.2.1 and 4.2.2 are identical in the sense that they describe exactly the same collections of probability distributions for $Z_{[1]}, \dots, Z_{[M]}$.

Unit j has both an observed covariate $\mathbf{x}_{[j]}$ and an unobserved covariate $u_{[j]}$. The model has two parts, a logit form linking treatment assignment $Z_{[j]}$ to the covariates $(\mathbf{x}_{[j]}, u_{[j]})$ and a constraint on $u_{[j]}$, namely,

$$\log\left(\frac{\pi_{[j]}}{1 - \pi_{[j]}}\right) = \kappa(\mathbf{x}_{[j]}) + \gamma u_{[j]}, \quad \text{with } 0 \leq u_{[j]} \leq 1, \qquad (4.2)$$

where $\kappa(\cdot)$ is an unknown function and γ is an unknown parameter.

Suppose units j and k have the same observed covariate, so $\mathbf{x}_{[j]} = \mathbf{x}_{[k]}$ and hence $\kappa(\mathbf{x}_{[j]}) = \kappa(\mathbf{x}_{[k]})$. In adjusting for $\mathbf{x}$, these two units might be matched or might be placed in the same stratum. Consider the ratio of the odds that units j and k receive the treatment

$$\frac{\pi_{[j]}(1 - \pi_{[k]})}{\pi_{[k]}(1 - \pi_{[j]})} = \exp\{\gamma(u_{[j]} - u_{[k]})\}. \qquad (4.3)$$

In other words, two units with the same $\mathbf{x}$ differ in their odds of receiving the treatment by a factor that involves the parameter γ and the difference in their unobserved covariates u.

The following proposition says that the model (4.2) is the same as the inequality (4.1). The proposition shows that the model (4.2) involves no additional assumptions beyond (4.1)—the logit form, the constraint on u, the absence of interaction terms linking $\mathbf{x}$ and u—these are not additional assumptions but rather are implied by (4.1).

Proposition 12 *With $e^\gamma = \Gamma \geq 1$, there is a model of the form (4.2) that describes the $\pi_{[1]}, \dots, \pi_{[M]}$ if and only if (4.1) is satisfied.*

Proof. Assume the model (4.2). Then $-1 \leq u_{[j]} - u_{[k]} \leq 1$, and so from (4.3),

$$\exp(-\gamma) \leq \frac{\pi_{[j]}(1 - \pi_{[k]})}{\pi_{[k]}(1 - \pi_{[j]})} \leq \exp(\gamma) \qquad \text{wherever} \quad \mathbf{x}_{[j]} = \mathbf{x}_{[k]}. \qquad (4.4)$$

In other words, if (4.2) holds, then (4.1) holds with $\Gamma = \exp(\gamma)$.

Conversely, assume the inequality (4.1) holds. For each value $\mathbf{x}$ of the observed covariate, find that unit k with $\mathbf{x}_{[k]} = \mathbf{x}$ having the smallest π, so

$$\pi_{[k]} = \min_{\{j:\mathbf{x}_{[j]}=\mathbf{x}\}} \pi_{[j]};$$

then set $\kappa(\mathbf{x}) = \log\{\pi_{[k]}/(1 - \pi_{[k]})\}$ and $u_{[k]} = 0$. If $\Gamma = 1$, then $\mathbf{x}_{[j]} = \mathbf{x}_{[k]}$ implies $\pi_{[j]} = \pi_{[k]}$, so set $u_{[j]} = 0$. If $\Gamma > 1$ and there is another unit j with the same value of $\mathbf{x}$, then set

$$u_{[j]} = \frac{1}{\gamma} \log\left(\frac{\pi_{[j]}}{1 - \pi_{[j]}}\right) - \frac{\kappa(\mathbf{x})}{\gamma} = \frac{1}{\gamma} \log\left(\frac{\pi_{[j]}(1 - \pi_{[k]})}{\pi_{[k]}(1 - \pi_{[j]})}\right). \qquad (4.5)$$

Now (4.5) implies the logit form in (4.2). Since $\pi_{[j]} \geq \pi_{[k]}$, it follows that $u_{[j]} \geq 0$. Using (4.1) and (4.5), it follows that $u_{[j]} \leq 1$. So the constraint on $u_{[j]}$ in (4.2) holds. ∎

The constraint on $u_{[j]}$ in (4.2) may be viewed in various ways. First, it may be seen simply as the formal expression of the statement (4.1) that subjects who appear similar in terms of $\mathbf{x}$ may differ in their odds of receiving the treatment by at most Γ. This faithful translation of the model into nontechnical terms is often helpful in discussing the results of a sensitivity analysis.

A second view of the constraint on $u_{[j]}$ in (4.2) relates it to the risk ratio inequality of Cornfield et al. (1959). In that inequality, the unobserved variable is a binary trait, $u = 0$ or $u = 1$. Of course, a binary trait satisfies the constraint in (4.2). Indeed, in later calculations, the relevant u's take the extreme values of 0 or 1. In other words, the constraint in (4.2) could, for most purposes, be replaced by the assumption that u is a binary variable taking values 0 or 1.

Finally, the constraint in (4.2) may be seen as a restriction on the scale of the unobserved covariate, a restriction needed if the numerical value of γ is to have meaning. This last view suggests consideration of other restrictions on the scale, for instance, restrictions that do not confine u to a finite range. As is seen later in Problem 9, it is not difficult to work with alternative restrictions on u, but the attractive nontechnical translation of the parameter Γ is then lost.

Throughout this chapter, the model (4.1) or its equivalent model (4.2) is assumed to hold.

4.2.3 The Distribution of Treatment Assignments After Stratification

As in §3.2.2, group units into S strata based on the observed covariate $\mathbf{x}$, with n_s units in stratum s, of whom m_s received the treatment. Under the equivalent models (4.1) or (4.2), the conditional distribution of the treatment assignment $\mathbf{Z} = (Z_{11}, \ldots, Z_{S,n_S})$ given $\mathbf{m}$ is no longer constant, as it was in §3.2.2 for a study free of hidden bias. Instead, the distribution is

$$\text{prob}(\mathbf{Z} = \mathbf{z} | \mathbf{m}) = \frac{\exp(\gamma \mathbf{z}^T \mathbf{u})}{\sum_{\mathbf{b} \in \Omega} \exp(\gamma \mathbf{b}^T \mathbf{u})} = \prod_{s=1}^{S} \frac{\exp(\gamma \mathbf{z}_s^T \mathbf{u}_s)}{\sum_{\mathbf{b}_s \in \Omega_s} \exp(\gamma \mathbf{b}_s^T \mathbf{u}_s)}, \quad (4.6)$$

where

$$\mathbf{z} = \begin{bmatrix} \mathbf{z}_1 \\ \vdots \\ \mathbf{z}_S \end{bmatrix}, \qquad \mathbf{u} = \begin{bmatrix} \mathbf{u}_1 \\ \vdots \\ \mathbf{u}_S \end{bmatrix},$$

and Ω_s is the set containing the $\binom{n_s}{m_s}$ different n_s-tuples with m_s ones and $n_s - m_s$ zeros. The easy steps leading from (4.2) to (4.6) form Problem 8.

Notice what (4.6) says. Given $\mathbf{m}$, the distribution of treatment assignments $\mathbf{Z}$ no longer depends of the unknown function $\kappa(\mathbf{x})$, but it continues to depend on the unobserved covariate u. In words, stratification on $\mathbf{x}$ was useful in that it eliminated part of the uncertainty about the unknown π's, specifically, the part due to $\kappa(\mathbf{x})$, but stratification on $\mathbf{x}$ was insufficient to render all treatment assignments equally probable, as they would be in an experiment.

There are two exceptions, two cases in which (4.6) gives equal probabilities $1/K$ to all K treatment assignments in Ω. If $\gamma = 0$, then the unobserved covariate u is not relevant to treatment assignment; see (4.2). If $u_{si} = u_{sj}$ for all $1 \leq i < j < n_s$ and for $s = 1, \ldots, S$, then units in the same stratum have the same value of u. In both cases, (4.6) equals $1/K$ for all $\mathbf{z} \in \Omega$. In other words, there is no hidden bias if u is either unrelated to treatment assignment or if u is constant for all units in the same stratum.

For any specific $(\gamma, \mathbf{u})$, the distribution (4.6) is a distribution of treatment assignments $\mathbf{Z}$ on Ω. If $(\gamma, \mathbf{u})$ were known, (4.6) could be used as a basis for permutation inference, as in Chapter 2. Since $(\gamma, \mathbf{u})$ is not known, a sensitivity analysis will display the sensitivity of inferences to a range of assumptions about $(\gamma, \mathbf{u})$. Specifically, for several values of γ, the sensitivity analysis will determine the most extreme inferences that are possible for $\mathbf{u}$ satisfying the constraint in (4.2), that is, for $\mathbf{u}$ in the N-dimensional unit cube $U = [0, 1]^N$. Specific procedures are discussed in the remaining sections of this chapter.

4.3 Matched Pairs

4.3.1 Sensitivity of Significance Levels: The General Case

With matched pairs, the model and methods have a particularly simple form. For this reason, matched pairs are discussed first and separately, though §4.3 could be formally subsumed within the more general discussion in §4.4.

With S matched pairs, each stratum $s, s = 1, \ldots, S$, has $n_s = 2$ units, one of which received the treatment, $1 = m_s = Z_{s1} + Z_{s2}$, so $Z_{s2} = 1 - Z_{s1}$. The model (4.6) describes S independent binary trials with unequal probabilities, namely,

$$\text{prob}(\mathbf{Z} = \mathbf{z}|\mathbf{m}) = \prod_{s=1}^{S} \left[\frac{e^{\gamma u_{s1}}}{e^{\gamma u_{s1}} + e^{\gamma u_{s2}}} \right]^{z_{s1}} \left[\frac{e^{\gamma u_{s2}}}{e^{\gamma u_{s1}} + e^{\gamma u_{s2}}} \right]^{1-z_{s1}}. \quad (4.7)$$

In a randomized experiment or in a study free of hidden bias, $\gamma = 0$ and every unit has an equal chance of receiving each treatment, so (4.7) is the uniform distribution on Ω. In contrast, if $\gamma > 0$ in (4.7), then the first unit in pair s is more likely to receive the treatment than the second if $u_{s1} > u_{s2}$.

As seen in Chapter 2, common test statistics for matched pairs are sign-score statistics, including Wilcoxon's signed rank statistic and McNemar's statistic. They have the form

$$T = t(\mathbf{Z}, \mathbf{r}) = \sum_{s=1}^{S} d_s \sum_{i=1}^{2} c_{si} Z_{si},$$

where c_{si} is binary, $c_{si} = 1$ or $c_{si} = 0$, and both $d_s \geq 0$ and c_{si} are functions of $\mathbf{r}$, and so are fixed under the null hypothesis of no treatment effect. In a randomized experiment, $t(\mathbf{Z}, \mathbf{r})$ is compared to its randomization distribution under the null hypothesis, but this is not possible here because the treatment assignments $\mathbf{Z}$ have distribution (4.7) in which $(\gamma, \mathbf{u})$ is unknown. Specifically, for each possible $(\gamma, \mathbf{u})$, the statistic $t(\mathbf{Z}, \mathbf{r})$ is the sum of S independent random variables, where the sth variable equals d_s with probability

$$p_s = \frac{c_{s1}\exp(\gamma u_{s1}) + c_{s2}\exp(\gamma u_{s2})}{\exp(\gamma u_{s1}) + \exp(\gamma u_{s2})}, \tag{4.8}$$

and equals 0 with probability $1 - p_s$. A pair is said to be concordant if $c_{s1} = c_{s2}$. If $c_{s1} = c_{s2} = 1$ then $p_s = 1$ in (4.8), while if $c_{s1} = c_{s2} = 0$ then $p_s = 0$, so concordant pairs contribute a fixed quantity to $t(\mathbf{Z}, \mathbf{r})$ for all possible $(\gamma, \mathbf{u})$.

Though the null distribution of $t(\mathbf{Z}, \mathbf{r})$ is unknown, for each fixed γ the null distribution is bounded by two known distributions. With $\Gamma = \exp(\gamma)$, define p_s^+ and p_s^- in the following way:

$$p_s^+ = \begin{cases} 0 & \text{if } c_{s1} = c_{s2} = 0, \\ 1 & \text{if } c_{s1} = c_{s2} = 1, \\ \frac{\Gamma}{1+\Gamma} & \text{if } c_{s1} \neq c_{s2}, \end{cases} \quad \text{and} \quad p_s^- = \begin{cases} 0 & \text{if } c_{s1} = c_{s2} = 0, \\ 1 & \text{if } c_{s1} = c_{s2} = 1, \\ \frac{1}{1+\Gamma} & \text{if } c_{s1} \neq c_{s2}. \end{cases}$$

Then using the constraint on $u_{[j]}$ in (4.2), it follows that $p_s^- \leq p_s \leq p_s^+$ for $s = 1, \ldots, S$. Define T^+ to be the sum of S independent random variables, where the sth variable takes the value d_s with probability p_s^+ and takes the value 0 with probability $1 - p_s^+$. Define T^- similarly with p_s^- in place of p_s^+. The following proposition says that, for all $\mathbf{u} \in U = [0, 1]^N$, the unknown null distribution of the test statistic $T = t(\mathbf{Z}, \mathbf{r})$ is bounded by the distributions of T^+ and T^-.

Proposition 13 *If the treatment has no effect, then for each fixed $\gamma \geq 0$,*

$$\text{prob}(T^+ \geq a) \geq \text{prob}\{T \geq a | \mathbf{m}\} \geq \text{prob}(T^- \geq a)$$

for all a and $\mathbf{u} \in U$.

For each γ, Proposition 13 places bounds on the significance level that would have been appropriate had $\mathbf{u}$ been observed. The sensitivity analysis for a significance level involves calculating these bounds for several values of γ.

The bounds in Proposition 13 are attained for two values of $\mathbf{u} \in U$, and this has two practical consequences. Specifically, the upper bound $\text{prob}(T^+ \geq a)$ is the distribution of $t(\mathbf{Z}, \mathbf{r})$ when $u_{si} = c_{si}$ and the lower bound $\text{prob}(T^- \geq a)$ is the distribution of $t(\mathbf{Z}, \mathbf{r})$ when $u_{si} = 1 - c_{si}$. The first consequence is that bounds in Proposition 13 are the best possible bounds: they cannot be improved unless additional information is given about the value of $\mathbf{u} \in U$. Second, the bounds are attained at values of $\mathbf{u}$ which perfectly predict the signs c_{si}. For conventional statistics like McNemar's statistic and Wilcoxon's signed rank statistic, this means that the bounds are attained for values of $\mathbf{u}$ that exhibit a strong, near perfect relationship with the response $\mathbf{r}$.

The bounding distributions of T^+ and T^- have easily calculated moments. For T^+, the expectation and variance are

$$E(T^+) = \sum_{s=1}^{S} d_s \, p_s^+ \quad \text{and} \quad \text{var}(T^+) = \sum_{s=1}^{S} d_s^2 \, p_s^+ (1 - p_s^+). \quad (4.9)$$

For T^- the expectation and variance are given by the same formulas with p_s^- in place of p_s^+. As the number of pairs S increases, the distributions of T^+ and T^- are approximated by Normal distributions, providing the number of discordant pairs increases with S and provided the d_s are reasonably well behaved, as they are for the McNemar and Wilcoxon statistics.

The method is illustrated in §4.3.2 for the McNemar test and in §4.3.3 for the Wilcoxon signed rank test.

4.3.2 Sensitivity Analysis for McNemar's Test: Simplified Formulas and an Example

In a study of the effects of smoking on lung cancer, Hammond (1964) paired 36,975 heavy smokers to nonsmokers on the basis of age, race, nativity, rural versus urban residence, occupational exposures to dusts and fumes, religion, education, marital status, alcohol consumption, sleep duration, exercise, severe nervous tension, use of tranquilizers, current health, history of cancer other than skin cancer, history of heart disease, stroke, or high blood pressure. Though the paired smokers and nonsmokers were similar in each of these ways, they may differ in many other ways. For instance, it has at times been suggested that there might be a genetic predisposition to smoke (Fisher 1958).

Of the $S = 36,975$ pairs, there were 122 pairs in which exactly one person died of lung cancer. Of these, there were 12 pairs in which the nonsmoker died of lung cancer and 110 pairs in which the heavy smoker died of lung cancer. In other words, 122 of the pairs are discordant for death from lung cancer. If this study were a randomized experiment, which it was not, or if the study were free of hidden bias, which we have little reason to believe, then McNemar's test would compare the 110 lung cancer deaths among smokers to a binomial distribution with 122 trials and probability of success $1/2$, yielding a significance level less than 0.0001. In the absence of hidden bias, there would be strong evidence that smoking causes lung cancer. How much hidden bias would need to be present to alter this conclusion?

Let $d_s = 1$ for all s, $Z_{si} = 1$ if the ith person in pair s is a smoker and $Z_{si} = 0$ if this person is a nonsmoker, and $c_{si} = 1$ if this person died of lung cancer and $c_{si} = 0$ otherwise. Then the sign-score statistic

$$T = t(\mathbf{Z}, \mathbf{r}) = \sum_{s=1}^{S} d_s \sum_{i=1}^{2} c_{si} Z_{si}$$

TABLE 4.1. Sensitivity Analysis for Hammond's Study of Smoking and Lung Cancer: Range of Significance Levels for Hidden Biases of Various Magnitudes.

Γ	Minimum	Maximum
1	< 0.0001	< 0.0001
2	< 0.0001	< 0.0001
3	< 0.0001	< 0.0001
4	< 0.0001	0.0036
5	< 0.0001	0.03
6	< 0.0001	0.1

is the number of smokers who died of lung cancer. A pair is concordant if $c_{s1} = c_{s2}$ and is discordant if $c_{s1} \neq c_{s2}$. No matter how the treatment, smoking, is assigned within pair s, if neither person died of lung cancer, then pair s contributes 0 to $t(\mathbf{Z}, \mathbf{r})$, while if both people died of lung cancer, then pair s contributes 1 to $t(\mathbf{Z}, \mathbf{r})$, so in either case, a concordant pair contributes a fixed quantity to $t(\mathbf{Z}, \mathbf{r})$. Removing concordant pairs from consideration subtracts a constant from $t(\mathbf{Z}, \mathbf{r})$ and does not alter the significance level. Therefore, set the concordant pairs aside before computing $T = t(\mathbf{Z}, \mathbf{r})$, leaving the 122 discordant pairs, so T is the number of discordant pairs in which the smoker died of lung cancer, $T = 110$; that is, T is McNemar's statistic.

With the concordant pairs removed, T^+ and T^- have binomial distributions with 122 trials and probabilities of success $p^+ = \Gamma/(1 + \Gamma)$ and $p^- = 1/(1 + \Gamma)$, respectively. Under the null hypothesis of no effect of smoking, for each $\gamma \geq 0$, Proposition 13 gives an upper and lower bound on the significance level, prob$\{T \geq 110 | \mathbf{m}\}$, namely, for all $\mathbf{u} \in U$,

$$\sum_{a=110}^{122} \binom{122}{a} (p^+)^a (1 - p^+)^{122-a} \geq \text{prob}\{T \geq 110 | \mathbf{m}\}$$

$$\geq \sum_{a=110}^{122} \binom{122}{a} (p^-)^a (1 - p^-)^{122-a} \tag{4.10}$$

In a randomized experiment or a study free of hidden bias, the sensitivity parameter γ is zero, so $p^+ = p^- = \frac{1}{2}$, the upper and lower bounds in (4.10) are equal, and both bounds give the usual significance level for McNemar's statistic. For $\gamma > 0$, (4.10) gives a range of significance levels reflecting uncertainty about $\mathbf{u}$.

Table 4.1 gives the sensitivity analysis for Hammond's data. For six values of Γ, the table gives the upper and lower bounds (4.10) on the significance level. For $\Gamma = 4$, one person in a pair may be four times as likely to smoke as the other because they have different values of the unobserved covariate u. In the case $\Gamma = 4$, the significance level might be less than 0.0001 or it might be as high as 0.0036, but for all $\mathbf{u} \in U$ the null hypothesis of no effect of smoking on lung cancer is not plausible. The null hypothesis of no effect begins to become plausible for at least some $\mathbf{u} \in U$ with $\Gamma = 6$. To attribute the higher rate of death from lung cancer to an unobserved covariate $\mathbf{u}$ rather than to an effect of smoking, that unobserved covariate would need to produce a sixfold increase in the odds of smoking, and it would need to be a near perfect predictor of lung cancer. As will be seen by comparison in later examples, this is a high degree of insensitivity to hidden bias: in many other studies, biases smaller than $\Gamma = 6$ could explain the association between treatment and response.

4.3.3 Sensitivity Analysis for the Signed Rank Test: Simplified Formulas and an Example

Wilcoxon's signed rank statistic for S matched pairs is computed by ranking the absolute differences $|r_{s1} - r_{s2}|$ from 1 to S and summing the ranks of the pairs in which the treated unit had a higher response than the matched control. In the notation of §4.3.1, d_s is the rank of $|r_{s1} - r_{s2}|$ with average ranks used for ties, and $c_{s1} = 1, c_{s2} = 0$ if $r_{s1} > r_{s2}$ or $c_{s1} = 0, c_{s2} = 1$ if $r_{s1} < r_{s2}$ so in both cases the pairs are discordant, and $c_{s1} = 0, c_{s2} = 0$ if $r_{s1} = r_{s2}$ so the pairs are concordant or tied.

If there are no ties among the absolute differences $|r_{s1} - r_{s2}|$ and no zero differences, then the expectation and variance (4.9) of T^+ have simpler forms

$$E(T^+) = \frac{p^+ S(S+1)}{2},$$

and

$$\mathrm{var}(T^+) = p^+(1 - p^+)\frac{S(S+1)(2S+1)}{6}, \tag{4.11}$$

where

$$p^+ = \frac{\Gamma}{1 + \Gamma};$$

see Problem 7. With $p^- = 1/(1 + \Gamma)$ in place of p^+, the same expressions give the expectation and variance of T^-. If $\gamma = 0$ so $e^\gamma = \Gamma = 1$, then $p^+ = p^- = \frac{1}{2}$, and these expressions for the moments of T^+ and T^- are the same as the usual formulas for the expectation and variance of Wilcoxon's statistic in a randomized experiment (Lehmann 1975, p. 128).

Ties and zero differences occur in practice, and in this case the general formula (4.9) may be used. Notice that ties do not alter the expectation of T^+ and T^-, but they alter the variance.

TABLE 4.2. Sensitivity Analysis for Lead in the Blood of Children: Range of Significance Levels for the Signed-Rank Statistic.

Γ	Minimum	Maximum
1	<0.0001	<0.0001
2	<0.0001	0.0018
3	<0.0001	0.0136
4	<0.0001	0.0388
4.25	<0.0001	0.0468
5	<0.0001	0.0740

Consider the example in Table 3.1 concerning lead in the blood of children whose parents work in a battery factory. The Wilcoxon signed rank statistic for the differences in children's lead levels is 527. As seen in §3.2.4, if the study were free of hidden bias, this would constitute strong evidence of an effect of parental exposure to lead on children's lead levels. Specifically, if the study were free of hidden bias, that is, if $\Gamma = 1$, the expectation of the signed rank statistic under the null hypothesis of no effect is 280 with variance 3130.63. The standardized deviate $(527 - 280)/\sqrt{3130.63}$ is 4.41, yielding an approximate one-sided significance level of less than 0.0001.

In Table 3.1, notice the several tied differences for which average ranks were used, and the one zero difference which is a concordant pair. Because of the ties and zero difference, the exact formula (4.9) was used, though in this case (4.11) gives nearly identical results; for instance, (4.9) gives 280.5 and 3132.25 for the null expectation and variance with deviate still equal to 4.41.

If the study were free of hidden bias, there would be strong evidence that parents' occupational exposures to lead increased the level of lead in their children's blood. The sensitivity analysis asks how this conclusion might be changed by hidden biases of various magnitudes. If Γ were 2, then matched children might differ in their odds of exposure to lead by a factor of 2 due to hidden bias. In this case, $E(T^+) = 373.33$, $E(T^-) = 186.67$, $\text{var}(T^+) = \text{var}(T^-) = 2782.78$, so the deviates are $(527 - 373.33)/\sqrt{2782.78} = 2.91$ and $(527 - 186.67)/\sqrt{2782.78} = 6.45$. The range of significance levels when $\Gamma = 2$ is therefore from less than 0.0001 to 0.0018. A hidden bias of size $\Gamma = 2$ is insufficient to explain the observed difference between exposed and control children.

Table 4.2 gives the sensitivity analysis for the significance levels from Wilcoxon's signed rank test, that is, the range of possible significance levels for various values of Γ. The table shows that to explain away the observed association between parental exposure to lead and child's lead level, a hidden bias or unobserved covariate would need to increase the odds of exposure by more than a factor of $\Gamma = 4.25$. The association cannot be attributed to small hidden biases, but it is somewhat more sensitive to bias than Hammond's study of heavy smokers in §4.3.3.

4.3.4 Sensitivity Analysis of the Hodges—Lehmann Point Estimate in Matched Pairs

In §3.2.4, the effect of parental exposure to lead on the level of lead in the child's blood was estimated to be 15 μg of lead per decaliter of whole blood. This estimate is based on two premises:

(i) the treatment has an additive effect τ, so a child's lead level is increased by τ as a consequence of parental exposure; and

(ii) the study is free of hidden bias.

If the effect is additive, then the adjusted responses $\mathbf{R} - \tau\mathbf{Z}$ satisfy the null hypothesis of no effect. The estimate of 15 $\mu g/dl$ is the Hodges–Lehmann estimate, obtained as the value $\hat{\tau}$ such that, if the signed rank statistic is computed after subtracting $\hat{\tau}$ from the responses of exposed children, then this statistic equals its expectation $E\{t(\mathbf{Z}, \mathbf{R} - \tau\mathbf{Z})\} = \bar{\bar{t}} = S(S + 1)/4$ under the null hypothesis of no effect in the absence of hidden bias. More precisely, as discussed in §2.7.2, to allow for the discreteness of the signed rank statistic, $\hat{\tau}$ is defined as

$$\hat{\tau} = \text{SOLVE}\{\bar{\bar{t}} = t(\mathbf{Z}, \mathbf{R} - \hat{\tau}\mathbf{Z})\}$$
$$= \frac{\inf\{\tau : \bar{\bar{t}} > t(\mathbf{Z}, \mathbf{R} - \tau\mathbf{Z})\} + \sup\{\tau : \bar{\bar{t}} < t(\mathbf{Z}, \mathbf{R} - \tau\mathbf{Z})\}}{2}. \quad (4.12)$$

If there is hidden bias, then the signed rank statistic computed from $\mathbf{R} - \tau\mathbf{Z}$ does not generally have expectation $S(S+1)/4$. If $E\{t(\mathbf{Z}, \mathbf{R} - \tau\mathbf{Z})\} \neq S(S+1)/4$, then there is no point in trying to find a $\hat{\tau}$ so that $t(\mathbf{Z}, \mathbf{R} - \hat{\tau}\mathbf{Z})$ is nearly $S(S + 1)/4$; rather, $\hat{\tau}$ should be found so $t(\mathbf{Z}, \mathbf{R} - \hat{\tau}\mathbf{Z})$ is close to $E\{t(\mathbf{Z}, \mathbf{R} - \tau\mathbf{Z})\}$. Under the model (4.1) or (4.2), the expectation of the signed rank statistic computed from $\mathbf{R} - \tau\mathbf{Z}$ is bounded by the expectations of T^+ and T^- in Proposition 13, that is, bounded by

$$\bar{\bar{t}}_{\min} = \frac{p^- S(S + 1)}{2} \quad \text{and} \quad \bar{\bar{t}}_{\max} = \frac{p^+ S(S + 1)}{2}, \quad (4.13)$$

where

$$p^- = \frac{1}{1 + \Gamma} \quad \text{and} \quad p^+ = \frac{\Gamma}{1 + \Gamma}.$$

In other words, for a given Γ, the expectation $\bar{\bar{t}} = E\{t(\mathbf{Z}, \mathbf{R} - \tau\mathbf{Z})\}$ is not known but is bounded by two known numbers, $\bar{\bar{t}}_{\min} \leq \bar{\bar{t}} \leq \bar{\bar{t}}_{\max}$. Moreover, these bounds are sharp in that they are attained for particular values of the unobserved covariate $\mathbf{u} \in U$. Consider calculating $\hat{\tau} = \text{SOLVE}\{\bar{\bar{t}} = t(\mathbf{Z}, \mathbf{R} - \hat{\tau}\mathbf{Z})\}$ for each $\bar{\bar{t}}$ in the interval $[\bar{\bar{t}}_{\min}, \bar{\bar{t}}_{\max}]$; this would produce the set of possible Hodges–Lehmann estimates. In fact, the minimum and maximum estimates in this set are easily determined. Since the signed rank

TABLE 4.3. Sensitivity Analysis for Lead in Children's Blood: Range of Hodges–Lehmann Estimates of Effect for Biases of Various Magnitudes.

Γ	Minimum	Maximum
1	15	15
2	10.25	19.5
3	8	23
4	6.5	25
5	5	26.5

statistic is effect increasing, the smallest $\hat{\tau} = \text{SOLVE}\ \{\bar{\bar{t}} = t(\mathbf{Z}, \mathbf{R} - \hat{\tau}\mathbf{Z})\}$ is found for $\bar{\bar{t}} = \bar{\bar{t}}_{\text{max}}$ and the largest $\hat{\tau}$ is found for $\bar{\bar{t}} = \bar{\bar{t}}_{\text{min}}$.

In short, the sensitivity analysis consists of calculating the range of possible Hodges–Lehmann estimates of an additive treatment effect for biases of various magnitudes. This is done by calculating $\hat{\tau} = \text{SOLVE}\ \{\bar{\bar{t}} = t(\mathbf{Z}, \mathbf{R} - \hat{\tau}\mathbf{Z})\}$ for $\bar{\bar{t}} = \bar{\bar{t}}_{\text{max}}$ and $\bar{\bar{t}} = \bar{\bar{t}}_{\text{min}}$ for several values of Γ.

Table 4.3 shows the sensitivity analysis for the data on lead in the blood of children. If there is no hidden bias, that is, if $\Gamma = 1$, then $\bar{\bar{t}}_{\text{max}} = \bar{\bar{t}}_{\text{min}} = S(S+1)/4 = 33(33+1)/4 = 280.5$, and the range of Hodges–Lehmann estimates of effect is the single number 15 $\mu g/dl$, namely, the usual Hodges–Lehmann estimate for a randomized experiment. If $\Gamma = 2$, then two subjects with the same observed covariates may differ by a factor of 2 in their odds of receiving the treatment. In this case, $\bar{\bar{t}}_{\text{max}} = p^{+}S(S+1)/2 = 2 \times 33(33+1)/6 = 374$ and $\bar{\bar{t}}_{\text{min}} = p^{-}S(S+1)/2 = 33(33+1)/6 = 187$, and solving $\hat{\tau} = \text{SOLVE}\ \{\bar{\bar{t}} = t(\mathbf{Z}, \mathbf{R} - \hat{\tau}\mathbf{Z})\}$ gives 10.25 and 19.5, respectively. In words, if $\Gamma = 2$, the estimated effect might be as small as 10.25 $\mu g/dl$ or as high as 19.5 $\mu g/dl$. Keep in mind that the median lead level in $\mu g/dl$ among controls is 16, so an effect of 10.25 is a 64% increase above the level found among controls. For $\Gamma = 5$, the estimated effect is between 5 and 26 $\mu g/dl$, though the smaller estimate does not differ significantly from zero; see Table 4.2.

To illustrate the computations in greater detail, consider the upper bound 19.5 for $\Gamma = 2$. As noted above, $\bar{\bar{t}}_{min} = 187$. By direct calculation with the signed rank statistic, $t(\mathbf{Z}, \mathbf{R} - 19.5\mathbf{Z}) = 193.5$, which is a bit too high, while $t(\mathbf{Z}, \mathbf{R} - 19.5001\mathbf{Z}) = 186$, which is a bit too low. Applying (4.12) with $\inf\{\tau : 187 > t(\mathbf{Z}, \mathbf{R} - \tau\mathbf{Z})\} = 19.5$ and $\sup\{\tau : 187 < t(\mathbf{Z}, \mathbf{R} - \tau\mathbf{Z})\} = 19.5$ yields the estimate 19.5.

4.3.5   Sensitivity Analysis for Confidence Intervals in Matched Pairs

This section discusses sensitivity analysis for a confidence interval for an additive effect τ. A $1 - \alpha$ confidence interval is the set of all values of τ

TABLE 4.4. Sensitivity Analysis for Lead in Children's Blood: Confidence Intervals for an Additive Effect for Biases of Various Magnitudes.

Γ	95% Confidence interval
1	(9.5, 20.5)
2	(4.5, 27.5)
3	(1.0, 32.0)
4	(−1.0, 36.5)
5	(−3.0, 41.5)

that are not rejected in an α-level test. For each $(\gamma, \mathbf{u})$, there is a two-sided α-level confidence interval derived from a test statistic $t(\mathbf{Z}, \mathbf{R} - \tau\mathbf{Z})$. For each fixed γ, the sensitivity analysis will report the union of these intervals as $\mathbf{u}$ ranges over U. Call this union the sensitivity interval. A value τ is in the interval if and only if there is some $\mathbf{u} \in U$ such that τ is not rejected in an α-level test.

Let T_τ^+ and T_τ^- be the bounding random variables in §4.3.1 computed from the adjusted responses $\mathbf{R} - \tau\mathbf{Z}$, and let $T_\tau = t(\mathbf{Z}, \mathbf{R} - \tau\mathbf{Z})$. Fix $\gamma \geq 0$ and suppose $\alpha/2 \geq \text{prob}(T_\tau^+ \geq a_1)$ and $\alpha/2 \geq \text{prob}(T_\tau^- \leq a_2)$. Then Proposition 13 implies $\alpha/2 \geq \text{prob}\{T_\tau \geq a_1|\mathbf{m}\}$ for all $\mathbf{u} \in U$, and $\alpha/2 \geq \text{prob}\{T_\tau \leq a_2|\mathbf{m}\}$, for all $\mathbf{u} \in U$. If $T_\tau \geq a_1$ or $T_\tau \leq a_2$, then τ is rejected at level α for all $\mathbf{u} \in U$, so τ is excluded from the sensitivity interval. When $\gamma = 0$, this yields the confidence interval in §2.6.2 for a randomized experiment, but as γ increases this interval becomes larger reflecting uncertainty about the impact of $\mathbf{u}$.

When using the Normal approximation to the distribution of T_τ^+ and T_τ^-, the endpoints of the 95% confidence interval are

$$\inf\left\{\tau : \frac{T_\tau - E(T_\tau^+)}{\sqrt{\text{var}(T_\tau^+)}} \leq 1.96\right\} \text{ and } \sup\left\{\tau : \frac{T_\tau - E(T_\tau^-)}{\sqrt{(T_\tau^-)}} \geq -1.96\right\}.$$

The procedure is illustrated in Table 4.4 using the signed rank test for the data on lead in children's blood. If the study were free of hidden bias, that is, if $\Gamma = 1$, the 95% confidence interval for the additive effect τ in $\mu g/dl$ would be (9.5, 20.5), as in §3.2.4. If $\Gamma = 2$, matched children might differ by a factor of two in their odds of exposure to lead due to differences in the unobserved covariate. In this case, the 95% confidence interval is longer, (4.5, 27.5), though the smallest plausible effect 4.5 is still 28% of the median lead level 16 $\mu g/dl$ among controls. For $\Gamma = 4$, slightly negative effects become just plausible, though large positive effects are also plausible. In comparing the tables, keep in mind that Table 4.4 describes a two-sided 95% confidence interval while Table 4.2 describes a one-sided significance test.

TABLE 4.5. Illustrative Computations for $\Gamma = 2$.

τ	$t(\mathbf{Z}, \mathbf{R} - \tau\mathbf{Z})$	$E(T_\tau^-)$	$\mathrm{var}(T_\tau^-)$	Deviate
27.4999	89.00	187.00	2609.00	-1.92
27.5000	86.00	187.00	2607.89	-1.98

τ	$t(\mathbf{Z}, \mathbf{R} - \tau\mathbf{Z})$	$E(T_\tau^+)$	$\mathrm{var}(T_\tau^+)$	Deviate
4.5000	476.50	374.00	2766.61	1.95
4.4999	479.00	374.00	2769.00	2.00

Table 4.5 contains some of the calculations leading to Table 4.4 for the case $\Gamma = 2$. The top half of Table 4.5 shows that 27.5 is the sup of all τ leading to a deviate greater than -1.96. The bottom half of the table shows that 4.5 is the inf of all τ leading to a deviate less than 1.96.

4.4 Sensitivity Analysis for Sign-Score Statistics

4.4.1 The General Method

As noted in §2.4.3, many common statistical tests are sign-score statistics. All sign-score statistics permit a sensitivity analysis similar to that in §4.3 for matched pairs. The purpose of §4.4 is to discuss this class of problems, first in general terms in §4.4.1, and then in specific situations in the later parts of §4.4. Sections 4.5 and 4.6 discuss larger classes of test statistics requiring computations that are just slightly more complex.

Recall that a sign-score statistic has the form

$$t(\mathbf{Z}, \mathbf{r}) = \sum_{s=1}^{S} d_s \sum_{i=1}^{n_s} c_{si} Z_{si} = \sum_{s=1}^{S} d_s B_s, \quad \text{where } B_s = \sum_{i=1}^{n_s} c_{si} Z_{si}, \quad (4.14)$$

c_{si} is binary, $c_{si} = 1$ or $c_{si} = 0$, d_s is nonnegative, $d_s \geq 0$, and both d_s and c_{si} are functions of $\mathbf{r}$. For binary responses $\mathbf{r}$, the signs are the responses themselves, $c_{si} = r_{si}$, and the scores are constant, $d_s = 1$; this yields Fisher's exact test for a 2×2 table, McNemar's statistic for matched pairs, and the Mantel–Haenszel statistic for a $2 \times 2 \times S$ table. In Wilcoxon's signed rank statistic for matched pairs, the signs, c_{si}, identify the unit in matched set i with the largest response and the score d_s is the rank of the difference between the larger and the smaller responses in a matched pair. In the median test, the signs c_{si} identify units with responses above the median response of the combined treated and control groups.

The main fact about sign-score statistics, namely Proposition 14, is similar to Proposition 13 for matched pairs. Specifically, Proposition 14 bounds the unknown distribution of $t(\mathbf{Z}, \mathbf{r})$ by two known distributions, namely, the

TABLE 4.6. Two-By-Two Table Associated With a Sign-Score Statistic.

	$Z_{si} = 1$	$Z_{si} = 0$	Total
$c_{si} = 1$	A	$c_{s+} - A$	c_{s+}
$c_{si} = 0$	$m_s - A$	$n_s - c_{s+} - m_s + A$	$n_s - c_{s+}$
Total	m_s	$n_s - m_s$	n_s

distributions at the most extreme $\mathbf{u}$'s $\in U$. Sign-score statistics are special in the following sense: The $\mathbf{u}$'s $\in U$ that provide the bounds may be determined immediately from the structure of the statistic itself.

Under the null hypothesis of no treatment effect, $\mathbf{r}$ is fixed, so d_s and c_{si} are fixed because they are functions of $\mathbf{r}$. Write $\mathbf{u}^+$ and $\mathbf{u}^-$ for the N-tuples with $u_{si}^+ = c_{si}$ and $u_{si}^- = 1 - c_{si}$, respectively. Fix γ, and let T^+ be the random variable $\sum_s d_s \sum_i c_{si} Z_{si}$ when $\mathbf{Z}$ has the distribution (4.6) with $\mathbf{u} = \mathbf{u}^+$, and define T^- similarly with $\mathbf{u} = \mathbf{u}^-$.

Proposition 14 *If the treatment has no effect and $T = t(\mathbf{Z}, \mathbf{r})$ is a sign-score statistic, then for each fixed $\gamma \geq 0$,*

$$\mathrm{prob}(T^+ \geq a) \geq \mathrm{prob}\{T \geq a | \mathbf{m}\} \geq \mathrm{prob}(T^- \geq a) \quad \text{for all } a \text{ and } \mathbf{u} \in U.$$

The proof is given in the appendix to this chapter.

As in §4.3, the sensitivity analysis consists of calculating the bounds, $\mathrm{prob}(T^+ \geq a)$ and $\mathrm{prob}(T^- \geq a)$, for a range of values of γ. For $\gamma = 0$, the bounds are equal, $\mathrm{prob}(T^+ \geq a) = \mathrm{prob}(T^- \geq a)$, and their common value is the usual significance level for a randomized experiment. As γ increases, the bounds move apart reflecting uncertainty about hidden biases. The general method is applied to particular cases in subsequent sections.

The proposition has two special cases that deserve explicit mention, because they arise frequently and because the bounding distributions have a familiar form. The extended hypergeometric distribution is most commonly obtained as the distribution of a binomial random variable conditionally given the sum of this variable and another independent binomial; see Johnson, Kotz, and Kemp (1992, §6.11) or Plackett (1981, §4.4.6). It is a distribution of the corner cell, A, in the 2×2 contingency table with fixed marginal total in Table 4.6, where $c_{s+} = \sum_{i=1}^{n_s} c_{si}$ and A must satisfy the inequality $\min(m_s, c_{s+}) \geq A \geq \max(0, m_s + c_{s+} - n_s)$ to ensure that every cell of the table has a nonnegative count. Specifically, under the extended hypergeometric distribution, $\mathrm{prob}(A \geq a)$ is given by

$$\Upsilon(n_s, m_s, c_{s+}, a, \Gamma) = \frac{\sum_{k=\max(a, m_s + c_{s+} - n_s)}^{\min(m_s, c_{s+})} \binom{c_{s+}}{k} \binom{n_s - c_{s+}}{m_s - k} \Gamma^k}{\sum_{k=\max(0, m_s + c_{s+} - n_s)}^{\min(m_s, c_{s+})} \binom{c_{s+}}{k} \binom{n_s - c_{s+}}{m_s - k} \Gamma^k}.$$

The corollary says that the extended hypergeometric distribution provides the sharp bounds in Proposition 14 when attention focuses on a single stratum s and $d_s = 1$, $d_{s'} = 0$ for $s' \neq s$.

Corollary 15 *If the treatment has no effect and $T = \sum_{i=1}^{n_s} c_{si} Z_{si}$ for a specific s, then for each fixed $\Gamma \geq 1$, and for all a and $\mathbf{u} \in U$,*

$$\Upsilon\left(n_s, m_s, c_{s+}, a, \Gamma\right) \geq \mathrm{prob}\{T \geq a | \mathbf{m}\} \geq \Upsilon\left(n_s, m_s, c_{s+}, a, \frac{1}{\Gamma}\right).$$

More generally, if $T = \sum_s \sum_i c_{si} Z_{si}$ —that is, if $d_s = 1$ for every s— then the bounds in Proposition 14 are given by the corresponding distributions of sums of independent extended hypergeometric distributions. Let $\overline{\overline{B}}_s$, $s = 1, \ldots, S$, be independent random variables with extended hypergeometric distributions $\Upsilon\left(n_s, m_s, c_{s+}, a, \Gamma\right)$ and let $\overline{B}_s$, $s = 1, \ldots, S$, be independent random variables with extended hypergeometric distributions $\Upsilon\left(n_s, m_s, c_{s+}, a, \frac{1}{\Gamma}\right)$.

Corollary 16 *If the treatment has no effect and $T = \sum_s \sum_i c_{si} Z_{si}$, then for each fixed $\Gamma \geq 1$, and for all a and $\mathbf{u} \in U$,*

$$\mathrm{prob}\left(\sum \overline{\overline{B}}_s \geq a\right) \geq \mathrm{prob}\left(T \geq a | \mathbf{m}\right) \geq \mathrm{prob}\left(\sum \overline{B}_s \geq a\right).$$

4.4.2 Matching with Multiple Controls Using Sign-Score Statistics

When controls are plentiful, it is common to match several controls to each treated unit, so each stratum s contains a single treated unit and one or more controls, $1 = \sum_i Z_{si}$ for $s = 1, \ldots, S$. In this case, the model (4.6) may be written

$$\mathrm{prob}(\mathbf{Z} = \mathbf{z} | \mathbf{m}) = \prod_{s=1}^{S} \frac{\exp\left(\gamma \sum_{i=1}^{n_s} z_{si} u_{si}\right)}{\sum_{i=1}^{n_s} \exp(\gamma u_{si})} \quad \text{for } \mathbf{z} \in \Omega, \tag{4.15}$$

since Ω_s contains just the n_s vectors $\mathbf{z}_s$ whose coordinates include a single one and $n_s - 1$ zeros. Compare (4.15) and (4.7). Also, B_s in (4.14) is a binary random variable. From (4.15), the $B_s, s = 1, \ldots, S$, are mutually independent and

$$p_s = \mathrm{prob}(B_s = 1 | \mathbf{m}) = \frac{\sum_{i=1}^{n_s} c_{si} \exp(\gamma u_{si})}{\sum_{i=1}^{n_s} \exp(\gamma u_{si})}. \tag{4.16}$$

This probability is bounded by its values at $\mathbf{u}^+$ and $\mathbf{u}^-$ in §4.4.1, namely,

$$p_s^- = \frac{c_{s+}}{c_{s+} + (n_s - c_{s+})\Gamma} \leq p_s \leq \frac{\Gamma c_{s+}}{\Gamma c_{s+} + n_s - c_{s+}} = p_s^+, \tag{4.17}$$

where $c_{s+} = \sum_i c_{si}$ and $\Gamma = e^\gamma$. Moreover, T^+ and T^- are the random variables that equal $\sum d_s B_s$ when, respectively, $p_s = p_s^+$ and $p_s = p_s^-$. It follows that the moments of T^+ are again given by formula (4.9), and the moments of T^- are given by the same formula with p_s^- in place of p_s^+.

If the outcome is binary, say 1 for survived and 0 for died, then the Mantel–Haenszel (1959) statistic is the usual test of the null hypothesis of no treatment effect; see §2.4.3. This statistic $T = t(\mathbf{Z}, \mathbf{r})$ is a sign-score statistic with $d_s = 1$ and $c_{si} = r_{si}$ in (4.14), so T is the number of treated subjects who survived. In (4.14), $B_s = 1$ if the treated subject in matched set s survived and $B_s = 0$ otherwise. The bounds on the approximate significance level are obtained by referring the standardized deviates,

$$\frac{|T - \sum p_s^+| - \frac{1}{2}}{\sqrt{\sum p_s^+ (1 - p_s^+)}} \quad \text{and} \quad \frac{|T - \sum p_s^-| - \frac{1}{2}}{\sqrt{\sum p_s^- (1 - p_s^-)}}, \qquad (4.18)$$

to tables of the standard normal distribution. When $\gamma = 0$, these two deviates are equal, and the square of either deviate equals the usual Mantel–Haenszel statistic.

In general, for any sign-score statistic used in matching with one or more controls, the deviates are

$$\frac{T - \sum d_s p_s^+}{\sqrt{\sum d_s^2 p_s^+ (1 - p_s^+)}} \quad \text{and} \quad \frac{T - \sum d_s p_s^-}{\sqrt{\sum d_s^2 p_s^- (1 - p_s^-)}}.$$

4.4.3   Sensitivity Analysis with Multiple Controls: The Example of Vitamin C and Cancer

Recall from Chapter 1 the study by Cameron and Pauling (1976) of vitamin C and advanced cancer. In this study, patients treated with vitamin C were each matched with 10 controls on the basis of gender, age, primary cancer, and histological tumor type. In this section, the Cameron and Pauling study helps to clarify what sensitivity analyses can and cannot do.

Focus on the 18 treated patients and their 180 matched controls having primary cancers of the colon and rectum. The subsequent randomized trial by Moertel et al. (1985) concerned patients with cancers of the colon and rectum. Cameron and Pauling (1976) used as a test statistic the number of matched sets in which the treated patient survived longer than the average survival among the 10 controls. This is the same as counting the number of treated patients living longer than the mean for all 11 patients in the set, and this count is a sign-score statistic in which $c_{si} = 1$ if subject (s, i) lived longer than the mean survival in set s and $c_{si} = 0$ otherwise. In fact, 16 of the 18 treated patients survived longer than the mean in their matched set.

Given the presence of some extreme responses and certain other features of the data, it is more appropriate to count the number of treated responses that exceed the median response in their matched set, though this too is 16

TABLE 4.7. Sensitivity Analysis for Vitamin C and Advanced Cancer.

Γ	p_s^+	p_s^-	Deviate at $\mathbf{u}^+$	Deviate at $\mathbf{u}^-$	Range of significance levels
1	0.45	0.45	3.46	3.46	$[0.0003,\ 0.0003]$
2	0.63	0.29	2.07	5.28	$[0.019, < 0.0001]$
3	0.71	0.22	1.38	6.62	$[0.082, < 0.0001]$

of the 18 matched sets. This statistic is also a sign-score statistic; in fact, it is a version of the standard median test, though here it is applied within matched sets and then combined across matched sets. The median test is a good test in large samples if the data are from a double exponential distribution and the treatment has a small effect; see Hájek, Sidák and Sen (1999, §4.1, p. 97) or Hettmansperger (1984, §3); however, see also the critical discussion of the median test by Freidlin and Gastwirth (2000). Here, $c_{si} = 1$ if subject (s, i) lived longer than the median survival in set s and $c_{si} = 0$ otherwise. Then for each s, $5 = \sum_i c_{si}$, because each set contains $n_s = 11$ subjects, and the sixth largest response is the median. Then in (4.17), $p_s^- = 5/(5 + 6\Gamma)$ and $p^+ = 5\Gamma/(5\Gamma + 6)$, and $p_s^- = p_s^+ = 5/11$ if there is no hidden bias in the sense that $\Gamma = 1$. Now expression (4.18) yields Table 4.7. The table suggests that the longer survival among patients receiving vitamin C would be highly significant in a randomized experiment, and is insensitive to small biases, but it becomes sensitive at about $\Gamma = 3$.

The sensitivity analysis just performed was based on the 16/18 treated subjects who lived longer than the typical subject in the matched set, and this closely parallels the analysis in the original study. One might wonder how the results might change with a different test statistic. In fact, with the same 18 matched sets but a different choice of test statistic, the results are far less sensitive, the upper bound on the significance level being 0.045 for $\Gamma = 10$; see Rosenbaum (1988) for detailed discussion. The test described in Table 4.7 made no use of the fact that most treated subjects had survival times much higher than the typical survival time in the subject's matched set, and this is the reason for the diverging findings from different test statistics. The test used for illustration in Rosenbaum (1988) yielded less sensitivity to bias in this example.

The first observation to make is that, with a suitable test, this study is insensitive to extremely large biases, $\Gamma = 10$, and yet the findings were contradicted by the randomized experiment by Moertel et al. (1985); see §1.2. This observation contains an important lesson about what sensitivity analyses can and cannot do. A sensitivity analysis can indicate the magnitude of the bias required to alter the qualitative conclusions of an observational study, but it cannot indicate what biases are present. Large biases have occurred in some studies. Chapters 6 through 9 are concerned with the collection of data that may indicate the presence or magnitude of hidden biases.

A second observation is of a technical nature, and it concerns the choice of test statistic. In this example, the choice of test statistic had a substantial impact on the sensitivity analysis. At present, little firm advice is available about the choice of test statistic for use in sensitivity analyses. Aside from a few special cases, statistical theory has emphasized the creation of test statistics that perform well in large samples against alternative hypotheses that are close to the null hypothesis, so-called local alternatives. The idea is that, in large samples, any reasonable test can detect a large departure from the null hypothesis, so tests should be designed to perform well against local alternatives. The situation is different in a sensitivity analysis. Large departures from the null hypothesis are of interest even in large samples, because only large departures are insensitive to hidden bias. An open area for research is the development of best tests for use in sensitivity analysis. For the most part, this chapter discusses traditional test statistics which, at least, are known to perform well in the absence of bias, that is, when $\Gamma = 1$.

4.4.4 Case-Referent Studies with Multiple Matched Referents

In the study of DES and vaginal cancer in Chapter 1, eight women with vaginal cancer, that is, eight cases, were each matched to four women without the disease, that is, four referents, and the cases and referents were compared in terms of their frequency of prenatal exposure to the treatment, DES. This study design, a case-referent study with several referents matched to each case, is quite common. If a disease or other outcome is rare, it may be difficult to locate additional cases, but referents may be plentiful; hence the tendency to match several referents to each case.

Note the difference between the study designs in §4.4.2 and 4.4.4. In matching with multiple controls in §4.4.2, each matched set contained one treated subject and several untreated controls, so $m_s = 1 = \sum_{i=1}^{n_s} Z_{si}$ for each s, but the design imposed no restrictions on the outcomes in a matched set. In a study in which cases are matched to several referents, matched set s may contain any number of treated subjects, that is, m_s can be $0, 1, \ldots,$ or n_s; however, each matched set contains exactly one case, so $1 = \sum_{i=1}^{n_s} R_{si}$. In §4.4.2, Ω_s contained n_s possible treatment assignments so the distribution of treatment assignments (4.6) had the simple form (4.15), but in a case-referent study, Ω_s contains $\binom{n_s}{m_s}$ possible treatment assignments, and (4.6) does not simplify further.

The null hypothesis of no treatment effect says that the binary indicator of a case of disease, R_{si}, is unaffected by the treatment Z_{si}, so R_{si} is a constant r_{si} that does not change with Z_{si}. The common test of no treatment effect is the Mantel–Haenszel statistic, which is a sign-score statistic with $d_s = 1$ and $c_{si} = r_{si}$, so the statistic (4.14) is the number of cases exposed to the treatment, namely, $T = \sum_{s=1}^{S} \sum_{i=1}^{n_s} r_{si} Z_{si} = \sum_{s=1}^{S} B_s$, where $B_s = \sum_{i=1}^{n_s} r_{si} Z_{si}$. Since each matched set contains one case in a case-referent study, B_s equals zero or one, as in §4.4.2.

Recall that in Proposition 14, the distribution of the test statistic T is bounded by the distributions of T^+ and T^- obtained by assuming $\mathbf{u} = \mathbf{u}^+$ or $\mathbf{u} = \mathbf{u}^-$. Since $T = \sum_{s=1}^{S} B_s$ in matched case-referent studies, these bounds may be obtained using the following bounds on $p_s = \text{prob}(B_s = 1|\mathbf{m})$ at $\mathbf{u} = \mathbf{u}^-$ or $\mathbf{u} = \mathbf{u}^+$

$$P_s^- = \frac{m_s}{m_s + (n_s - m_s)\Gamma} \leq p_s \leq \frac{m_s \Gamma}{m_s \Gamma + (n_s - m_s)} = P_s^+ \quad \text{for all } \mathbf{u} \in U.$$
(4.19)

For the derivation of (4.19), see Problem 6. If there is no treatment effect and no hidden bias, so $\Gamma = 1$, then $p_s = P_s^+ = P_s^- = m_s/n_s$; this says the chance the case was exposed to the treatment equals the proportion of subjects in the matched set who were exposed to the treatment. Note that (4.17) and (4.19) are similar in form but different in detail. The large sample approximations to the bounds on the significance levels for the Mantel–Haenszel statistic are obtained by referring the deviates,

$$\frac{|T - \sum P_s^+| - \frac{1}{2}}{\sqrt{\sum P_s^+(1 - P_s^+)}} \quad \text{and} \quad \frac{|T - \sum P_s^-| - \frac{1}{2}}{\sqrt{\sum P_s^-(1 - P_s^-)}},$$
(4.20)

to tables of the standard normal distribution. As with (4.18), these two deviates are equal if hidden bias is absent, in the sense that $\Gamma = 1$, and in this case their square is the usual Mantel–Haenszel statistic. These computations are performed for five studies in §4.4.5. This illustrates the computations and gives an indication of the varied results that a sensitivity analysis may produce.

4.4.5 Matched Case-Referent Studies: Five Examples

Recall from Chapter 1 that in the case-referent study by Herbst, Ulfelder, and Poskanzer (1971) of DES and vaginal cancer, each of eight cases of vaginal cancer was matched to four referents using day of birth and type of service, ward or private. Table 4.8 contains the data together with some of the calculations for the sensitivity analysis.

In Table 4.8, there are $S = 8$ matched sets, $s = 1, \ldots, 8$, each containing one case and four referents. In seven matched sets, exactly one patient had in utero exposure to DES, that is, $m_s = 1$, but in set $s = 5$ there were no exposures to DES, $m_s = 0$. In all of the seven matched sets, it was the case who was exposed to DES, that is, $B_s = 1$ for $s \neq 5$. The relationship between DES and vaginal cancer appears to be extremely strong, though

TABLE 4.8. DES and Vaginal Cancer: Data and Computations.

s	B_s	m_s	$P_s^- = P_s^+$ for $\Gamma = 1$	P_s^- for $\Gamma = 2$	P_s^+ for $\Gamma = 2$
1	1	1	0.20	0.11	0.33
2	1	1	0.20	0.11	0.33
3	1	1	0.20	0.11	0.33
4	1	1	0.20	0.11	0.33
5	0	0	0.00	0.00	0.00
6	1	1	0.20	0.11	0.33
7	1	1	0.20	0.11	0.33
8	1	1	0.20	0.11	0.33
Total	7	7	1.40	0.78	2.33

of course there are only eight cases. The one matched set with no exposed patients is concordant, and it could be removed without changing the value of the statistics in (4.20).

To test the null hypothesis of no effect, suppose for the moment that DES does not cause vaginal cancer. Consider one of the seven matched sets with $n_s = 5$ patients of whom $m_s = 1$ was exposed to DES. If the study were free of hidden bias, so $\Gamma = 1$, then each of the five patients has the same chance of being exposed, so the chance that the one exposed patient is the case is 1/5 or 0.20, as in Table 4.8. This would lead us to expect $1.40 = 7 \times 0.20$ cases to be exposed to DES, though, in fact, $T = 7$ cases were exposed, with a variance of $7 \times 0.20 \times 0.80 = 1.12$, and a deviate of $(|7 - 1.4| - \frac{1}{2})/\sqrt{1.12} = 4.82$, whose square $4.82^2 = 23.2$ is the usual Mantel–Haenszel statistic. If the study were free of hidden bias, there would be strong evidence that DES causes vaginal cancer. If hidden bias were present to the extent that matched subjects might differ in their odds of exposure to DES by a factor of two, so $\Gamma = 2$, then the chance that the case was exposed to DES might be as low as $m_s/\{m_s + \Gamma(n_s - m_s)\} = 1/(1+2\times4) = 0.11$ or as high as $\Gamma m_s/\{\Gamma m_s + (n_s - m_s)\} = 2/(2+4) = 0.33$, so the expected number of exposed cases might be as low as $7 \times 0.11 = 0.78$ with variance $7 \times 0.11 \times 0.89 = 0.69$, or as high as $7 \times 0.33 = 2.33$ with variance $7 \times 0.33 \times 0.67 = 1.56$, whereas seven exposed cases were observed. For $\Gamma = 2$, the deviates in (4.20) are

$$\frac{(|7 - 2.33| - \frac{1}{2})}{\sqrt{1.56}} = 3.34 \quad \text{and} \quad \frac{(|7 - 0.78| - \frac{1}{2})}{\sqrt{0.69}} = 6.88,$$

yielding a range of significance levels from less than 0.0001 to at most 0.0004. A hidden bias of magnitude $\Gamma = 2$ cannot reasonably explain the strong association seen between DES and vaginal cancer. Table 4.9 gives results for other values of Γ. Only beyond $\Gamma = 7$ is hidden bias a plausible

TABLE 4.9. Sensitivity Analysis for DES and Vaginal Cancer.

Γ	Deviate$^-$	Deviate$^+$	Range of signifance levels
1	4.82	4.82	< 0.0001
2	6.88	3.34	< 0.0001 to 0.0004
4	9.78	2.27	< 0.0001 to 0.012
6	12.00	1.77	< 0.0001 to 0.038
7	12.96	1.61	< 0.0001 to 0.054

explanation of the association between DES and vaginal cancer. The hidden bias would need to create a sevenfold increase in the odds of exposure to DES.

Table 4.10 compares the sensitivity of five matched case-referent studies. The purpose of the table is to gain some insight into the variety of results a sensitivity analysis may produce. The study of DES and vaginal cancer by Herbst et al. (1976) has just been discussed. The study by Mack et al. (1976) compared each of $S = 63$ cases of endometrial cancer to four referents matched for age and marital status. Cases and referents, who had been drawn from a large retirement community, were compared with respect to use of estrogens. The data are given in the detailed form needed for computations in Breslow and Day (1980, §5.3).

Jick et al. (1973) studied coffee consumption and myocardial infarction in white patients aged 40 to 69 years who were not alcoholic. Matching on age, gender, history of myocardial infarction, smoking, time of entry into the study, and hospital, they formed 27 matched pairs and 88 matched sets with two referents. This portion of the study included only subjects who either drank zero cups of coffee per day or at least six cups of coffee per day.

The study by Kelsey and Hardy (1975) examined the relationship between acute herniated lumbar disc—a back injury—and driving a truck as an occupation. They matched cases of herniated lumbar disc in New Haven to referents of the same gender, age, and hospital service or radiologist's office. Each case was matched to a single referent, and there were $S = 128$ matched pairs.

The study by Trichopoulos et al. does not appear to have been published. It is described only briefly by Miettinen (1969). It concerned prior induced abortion as a risk factor for subsequent ectopic pregnancy. A total of 18 cases were each matched to five referents.

Table 4.10 gives the upper bound on the significance level, the lower bound being highly significant in all cases in the table. The table shows that the five studies are quite different in their sensitivity to hidden bias.

TABLE 4.10. Comparison of the Sensitivity of Five Matched Case Control Studies: Upper Bounds on the Significance Level for Several Values of Γ.

Γ	1	2	3	4
DES and vaginal cancer Herbst et al. (1971), $n_s = 5$	< 0.0001	0.0004	0.0038	0.012
Endometrial cancer Mack et al. (1976), $n_s = 5$	< 0.0001	0.0004	0.013	0.068
Coffee and MI Jick et al. (1973), $n_s = 2$ or $n_s = 3$	0.0038	> 0.2		
Ectopic pregnancy Trichopoulos, et al., as in Miettinen (1969), $n_s = 5$	0.0001	0.0066	0.0301	0.0668
Herniated disc Kelsey and Hardy (1975), $n_s = 2$	0.0057	0.093	0.24	

Least sensitive is the study of DES and vaginal cancer, despite its small size, because of the strong association between treatment and outcome seen in the data in Table 4.8. To attribute this association to an unobserved covariate, one must postulate a covariate with a dramatic relationship with both exposure to DES and vaginal cancer.

Most sensitive is the study of coffee and myocardial infarction, where the upper bound on the significance level exceeds the conventional 0.05 for all $\Gamma \geq 1.3$. An unobserved covariate strongly related to myocardial infarction but only weakly related to coffee consumption could explain the observed association, despite the small significance level of 0.0038 for the usual Mantel–Haenszel statistic. Coffee may, in fact, cause myocardial infarctions, and this study may be free of hidden biases; however, fairly small biases, too small to easily detect, could readily explain the observed association.

The studies of endometrial cancer and of ectopic pregnancy are about equally sensitive to bias. The former study is more than three times larger, but the latter exhibits a stronger relationship between treatment and outcome.

Sensitivity to hidden bias is an aspect of the conclusions of an observational study, an aspect relevant to causal interpretation. However, sensitivity to hidden bias is not a fault of a study or its authors, as insensitivity to bias is not an accomplishment.

The study of herniated lumbar discs is different from the others in its structure, and a slightly different sensitivity analysis was performed. That study distinguished three types of cases:

(i) surgical cases, in which records indicate the herniated disc was seen by a surgeon during surgery;

(ii) probable cases, having the same symptoms as surgical cases but without an explicit record that a surgeon saw the herniated disc; and

(iii) possible cases, having some but not all of the symptoms of a herniated disc.

To give greatest weight to the surgical cases, numerical scores were assigned to each matched pair, specifically score $d_s = 3$ was assigned to pairs with a surgical case, score $d_s = 2$ was assigned to probable cases, and score $d_s = 1$ was assigned to possible cases. Set $c_{si} = 1$ for cases and $c_{si} = 0$ for referents. Then the sign-score statistic $T = \sum_s d_s \sum_i c_{si} Z_{si}$ is the total of the case scores for matched sets in which the case was exposed to the treatment, specifically, the total of the case scores for matched pairs in which the case was a truck driver. The upper bounds on the significance level are obtained

TABLE 4.11. Allopurinol and Rash.

			Z_{1i}		$n_1 = 719$
Males, $s = 1$			Allopurinol Other		$m_1 = 38$
	r_{1i}	Rash cases	5	36	$r_{1+} = 41$
		Noncases	33	645	
			Z_{2i}		$n_2 = 605$
Females, $s = 2$			Allopurinol Other		$m_2 = 29$
	r_{2i}	Rash cases	10	58	$r_{2+} = 68$
		Noncases	19	518	

using the deviates

$$\frac{|T - \sum_s d_s P_s^+| - \frac{1}{2}}{\sqrt{\sum d_s^2 P_s^+ (1 - P_s^+)}} \quad \text{and} \quad \frac{|T - \sum_s d_s P_s^-| - \frac{1}{2}}{\sqrt{\sum d_s^2 P_s^- (1 - P_s^-)}}, \quad (4.21)$$

which generalize the Mantel–Haenszel deviates in (4.20) to the case where $d_s \neq 1$ for some s. (The continuity correction should only be used if the scores d_s are integers, as is true in this example.)

4.4.6   The $2 \times 2 \times S$ Contingency Table

Table 4.11 concerns a case-referent study of the possible effects of the drug allopurinol as a cause of rash (Boston Collaborative Drug Surveillance Program 1972). The table has two strata, males and females. Here, $r_{si} = 1$ for a case of rash, $r_{si} = 0$ for a referent, $Z_{si} = 1$ for use of allopurinol, $Z_{si} = 0$ otherwise, and $r_{s+} = \sum_i r_{si}$ is the number of cases in stratum s. The Mantel–Haenszel statistic, as discussed in §2.4.3, is based on the number of cases who used allopurinol, namely $T = \sum_s \sum_i Z_{si} r_{si} = 5 + 10 = 15$. If allopurinol had no effect, then in a randomized experiment or in the absence of hidden bias, T would have expectation 5.43 and variance 4.70, yielding a deviate of $(|15 - 5.43| - \frac{1}{2})/\sqrt{4.70} = 4.19$, with an approximate one-sided significance level less than 0.0001. In a randomized experiment or in the absence of hidden bias, Table 4.11 would constitute strong evidence that allopurinol causes rash.

The procedure for conducting a sensitivity analysis for the Mantel–Haenszel statistic is as follows. The exact bounds are provided by Corollary 16, and the large sample approximation that follows is developed in the appendix, §4.7.4. Determine the unique root $\tilde{E}_s$ of the quadratic equation

$$\tilde{E}_s^2 (\Gamma - 1) - \tilde{E}_s \{(\Gamma - 1)(m_s + r_{s+}) + n_s\} + \Gamma r_{s+} m_s = 0, \quad (4.22)$$

with $\max(0, r_{s+} + m_s - n_s) \leq \tilde{E}_s \leq \min(r_{s+}, m_s)$, calculate

$$\tilde{V}_s = \left(\frac{1}{\tilde{E}_s} + \frac{1}{r_{s+} - \tilde{E}_s} + \frac{1}{m_s - \tilde{E}_s} + \frac{1}{n_s - r_{s+} - m_s + \tilde{E}_s} \right)^{-1}, \quad (4.23)$$

and refer the deviate $(|T - \sum \tilde{E}_s| - \frac{1}{2})/\sqrt{\sum \tilde{V}_s}$ to tables of the standard normal distribution. Repeat with Γ replaced by $1/\Gamma$ to determine the other bound on the significance level.

To illustrate, for $\Gamma = 2$, $s = 1$, equation (4.22) is

$$\tilde{E}_1^2 (2 - 1) - \tilde{E}_1 \{(2 - 1)(38 + 41) + 719\} + 2 \times 41 \times 38 = 0$$

or

$$\tilde{E}_1^2 - 789 \tilde{E}_1 + 3,116 = 0,$$

with roots 794.08 and 3.92, of which the correct root is $\tilde{E}_1 = 3.92$ since $\max(0, 41 + 38 - 719) = 0 \leq 3.92 \leq 38 = \min(41, 38)$. Substituting $\tilde{E}_1 = 3.92$ into (4.23) yields $V_1 = 3.20$. In the same way, $\tilde{E}_2 = 5.66$, $V_2 = 4.21$, and $(|T - \sum \tilde{E}_s| - \frac{1}{2})/\sqrt{\sum \tilde{V}_s} = (|15 - 9.59| - \frac{1}{2})/\sqrt{7.41} = 1.80$, giving an upper bound on the one-sided significance level of 0.036.

Computed in this way, the upper bounds on the significance levels from Table 4.11 for $\Gamma = 1, 2$, and 3 are 0.0001, 0.036, and 0.30. The study is insensitive to a bias that would double the odds of exposure to allopurinol but sensitive to a bias that would triple the odds. It is more sensitive to bias than three of the five studies in §4.4.5 and less sensitive than the other two.

When $\mathbf{u} = \mathbf{u}^+$ and the treatment has no effect, expressions (4.22) and (4.23) yield, respectively, large sample approximations to the expectation and variance of the upper left corner cell count in the sth 2×2 table, namely $\sum_i Z_{si} r_{si}$. A detailed derivation of (4.22) and (4.23) is given in the Appendix to this chapter.

The approximate expectation (4.22) and variance (4.23) are appropriate when each marginal total of table s is large, that is, when m_s, $n_s - m_s$, r_{s+}, and $n_s - r_{s+}$ are each large. If table s has a small marginal total, the exact moments of $\sum_i Z_{si} r_{si}$ may be used in place of the approximations. The exact expectation and variance are

$$\overline{E}_s = \frac{\sum_{t=\max(0, m_s + r_{s+} - n_s)}^{\min(m_s, r_{s+})} t \binom{r_{s+}}{t} \binom{n_s - r_{s+}}{m_s - t} \Gamma^t}{\sum_{t=\max(0, m_s + r_{s+} - n_s)}^{\min(m_s, r_{s+})} \binom{r_{s+}}{t} \binom{n_s - r_{s+}}{m_s - t} \Gamma^t} \qquad (4.24)$$

and

$$\overline{V}_s = \frac{\sum_{t=\max(0, m_s + r_{s+} - n_s)}^{\min(m_s, r_{s+})} t^2 \binom{r_{s+}}{t} \binom{n_s - r_{s+}}{m_s - t} \Gamma^t}{\sum_{t=\max(0, m_s + r_{s+} - n_s)}^{\min(m_s, r_{s+})} \binom{r_{s+}}{t} \binom{n_s - r_{s+}}{m_s - t} \Gamma^t} - \overline{E}_s^2$$

TABLE 4.12. Comparison of Results Using Exact and Approximate Moments of T.

Γ		Expectation	Variance	Deviate
1	Exact	5.43	4.70	4.19
	Approximate	5.43	4.69	4.19
2	Exact	9.60	7.44	1.80
	Approximate	9.59	7.41	1.80
3	Exact	12.97	9.17	0.51
	Approximate	12.94	9.13	0.52

$$= \frac{\Gamma m_s r_{s+} - \{n_s - (m_s + r_{s+})(1 - \Gamma)\}\overline{E}_s}{(1 - \Gamma)} - \overline{E}_s^2. \qquad (4.25)$$

As an example of a table with small counts, suppose $n_s = 8$, $m_s = 4$, and $r_{s+} = 3$. For $\Gamma = 2$, the exact moments in this hypothetical table are $\overline{E}_s = 1.87$ and $\overline{V}_s = 0.52$, while the approximations are $\tilde{E}_s = 1.82$ and $V_s = 0.45$. In this case, the exact expectation in (4.24) is the sum of four terms, namely, from $t = 0 = \max(0, 4 + 3 - 8)$ to $t = \min(4, 3) = 3$. (Expression (4.25) is derived from an observation of Johnson, Kotz, and Kemp (1992, p. 280, expression (6.160)).)

Returning to the case-referent study of allopurinol and rash, Table 4.12 compares the calculations based on the exact and approximate moments of T. In this example, the approximation is quite satisfactory.

The sensitivity analysis for binary responses has an intimate connection with setting confidence limits for an odds ratio in a $2 \times 2 \times S$ contingency table, so a program such as Cytel Corporation's StatXact can do certain sensitivity calculations. This link has not been stressed in this chapter because it does not apply with outcomes that are not binary.

4.4.7 Comparing Rates in a Treated Group to a Population of Controls

Some observational studies compare a treated group of finite size to a control group that is, in effect, infinite in size, that is, a control group in which sampling variability is negligible. This is particularly common in studies in occupational health, where the mortality in an industry is often compared to mortality rates for the nation as a whole. For instance, in a study of the hazards of asbestos, Nicholson, Selikoff, Seidman, Lilis, and Formby (1979) studied a treated group consisting of 544 male miners of chrysotile asbestos at the Thetford Mines in Quebec. All had been miners for at least 20 years as of 1961, and their mortality between 1961 and 1977 is under study.

These 544 miners were observed for a total of 7408 years; that is, there were 7408 person-years of observation. The control group was constructed from the mortality experience for all males in Canada. By reweighting national mortality rates that were specific to age, gender, and cause of death, the investigators computed mortality rates for a constructed population of Canadians having the same distribution of ages and the same number of years of observation as the 544 miners. For instance, they found that the constructed population would have a mortality rate of 0.01232 from noninfectious pulmonary diseases other than cancer, so 544 individuals selected at random from this constructed population would be expected to have $0.01232 \times 544 = 6.70$ deaths from this disease. Because of the enormous size of the population of Canada, this mortality rate is effectively free of sampling variability; however, since miners may differ from the typical Canadian in terms of covariates other than age, the rate may be affected by hidden biases. In fact, the 544 miners experienced 30 deaths from noninfectious pulmonary diseases other than cancer, or $30/6.70 = 4.48$ times more than expected. The ratio $30/6.70 = 4.48$ is called the *standardized mortality ratio*. In the absence of hidden bias, the usual test of no treatment effect views the 30 deaths as either a binomial count with probability 0.01232 and sample size 544 or as a Poisson count with expectation 6.70; see, for instance, Armitage (1977, p. 389), Gastwirth (1988, §10), Mosteller and Tukey (1977, §11E), or Hakulinen (1981). The binomial test may be viewed as asking whether the 544 miners look like a random sample of size 544 from an infinite population with rate 0.01232.

A small change in the method in §4.4.6 for 2×2 tables addresses the infinite sample size in the control group. The large sample argument in §4.4.6 assumed that the margins, $m_s, n_s - m_s, r_{s+}, n_s - r_{s+}$, of each 2×2 table were large, with all margins increasing at about the same rate. When comparing 544 miners to a reweighting of all of Canada, a more realistic asymptotic argument assumes that $m_s = 544$ is fixed, $n_s \to \infty$, with $r_{s+}/n_s \to \zeta_s = 0.01232$. At $\mathbf{u} = \mathbf{u}^+$ under the null hypothesis of no effect, $\sum_i Z_{si} r_{si}$ is asymptotically binomial with sample size m_s and probability

$$\nu_s = \frac{\zeta_s \Gamma}{(1 - \zeta_s) + \zeta_s \Gamma};$$

see the appendix, §4.7.4, for proof. If there is no hidden bias, then $\Gamma = 1$ and $\nu_s = \zeta_s$, leading to the usual binomial distribution for $\sum_i Z_{si} r_{si}$. With $T = \sum_s \sum_i Z_{si} r_{si}$, the deviate is

$$\frac{|T - \sum m_s \nu_s| - \frac{1}{2}}{\sqrt{\sum m_s \nu_s (1 - \nu_s)}}.$$

For the bound at $\mathbf{u} = \mathbf{u}^-$, the same calculation is performed with $1/\Gamma$ in place of Γ; see the appendix, §4.7.5. Gastwirth and Greenhouse (1987)

TABLE 4.13. Sensitivity Analysis for the Asbestos Miners' Standardized Mortality Ratio.

Γ	ν_1	$\nu_1 m_1$	Deviate
1	0.01232	6.70	8.86
2	0.02434	13.24	4.52
3	0.03607	19.62	2.27
4	0.04752	25.85	0.73

perform a similar calculation starting from the inequality of Cornfield et al. (1959).

In the example, there is one stratum, $S = 1$, and $m_1 = 544$, $\zeta = 0.01232$. Table 4.13 gives the sensitivity analysis for the upper bound on the significance level. In the absence of hidden bias, $\Gamma = 1$, 6.70 deaths were expected, the deviate is 8.86 with significance level less than 0.0001, and the standardized mortality ratio is $30/6.70 = 4.5$. A bias of magnitude $\Gamma = 2$ could raise the expected number of deaths to 13.24, though the observed 30 deaths are significantly more than 13.24, yielding a deviate of at least 4.52, a significance level less than 0.0001, and a standardized mortality ratio of at least $30/13.24 = 2.3$. This study is less sensitive to bias than the studies of allopurinol in §4.4.6 or coffee in §4.4.5, but it is more sensitive than the studies of smoking in §4.3.2 or lead exposure in §4.3.3.

4.4.8 Other Sign-Score Statistics: Censored Data, Median Tests

All sign-score statistics permit a sensitivity analysis based on Proposition 14. Sections 4.3 and 4.4 have presented details and examples for methods that are widely used. This final part of §4.4 briefly mentions several other sign-score statistics that are useful in particular situations.

(i) *Censored Survival Data in Matched Pairs.* When the outcome is the time until a particular event occurs, it may happen that for some individuals the event has not yet occurred at the time the data are analyzed, in which case the outcome is censored. This is often true in studies of human survival, in which some individuals are alive at the end of the study. It is also often true in studies that compare the effectiveness of various punishments for criminal acts, in which the outcome is the time until the act is repeated. O'Brien and Fleming (1987) propose and evaluate their Prentice–Wilcoxon statistic for censored survival times in matched pairs. This statistic may be written as a sign-score statistic, so the sensitivity analysis follows directly from the general procedure in §4.4.2 with each $n_s = 2$.

(ii) *Median, Quantile, and Sign Tests.* In comparing treated and control groups, the median test combines the groups, finds the median of the combination, and asks how many treated subjects have responses above the combined median. Though inefficient for responses obtained from a Normal distribution, the median test has good properties in large samples for data from a distribution more prone to extreme observations, specifically the double exponential distribution; see Hájek, Sidák and Sen (1999, §4.1, p. 97) or Hettmansperger (1984, §3) for detailed discussion. The median test is usually attributed to G. Brown and A. Mood. See also §5.3.

Other quantiles are sometimes used in place of the median. A version of the median test was used in §4.4.3 for matching with multiple controls. In pair matching, the median test becomes the sign test. The median test, the stratified or matched median test, the sign test, and other quantile tests are all sign-score tests.

4.5 Matching with Multiple Controls and Continuous Responses

4.5.1 A Simple, Effective Research Design

Matching with continuous responses and multiple controls is a simple research design that is often more effective than other simple designs, such as pair matching or stratification, and yet is it not widely used. Several of the statistical methods for matching with multiple controls, such as the aligned rank test of Hodges and Lehmann (1962), are also simple to use and highly effective, and yet they too are much less widely known than their close relatives, the signed rank test and the rank sum test. Before discussing procedures in detail, some motivation for this good choice of design is needed.

In matching with multiple controls, each treated subject is matched to one or more controls, so $m_s = 1$, $n_s \geq 2$, for $s = 1, \ldots, S$. When controls are easily available and inexpensive, the use of more than one control for each treated subject often increases the precision of estimates and the power of tests. At the same time, matching can ensure that treated and control groups are balanced with respect to observed covariates, while entirely excluding from the comparison potential controls who are unlike virtually all treated subjects.

For example, in a study of effective management of nursing at hospitals, Smith (1997) compared three versions of matching: matching with 1 control hospital, $n_s = 2$, matching with 8 control hospitals, $n_s = 9$, and matching with 15 control hospitals, $n_s = 16$. In this one particular data set, Smith (1997, p347) found that matching with 8 controls yielded substantially smaller standard errors than matching with 1 control, but the bias in observed covariates was not greatly increased. On the other hand, Smith also found that matching with 15 controls was only slightly better than matching with 8 controls in terms of standard errors, and was much worse in terms of comparability on observed covariates. See also Aiken, Smith, and Lake (1994) for results linking nursing management and hospital mortality.

Matching with a variable number of controls, so that n_s varies with s, can remove substantially more bias in observed covariates than matching with a fixed number of controls, when studies of equal total sample size $\sum n_s$ are compared. Moreover, the loss of precision from letting n_s vary with s is often quite small. See Ming and Rosenbaum (2000) for detailed discussion. The optimal design for an observational study is a "full matching," which is a generalization of matching with multiple controls (Rosenbaum 1991c, Gu and Rosenbaum 1993). These issues are discussed in some detail in Chapter 10.

4.5.2   General Method

In the case of sign-score statistics, such as Wilcoxon's signed rank test or the Mantel–Haenszel test, it was possible to bound the distribution of the test statistic, T, by two known distributions for random variables T^+ and T^-. This is not possible in general. However, for matching with multiple controls, it is possible to find two approximate bounds for tail probabilities, the approximation being quite good when the number of matched sets S is large. The general method is developed by Gastwirth, Krieger, and Rosenbaum (2000). The underlying idea is quite simple. If, as a consequence of the central limit theorem, T is asymptotically Normal as $S \to \infty$, then its limiting Normal distribution is characterized by its expectation and variance. If one cannot find a distribution that provides an upper bound on the chance that $T \geq k$, perhaps instead one can find the limiting Normal distribution that maximizes the chance that $T \geq k$. Here, k is some critical value in the upper tail of the distribution, so it is above the expectation of T. As intuition suggests, the limiting Normal distribution that attaches the greatest chance to $T \geq k$ has the largest possible expectation for T, and among distributions with this largest possible expectation, it is the distribution with the largest variance. The demonstration that this intuition is correct is straightforward, but lengthy, and the interested reader should turn to Gastwirth, Krieger and Rosenbaum (2000) for the proof and for minor regularity conditions. A key aspect of the proof is that, as $S \to \infty$, maximizing the expectation is more important than increasing the variance, whereas for finite S this may not be the case. Here, the simple procedure is described, and then discussed for two common tests, the stratified Wilcoxon rank sum test and the aligned rank test of Hodges and Lehmann (1962). Tests are discussed first, then inverted to obtain confidence intervals, and the device of Hodges and Lehmann (1963) yields point estimates.

In matching with multiple controls, $1 = m_s = \sum_{i=1}^{n_s} Z_{si}$ for each s and the distribution of treatment assignments is given by (4.15). Let q_{si} be a fixed score associated with the ith subject in stratum s, and consider the distribution of the statistic $T = \sum_{s=1}^{S} \sum_{i=1}^{n_s} Z_{si} q_{si}$. The expectation and variance of T are then:

$$E\left(T|\mathbf{m}\right) = \sum_{s=1}^{S} \frac{\sum_{i=1}^{n_s} q_{si} \exp\left(\gamma u_{si}\right)}{\sum_{i=1}^{n_s} \exp\left(\gamma u_{si}\right)}$$

and

$$var\left(T|\mathbf{m}\right) = \left\{ \sum_{s=1}^{S} \frac{\sum_{i=1}^{n_s} q_{si}^2 \exp\left(\gamma u_{si}\right)}{\sum_{i=1}^{n_s} \exp\left(\gamma u_{si}\right)} \right\} - \left\{ E\left(T|\mathbf{m}\right) \right\}^2,$$

and these are unknown because u_{si} is unknown.

For notational convenience, within each matched set, renumber the n_s subjects, $i = 1, \ldots, n_s$, so the q_{si} are in nondecreasing order, $q_{s,n_s} \geq \ldots \geq q_{s1}$. Let a be an integer, $0 < a < n_s$, and consider the contribution $\sum_{i=1}^{n_s} Z_{si} q_{si}$ to T from matched set s when $u_{s1} = 0, \ldots, u_{sa} = 0, u_{s,a+1} = 1, \ldots, u_{s,n_s} = 1$. At this value of the u_{si}, the expectation and variance of $\sum_{i=1}^{n_s} Z_{si} q_{si}$ are, respectively

$$\mu_{sa} = \frac{\sum_{i=1}^{a} q_{si} + \Gamma \sum_{i=a+1}^{n_s} q_{si}}{a + \Gamma\left(n_s - a\right)}$$

and

$$\nu_{sa}^2 = \frac{\sum_{i=1}^{a} q_{si}^2 + \Gamma \sum_{i=a+1}^{n_s} q_{si}^2}{a + \Gamma\left(n_s - a\right)} - \mu_{sa}^2.$$

Let the maximum of these a expectations be

$$\mu_s = \max_{0<a<n_s} \mu_{sa},$$

and let $A_s = \{a : \mu_{sa} = \mu_s\}$ be the set of values of a giving rise to this maximum expectation. Proposition 9 in the appendix shows that μ_s is the maximum expectation of $\sum_{i=1}^{n_s} Z_{si} q_{si}$. Let

$$\nu_s^2 = \max_{a \in A_s} \nu_{sa}^2,$$

so ν_s^2 is the maximum variance among a's giving rise to the maximum expectation. If $t > \sum \mu_s$, then approximate the upper bound on the tail probability, $prob\left(T \geq t|\mathbf{m}\right)$, by

$$1 - \Phi\left(\frac{t - \sum_{s=1}^{S} \mu_s}{\sqrt{\sum_{s=1}^{S} \nu_s^2}}\right),$$

where $\Phi\left(\cdot\right)$ is the standard Normal cumulative distribution. Under regularity conditions, Gastwirth, Krieger, and Rosenbaum (2000) show that this approximate upper bound converges to the correct, sharp upper bound on the tail probability as the number of matched sets increases, $S \to \infty$. If $\sum \mu_s \geq t$, then the upper bound on prob $(T \geq t|\mathbf{m})$ is at least $\frac{1}{2}$. The upper bound on the opposite tail area, prob $(T \leq t|\mathbf{m})$, is obtained by minimizing the expected contributions to T, and maximizing the variance among expected contributions that produce the same minimum expectation; see Problem 11.

Under the model of an additive treatment effect, $r_{Tsi} = r_{Csi} + \tau$, the hypothesis $H_0 : \tau = \tau_0$ is tested by applying the above procedure to the adjusted responses, $R_{si} - \tau_0 Z_{si}$. A confidence interval is formed by testing each τ_0 and retaining the values not rejected by the test.

4.5.3 Stratified Wilcoxon Rank Sum Test

In the stratified Wilcoxon rank sum test, q_{si} is the rank of R_{si} when the responses of the n_s subjects in matched set s are ranked from 1 to n_s, with average ranks for ties. Then T is the sum of the ranks for the S treated subjects. With matched pairs, $n_s = 2$, the stratified rank sum test is equivalent to the sign test, which is inefficient for Normal data, but fairly efficient for some distributions with longer tails. For Normal data, the performance of the stratified rank sum test improves as n_s increases; see Hodges and Lehmann (1962).

When there are no ties, the ranks are just $1, \ldots, n_s$, and μ_s and ν_s^2 can be tabulated, saving some arithmetic. Table 4.14 is from Gastwirth, Krieger, and Rosenbaum (2000). With this table, the calculations are elementary arithmetic. If one is conducting a sensitivity analysis with $\Gamma = 2$, and matched set s is an untied matched triple, one treated and two controls, $n_s = 3$, then from the table, $\mu_s = \frac{9}{4}$ and $\nu_s^2 = \frac{11}{16}$. Summing S such terms gives the required expectation and variance. If $\sum \mu_s \geq T = \sum_{s=1}^{S} \sum_{i=1}^{n_s} Z_{si} q_{si}$ then the upper bound on the one-sided significance level is at least $\frac{1}{2}$; otherwise,

$$\frac{T - \sum \mu_s}{\sqrt{\sum \nu_s^2}}$$

is compared to the standard Normal distribution to approximate the upper bound on the significance level. The minimum Hodges–Lehmann estimate is obtained by computing T from $R_{si} - \tau_0 Z_{si}$, and finding the value τ_0 that equates T to the maximum expectation $\sum_{s=1}^{S} \mu_s$.

TABLE 4.14. Extreme Moments for the Stratified Rank Sum Statistic.

n_s	Γ	μ_s	ν_s^2
2	1	3/2	1/4
	2	5/3	2/9
	3	7/4	3/16
	4	9/5	4/25
3	1	2	2/3
	2	9/4	11/16
	3	12/5	16/25
	4	5/2	7/12
4	1	5/2	5/4
	2	17/6	41/36
	3	3	4/3
	4	22/7	62/49
5	1	3	2
	2	24/7	96/49
	3	11/3	16/9
	4	42/11	194/121

4.5.4 Aligned Rank Test

In the aligned rank test of Hodges and Lehmann (1962), ranks q_{si} from 1 to $N = n_1 + \ldots + n_S$ are assigned to the N aligned responses, $r_{si} - (1/n_s)\sum_{j=1}^{n_s} r_{sj}$, formed by subtracting the mean of the n_s responses in each matched set s. In matched pairs, the aligned rank test is similar to Wilcoxon's signed rank test, and is more efficient with Normal data than the stratified rank sum test; see Lehmann (1975) for detailed discussion. Because the ranks within one matched set now depend on the responses in other matched sets, it is no longer possible to tabulate μ_s and ν_s^2, and a little more arithmetic is required. When testing the hypothesis $H_0 : \tau = \tau_0$ under the model of an additive treatment effect, $r_{Tsi} = r_{Csi} + \tau$, one computes adjusted responses, $R_{si} - \tau_0 Z_{si}$, which equal r_{Csi} under the null hypothesis, and then aligns these adjusted responses. A numerical example is given by Gastwirth, Krieger, and Rosenbaum (2000).

4.6   Sensitivity Analysis for Comparing Two Unmatched Groups

4.6.1   Why Is There a Difference Between Sum Statistics and Sign-Score Statistics?

This section discusses the sensitivity analysis for statistics T, such as the rank sum statistic, Gehan's statistic, or the log-rank statistic, used to compare unmatched treated and control groups. Here, there is a single stratum, $S = 1$, so the subscript s is omitted, and $m = \sum Z_i$. As defined in Chapter 2, a sum statistic has the form

$$T = \sum_{i=1}^{n} Z_i q_i, \tag{4.26}$$

where the q_i are functions of the responses and so are fixed under the null hypothesis of no treatment effect. For instance, in Wilcoxon's rank sum statistic, the q_i are the ranks of the responses r_i, and T is the sum of the ranks in the treated group.

The sensitivity analysis for sign-score statistics in §4.2 and 4.3 and for sum statistics in the current section are similar in certain ways and different in others. They are identical in interpretation and similar in their broad outlines, but they differ in an important detail of implementation. In Proposition 14, the distribution of the test statistic T is bounded at all points "a" by the distributions of two other random variables, T^+ and T^-, which are obtained from the distribution (4.6) at two points $\mathbf{u}^+$ and $\mathbf{u}^-$. In contrast, with sum statistics, there is a point $\mathbf{u}^+ \in U$ that provides an upper bound on the probability $\text{prob}(T \geq a|m)$; however, the point $\mathbf{u}^+$ changes as "a" changes. This distinction has a noticeable effect on calculations, but no effect on the interpretation of the sensitivity analysis.

4.6.2   Sensitivity Analysis for Sum Statistics: The General Method

This section obtains bounds on the behavior of a sum statistic $T = \sum Z_i q_i$ under the null hypothesis of no treatment effect. Specifically, bounds are obtained for the tail area or significance level $\text{prob}(T \geq a|m)$ and for the expectation $E(T)$. These are the basis for the sensitivity analysis for significance levels, confidence intervals, and point estimates. For this purpose, in §4.6.2, assume that the null hypothesis of no treatment effect holds, and that $\Gamma \geq 1$. Renumber the n subjects so that $q_1 \geq q_2 \geq \ldots \geq q_n$. Since no quantity depends on the numbering of the n subjects, this is just a notational convenience.

The maximum value of $\text{prob}(T \geq a|m)$ from model (4.6) with $\mathbf{u} \in U$ is found at point $\mathbf{u}$ of the form $u_1 = 1, u_2 = 1, \ldots, u_k = 1, u_{k+1} =$

$0, u_{k+2} = 0, \ldots, u_n = 0$, that is, at a point whose coordinates consist of k ones followed by $n - k$ zeros for some number k. Let U^+ be the set of all such $\mathbf{u}$, for $k = 0, 1, \ldots, n$. Similarly, the minimum value of $\text{prob}(T \geq a|m)$ is found at a point whose coordinates are k zeros followed by $n - k$ ones for some k. Write U^- for the set of all such points. The value of k may change as "a" is changed. The maximum and minimum values of $E(T)$ are also found at points of, respectively, U^+ and U^-. These facts are proved in the appendix to this chapter, specifically in §4.7.3.

For a given $\mathbf{u}$, let $\mu_{\mathbf{u}}$ and $\sigma_{\mathbf{u}}$ be the expectation and standard deviation of T. Exact expressions and easily calculated approximations to $\mu_{\mathbf{u}}$ and $\sigma_{\mathbf{u}}$ are discussed shortly. The bounds on $E(T)$ are then

$$\min_{\mathbf{u} \in U^-} \mu_{\mathbf{u}} \leq E(T|m) \leq \max_{\mathbf{u} \in U^+} \mu_{\mathbf{u}}. \tag{4.27}$$

Large sample approximations to the bounds on $\text{prob}(T \geq a|m)$ are obtained by referring the standardized deviates,

$$\max_{\mathbf{u} \in u^-} \frac{a - \mu_{\mathbf{u}}}{\sigma_{\mathbf{u}}} \quad \text{and} \quad \min_{\mathbf{u} \in u^+} \frac{a - \mu_{\mathbf{u}}}{\sigma_{\mathbf{u}}}, \tag{4.28}$$

to tables of the standard normal distribution. The effort required to compute the maximum and minimum in (4.28) is of the same order of magnitude as computing the rank sum statistic $2n$ times, an easy task with a computer.

4.6.3 *Expectation and Variance of T*

To use (4.27) and (4.28) for inference, one needs the moments $\mu_{\mathbf{u}}$ and $\sigma_{\mathbf{u}}$ of T when $\mathbf{u}$ has k coordinates equal to 1 and $n - k$ coordinates equal to 0. The current section discusses exact moments and the next section discusses large sample approximations. As is often true with continuous approximations to discrete distributions, the exact calculations use only elementary probability theory and somewhat more arithmetic, while the approximations use asymptotic theory and less arithmetic. The exact calculations are easily programmed in, say, S+, while the approximations involve solving a quadratic equation and can be done in a spreadsheet. Although the approximations perform well and are sometimes convenient, the formulas provide no insight, so the sections describing the approximations are marked with an asterisk indicating these sections may be skipped.

Fix a $\mathbf{u}$ with k coordinates equal to 1 and $n - k$ coordinates equal to 0. For this $\mathbf{u}$, write

$$\varrho_i = \text{prob}\,(Z_i = 1|m) = E\,(Z_i|m)$$

and

$$\varrho_{ij} = \text{prob}\,(Z_i = 1, Z_j = 1|m) = E\,(Z_i Z_j|m).$$

Notice carefully that $\varrho_{ii} = \varrho_i$. In words, given that m subjects were treated, the chance that subject i was treated is ϱ_i and the chance that both subject i and subject j were treated is ϱ_{ij}. Expressions for ϱ_i and ϱ_{ij} are given soon. Now $\mu_{\mathbf{u}} = E\left(\sum Z_i q_i\right) = \sum q_i \varrho_i$. Also,

$$\operatorname{cov}\left(Z_i, Z_j | m\right) = E\left(Z_i Z_j | m\right) - E\left(Z_i | m\right) E\left(Z_j | m\right) = \varrho_{ij} - \varrho_i \varrho_j,$$

so

$$\sigma_{\mathbf{u}}^2 = \operatorname{var}\left(\sum Z_i q_i\right) = \sum q_i q_j \operatorname{cov}\left(Z_i, Z_j | m\right) = \sum q_i q_j \left(\varrho_{ij} - \varrho_i \varrho_j\right).$$

In other words, the needed moments are simple functions of the probabilities, ϱ_i and ϱ_{ij}.

To find $\varrho_i = \operatorname{prob}\left(Z_i = 1 | m\right)$, the probability (4.6) with $S = 1$ must be summed over all $\mathbf{z} \in \Omega$ such that $z_i = 1$. Remember that $u_i = 1$ for k subjects and $u_i = 0$ for the remaining $n - k$ subjects. Write

$$\zeta\left(n, m, k\right) = \sum_{\mathbf{z} \in \Omega} \exp\left(\gamma \mathbf{z}^T \mathbf{u}\right) = \sum_{a=\max(0, m+k-n)}^{\min(m,k)} \binom{k}{a}\binom{n-k}{m-a}\Gamma^a$$

and *define* $\zeta\left(n, m, k\right) = 0$ if either $m < 0$ or $k < 0$. Now $\zeta\left(n, m, k\right)$ is the denominator of (4.6) with $S = 1$. Moreover,

$$\varrho_i = \frac{\exp\left(\gamma u_i\right)\zeta\left(n-1, m-1, k-u_i\right)}{\zeta\left(n, m, k\right)}.$$

In parallel, $\varrho_{ij} = \operatorname{prob}\left(Z_i = 1, Z_j = 1 | m\right)$ is the sum of the probability (4.6) with $S = 1$ over all $\mathbf{z} \in \Omega$ such that $z_i = 1$ and $z_j = 1$. If $i = j$, then $\varrho_{ii} = \varrho_i$, and if $i \neq j$ then

$$\varrho_{ij} = \frac{\exp\left\{\gamma\left(u_i + u_j\right)\right\}\zeta\left(n-2, m-2, k-u_i-u_j\right)}{\zeta\left(n, m, k\right)}.$$

Because each u_i is either 0 or 1, there are only two values of ϱ_i to be computed, and only three values of ϱ_{ij} to be computed.

4.6.4   *Large Sample Approximations to the Expectation and Variance of T

Large sample approximations to $\mu_{\mathbf{u}}$ and $\sigma_{\mathbf{u}}$ are now given when $\mathbf{u}$ has k coordinates equal to 1 and $n - k$ coordinates equal to zero. The approximations are developed in the appendix, §4.7.4; they involve approximations $\tilde{E}$ and $\tilde{V}$ to the expectation and variance of $\mathbf{Z}^T \mathbf{u}$, where $\mathbf{Z}^T \mathbf{u}$ is the number of treated subjects with $u = 1$. As in §4.4.6, the calculation involves solving the quadratic equation

$$\tilde{E}^2(\Gamma - 1) - \tilde{E}\{(\Gamma - 1)(m + k) + n\} + \Gamma k m = 0 \qquad (4.29)$$

for the root $\tilde{E}$ with $\max(0, k + m - n) \leq \tilde{E} \leq \min(k, m)$. If $k = 0$, then $\tilde{E} = 0$, but if $k = n$, then $\tilde{E} = m$, and in either case let $\tilde{V} = 0$; otherwise, let

$$\tilde{V} = \left(\frac{1}{\tilde{E}} + \frac{1}{k - \tilde{E}} + \frac{1}{m - \tilde{E}} + \frac{1}{n - k - m + \tilde{E}} \right)^{-1}. \qquad (4.30)$$

Then calculate the mean and variance of the scores q_i separately for subjects with $u_i = 1$ and $u_i = 0$,

$$\bar{q}_1 = \frac{1}{k} \sum_{i:u_i=1} q_i, \quad w_1 = \frac{1}{k-1} \sum_{i:u_i=1} (q_i - \bar{q}_1)^2,$$

$$\bar{q}_0 = \frac{1}{n-k} \sum_{i:u_i=0} q_i, \quad w_0 = \frac{1}{n-k-1} \sum_{i:u_i=0} (q_i - \bar{q}_0)^2,$$

where we define $\bar{q}_1 = 0$ if $k = 0$, $\bar{q}_0 = 0$ if $k = n$, $w_1 = 0$ if $k \leq 1$, and $w_0 = 0$ if $k \geq n - 1$.

Finally, if $k \neq 0$ and $k \neq n$, then approximate $\mu_{\mathbf{u}}$ by

$$\tilde{E}\bar{q}_1 + (m - \tilde{E})\bar{q}_0 \qquad (4.31)$$

and $\sigma_{\mathbf{u}}$ by

$$\sqrt{(w_1 - w_0)\tilde{E} - (\tilde{E}^2 + \tilde{V})\left(\frac{w_1}{k} + \frac{w_0}{n-k}\right) + \frac{m(n-k-m+2\tilde{E})w_0}{n-k} + \tilde{V}(\bar{q}_1 - \bar{q}_0)^2}; \qquad (4.32)$$

otherwise, if $k = 0$ or $k = n$,

$$\mu_{\mathbf{u}} = \frac{m}{n} \sum q_i \text{ and } \sigma_{\mathbf{u}} = \sqrt{\frac{m(n-m)}{n(n-1)} \sum (q_i - \bar{q})^2}, \text{ where } \bar{q} = \frac{1}{n} \sum q_i. \qquad (4.33)$$

In these calculations, $\tilde{E}$ and $\tilde{V}$ approximate the expectation and variance of $\mathbf{Z}^{\mathrm{T}}\mathbf{u}$, that is, of the number of treated subjects with $u_i = 1$. If the exact expectation $\overline{E}$ and variance $\overline{V}$ of $\mathbf{Z}^{\mathrm{T}}\mathbf{u}$ were used in place of $\tilde{E}$ and $\tilde{V}$, then (4.31) and (4.32) would give the exact moments of T. While the exact moments $\overline{E}$ and $\overline{V}$ can be used, their computation involves somewhat more effort; moreover, the approximations $\tilde{E}$ and $\tilde{V}$ perform well and are widely used in other contexts. Exact and approximate moments are compared in an example in §4.6.5 and they are discussed in detail in §4.7.3.

4.6.5 *Simplified Formulas for the Rank Sum Test When There Are No Ties

In the absence of ties among the responses, Wilcoxon's (1945) rank sum statistic T involves ranking all n responses from 1 to n and summing the ranks in the treated group. With the n subjects arranged as in §4.6.2, the rank sum is $T = \sum Z_i q_i$ with $q_1 = n, q_2 = n - 1, \ldots, q_n = 1$. This leads to simplifications of the formulas in §4.6.5. In particular, using familiar arguments about sums of integers and sums of squares of integers (Lehmann 1975, p. 329), it follows that for $\mathbf{u} \in U^+$ having k ones followed by $n - k$ zeros

$$\bar{q}_0 = \frac{n - k + 1}{2}, \quad w_0 = \frac{(n - k)(n - k + 1)}{12},$$

$$\bar{q}_1 = \frac{k + 1}{2} + n - k, \quad w_1 = \frac{k(k + 1)}{12}.$$

Substituting this in (4.31) gives

$$\frac{\tilde{E}n + m(n - k + 1)}{2}, \tag{4.34}$$

as the approximation to $\mu_{\mathbf{u}}$, and similarly (4.32) gives

$$\sqrt{\frac{\{(n+1)(2k-n)+2m(n-k+1)\}\tilde{E}-(n+2)\tilde{E}^2}{12} + \frac{m(n-k-m)(n-k+1)}{12} + \frac{\tilde{V}(n-1)(n+\frac{2}{3})}{4}} \tag{4.35}$$

as the approximation to $\sigma_{\mathbf{u}}$.

The same formulas may by used for an element, say $\mathbf{u}^*$, of U^-, by proceeding as follows. To approximate $\mu_{\mathbf{u}^*}$ and $\sigma_{\mathbf{u}^*}$ for $\mathbf{u}^* \in U^-$, determine $\mathbf{u} = 1 - \mathbf{u}^*$, so $\mathbf{u} \in U^+$, replace Γ by $1/\Gamma$, and apply formulas (4.34) and (4.35). This works because the distribution (4.6) with $(\Gamma, \mathbf{u}^*)$ is the same as with $(1/\Gamma, \mathbf{u})$; see Appendix §4.7.4, for detailed discussion.

4.6.6 An Example: Chromosome Damage from Contaminated Fish

Skerfving, Hansson, Mangs, Lindsten, and Ryman (1974) studied 23 subjects who had eaten large quantities of fish contaminated with methylmercury. Each of the 23 exposed subjects had eaten at least three meals a week of contaminated fish for more than three years. The control group consisted of 16 subjects who did not regularly consume contaminated fish and who ate far less fish of all kinds. Table 4.15 gives their data on two outcome measures, the amount of mercury found in the subject's blood, recorded in ng/g, and the percent of cells exhibiting a particular chromo-some abnormality called C_u cells, specifically, asymmetrical or incomplete symmetrical chromosome aberrations as recorded in cells cultured between 48 and 120 hours. Although the original study examined several types of chromosome abnormalities, the discussion here considers only this specific type of abnormality.

The table shows that subjects who ate contaminated fish had much higher levels of mercury in their blood and somewhat higher frequencies of chromosome aberrations. The data contain some extreme observations, for instance, subject 26 with a mercury level of 1100 ng/g and subject 24 with 9.5% of cells exhibiting chromosome aberrations. There are ties among the responses, particularly the subjects with 0% chromosome aberrations.

Consider first the level of mercury found in the blood. There are four pairs of tied observations. For instance, subjects #18 and #29 both had mercury levels of 70 ng/g, so instead of giving ranks 23 and 24 to these observations, they are both given the average rank of 23.5 in Wilcoxon's rank sum statistic, T. The sum of the ranks in the exposed group for level of mercury is $T = 642$. In the absence of hidden bias, that is, with $\Gamma = 1$, the calculations in §4.6.2 yield an approximate expectation and variance for T of 460 and 1202.92 under the null hypothesis of no treatment effect, so the deviate is $(642 - 460)/\sqrt{1202.92} = 5.25$, with significance level < 0.0001. If there were no hidden bias, there would be strong evidence that eating fish contaminated with methylmercury causes an increase in mercury levels in the blood.

Now, suppose $\Gamma = 2$, so that one subject may be twice as likely as another to eat contaminated fish because they differ with respect to an unobserved covariate for which adjustments are required. The largest significance level for all $\mathbf{u} \in U$ is approximated using the deviate

$$\min_{\mathbf{u} \in U^+} \frac{T - \mu_{\mathbf{u}}}{\sigma_{\mathbf{u}}} \tag{4.36}$$

which equals 4.40 if the approximate expectation and variance in §4.6.2 are used. This says that a hidden bias of magnitude $\Gamma = 2$ could not begin to explain the higher levels of mercury found in the blood of subjects who ate contaminated fish; no covariate that doubled the odds of eating contaminated fish could reasonably be expected to produce the observed results.

The minimum in (4.36) is calculated directly, evaluating $(T - \mu_{\mathbf{u}})/\sigma_{\mathbf{u}}$ for $k = 0, 1, \ldots, 39$ and taking the minimum. In this case, the minimum occurs at $k = 21$, that is, at a $\mathbf{u} = (1, 1, \ldots, 1, 0, 0, \ldots, 0)$ containing $k = 21$ ones followed by $n - k = 39 - 21 = 18$ zeros. The neighborhood of the minimum is quite flat. For $k = 20, 21$, and 22, the deviates $(T - \mu_{\mathbf{u}})/\sigma_{\mathbf{u}}$ are, respectively, 4.4035, 4.3985, and 4.3987. If exact expectations and variances are used in place of the approximations, the minimum still occurs at $k = 21$ and the deviate is 4.33. There is no need to calculate the maximum deviate in (4.28),

TABLE 4.15. Methymercury in Fish and Human Chromosome Damage.

Id.		%Cu cells	Mercury in blood
1	Control	2.7	5.3
2	Control	0.5	15
3	Control	0	11
4	Control	0	5.8
5	Control	5	17
6	Control	0	7
7	Control	0	8.5
8	Control	1.3	9.4
9	Control	0	7.8
10	Control	1.8	12
11	Control	0	8.7
12	Control	0	4
13	Control	1	3
14	Control	1.8	12.2
15	Control	0	6.1
16	Control	3.1	10.2
17	Exposed	0.7	100
18	Exposed	4.6	70
19	Exposed	0	196
20	Exposed	1.7	69
21	Exposed	5.2	370
22	Exposed	0	270
23	Exposed	5	150
24	Exposed	9.5	60
25	Exposed	2	330
26	Exposed	3	1,100
27	Exposed	1	40
28	Exposed	3.5	100
29	Exposed	2	70
30	Exposed	5	150
31	Exposed	5.5	200
32	Exposed	2	304
33	Exposed	3	236
34	Exposed	4	178
35	Exposed	0	41
36	Exposed	2	120
37	Exposed	2.2	330
38	Exposed	0	62
39	Exposed	2	12.8

since we know from the case of $\Gamma = 1$ that the minimum significance level will be less than 0.0001.

For $\Gamma = 5, 20$, and 35, the minimum deviates are 3.48, 2.43, and 2.10, respectively, with approximate significance levels 0.0003, 0.0075, and 0.0179. Even if one subject were 35 times more likely than another to eat contaminated fish—an extreme departure from randomization—this would not explain the high levels of mercury in the exposed group. This is a high level of insensitivity to hidden bias, a higher level than encountered in previous examples.

More important than the level of mercury is the possible damage to chromosomes. The rank sum for the percentage of abnormal cells is 551.5, with deviate 2.65 and significance level 0.004 in the absence of hidden bias, that is, with $\Gamma = 1$. A difference of this magnitude would provide strong evidence of an effect in a randomized experiment. For $\Gamma = 2$ and 3, the deviates are 1.78 and 1.28, respectively, with upper bounds on the significance levels 0.0375 and 0.10. An unobserved covariate that tripled the odds of eating contaminated fish could explain the difference in the frequency of chromosome damage in exposed and control groups. For chromosome aberrations, the study is insensitive to small biases, but sensitive to biases of moderate size.

The percentages of abnormal cells have many ties. Twelve subjects have no abnormal cells, so all twelve are given the average rank of 6.5. There are many other ties as well. If average ranks are used in computing the rank sum statistic T but are ignored in computing the expectation and variance, that is, if the simpler formulas (4.34) and (4.35) are used, the minimum deviates for $\Gamma = 2$ and 3 are 1.75 and 1.26 instead of the 1.78 and 1.28 given above. Even with many ties, the moment formulas that ignored ties gave similar results. If instead the exact expectation and variance of T are used, then the deviates for $\Gamma = 2$ and 3 are 1.74 and 1.23, a negligible change.

4.6.7 Hodges–Lehmann Point Estimates of an Additive Effect

This section conducts a sensitivity analysis for a Hodges–Lehmann (1963) estimate of an additive effect obtained from the rank sum test. The approach is similar to the sensitivity analysis in §4.3.5 for matched pairs. The model of an additive treatment effect makes little sense for coarse data with many ties, so in building point and interval estimates for an additive effect, it is assumed that there are no ties, and the moment formulas in §4.6.5 are used.

If the treatment has an additive effect τ, then the adjusted responses $\mathbf{R} - \tau\mathbf{Z}$ satisfy the null hypothesis of no effect. The Hodges–Lehmann estimate of τ is the value $\hat{\tau}$ such that $\mathbf{R} - \hat{\tau}\mathbf{Z}$ appears to precisely satisfy the null hypothesis of no effect, in the sense that the rank sum statistic $T = t(\mathbf{Z}, \mathbf{R} - \hat{\tau}\mathbf{Z})$ computed from $\mathbf{R} - \hat{\tau}\mathbf{Z}$ equals its null expectation, say $\bar{t}$. As with the signed rank statistic in §4.3.5; the rank sum statistic $t(\mathbf{Z}, \mathbf{R} - \tau\mathbf{Z})$ declines in discrete steps as τ increases, so an equation of the form $t(\mathbf{Z}, \mathbf{R} - \hat{\tau}\mathbf{Z}) = \bar{\bar{t}}$ may not have an exact solution. As a result, $\hat{\tau}$ is defined as the average of the smallest value that is too large and the largest value that is too small, that is, $\hat{\tau}$ is defined by $\hat{\tau} = \text{SOLVE}\{\bar{\bar{t}} = t(\mathbf{Z}, \mathbf{R} - \hat{\tau}\mathbf{Z})\}$ in (4.12).

When $\Gamma \neq 1$, the null expectation $\bar{\bar{t}}$ of the rank sum statistic is $\mu_{\mathbf{u}}$ which is not known because $\mathbf{u}$ is not known. However, bounds on $\bar{\bar{t}}$ are available from (4.27),

$$\bar{\bar{t}}_{\min} = \min_{\mathbf{u} \in U^-} \mu_{\mathbf{u}} \leq \bar{\bar{t}} \leq \max_{\mathbf{u} \in U^+} \mu_{\mathbf{u}} = \bar{\bar{t}}_{\max}, \text{ for all } \mathbf{u} \in U. \tag{4.37}$$

Since $t(\mathbf{Z}, \mathbf{R} - \tau\mathbf{Z})$ declines as τ increases, the maximum $\hat{\tau}$ is found as the solution $\text{SOLVE}\{\bar{\bar{t}}_{\min} = t(\mathbf{Z}, \mathbf{R} - \hat{\tau}\mathbf{Z})\}$ while the minimum $\hat{\tau}$ is found as $\text{SOLVE}\{\bar{\bar{t}}_{\max} = t(\mathbf{Z}, \mathbf{R} - \hat{\tau}\mathbf{Z})\}$.

The procedure is as follows. For each fixed Γ, compute $\bar{\bar{t}}_{\max}$ and $\bar{\bar{t}}_{\min}$ using (4.34) in §4.6.5. With these fixed values, compute $\text{SOLVE}\{\bar{\bar{t}}_{\max} = t(\mathbf{Z}, \mathbf{R} - \hat{\tau}\mathbf{Z})\}$ and $\text{SOLVE}\{\bar{\bar{t}}_{\min} = t(\mathbf{Z}, \mathbf{R} - \hat{\tau}\mathbf{Z})\}$ which give the range of possible Hodges–Lehmann estimates given uncertainty about the value of the unobserved covariate $\mathbf{u}$.

4.6.8 An Example: Point Estimates of the Increase in Chromosome Abnormalities

Continuing the example from §4.6.6, this section examines point estimates of the increase in chromosome abnormalities. If there is no hidden bias, that is, if $\Gamma = 1$ then the null expectation of the rank sum statistic is $\bar{\bar{t}} = 460 = m(n+1)/2 = 23(39+1)/2$. In fact, the rank sum statistic T for the proportion of cells with abnormalities is far higher, namely, $T = 551.5$. The Hodges–Lehmann estimate is the value $\hat{\tau}$ which, when subtracted from the responses for the 23 exposed subjects, gives a rank sum as close as possible to 460. Now, if 1.7 is subtracted from the 23 exposed proportions in Table 4.15 before the rank sum is computed, then $T = 464.5$ which is just slightly too high, but if 1.70001 is subtracted, then $T = 458$ which is just slightly too low; so the Hodges–Lehmann estimate is $\hat{\tau} = 1.7$. Were this a randomized experiment, eating contaminated fish would be estimated to increase the percentage of abnormalities by the additive effect 1.7%.

If there were a hidden bias of magnitude $\Gamma = 2$, then one subject might be twice as likely as another to eat contaminated fish because they have differing values of the unobserved covariate. In this case, the expectation of the rank sum statistic in the absence of a treatment effect, namely $\bar{\bar{t}}$, is not known because it depends on $\mathbf{u}$. Still, for $\mathbf{u} \in U$, the expectation $\bar{\bar{t}}$ is at most $\bar{\bar{t}}_{\max} = 491.57$ and at least $\bar{\bar{t}}_{\min} = 428.43$. The maximum $\bar{\bar{t}}_{\max}$ is found at $\mathbf{u} = (1, 1, \ldots, 1, 0, 0, \ldots, 0)$ with $k = 20$ ones followed by $n - k = 39 - 20 = 19$ zeros, while $\bar{\bar{t}}_{\min}$ found at $\mathbf{u}^* = (0, 0, \ldots, 0, 1, 1, \ldots, 1)$ with $k = 19$ zeros followed by $n - k = 39 - 19 = 20$ ones. These bounds on $\bar{\bar{t}}$ are calculated using (4.37) and (4.34), together with the observation in

TABLE 4.16. Sensitivity Analysis for Methylmercury Data.

Γ	Range of estimates of effect	Range of significance levels
1	1.7	0.004
2	1.0 to 2.0	0.0001 to 0.0375
3	0.7 to 2.2	< 0.0001 to 0.1

§4.6.5 about the relationship between the max over U^+ and the min over U^-.

The next step is to calculate $\text{SOLVE}\{\bar{\bar{t}}_{\max} = t(\mathbf{Z}, \mathbf{R} - \hat{\tau}\mathbf{Z})\}$ with $\bar{\bar{t}}_{\max} =$ 491.57. Now $t(\mathbf{Z}, \mathbf{R} - 0.9999\mathbf{Z}) = 497$ which is a little too large, but $t(\mathbf{Z}, \mathbf{R} - 1\mathbf{Z}) = 490.5$ which is a little too small; hence $1.0 = \text{SOLVE}\{491.57 = \tau(\mathbf{Z}, \mathbf{R} - \hat{\tau}\mathbf{Z})\}$. In words, if 1% is subtracted from the percentages of abnormalities in the exposed group, then the rank sum just about equals the maximum possible expectation for $\Gamma = 2$. In the same way, $\text{SOLVE}\{\bar{\bar{t}}_{\min} = t(\mathbf{Z}, \mathbf{R} - \hat{\tau}\mathbf{Z})\} = \text{SOLVE}\{428.43 = t(\mathbf{Z}, \mathbf{R} - \hat{\tau}\mathbf{Z})\} = 2.0$. Because we do not know $\mathbf{u}$, we cannot calculate a single Hodges–Lehmann estimate; however, if $\Gamma = 2$, the largest possible estimate is 2% and the smallest is 1%, all of these values being significantly different from 0% as seen in §4.6.6.

Repeating this for $\Gamma = 3$ gives $\bar{\bar{t}}_{\max} = 509.28, \text{SOLVE}\{509.28 = t(\mathbf{Z}, \mathbf{R} - \hat{\tau}\mathbf{Z})\} = 0.7, \bar{\bar{t}}_{\min} = 410.72$, and $\text{SOLVE}\{410.72 = t(\mathbf{Z}, \mathbf{R} - \hat{\tau}\mathbf{Z})\} = 2.2$. So the range of Hodges–Lehmann estimates for $\Gamma = 3$ is from 0.7 to 2.2%. However, from §4.6.6, not all of these estimates differ significantly from 0%. Table 4.16 summarizes these calculations.

Although the null hypothesis of no treatment effect begins to be plausible at about $\Gamma = 3$, the estimated effect is still at least 0.7%, which is not a small number given the levels of abnormalities found in the control group. Going beyond the table, the smallest estimated effect equals zero when $\Gamma = 8.9$.

The model of an additive effect is not fully consistent with the data in Table 4.15, because the exposed group has percentages of abnormalities that are not only higher but also more variable. An alternative to an analysis of the percentages R_{si} is an analysis of transformed percentages, say $\log(1 + R_{si})$ or $\sqrt{R_{si}}$, since these transformations tend to reduce the dispersion of larger values. An alternative approach is discussed in §5.3.

4.6.9 Confidence Intervals for an Additive Effect

Confidence intervals for τ are obtained by inverting the rank sum test. Fix $\Gamma \geq 1$. With Γ fixed, for each $\mathbf{u} \in U$, the distribution (4.6) yields a confidence interval for τ. Since $\mathbf{u}$ is unknown, the sensitivity interval is defined to be the union of these confidence intervals, the union taken over all $\mathbf{u} \in U$. A value τ is in the sensitivity interval if τ is plausible for some $\mathbf{u} \in U$, and τ is not in the sensitivity interval if it is rejected for every $\mathbf{u} \in U$. If $\Gamma = 1$, then the sensitivity interval is the single confidence interval commonly obtained from the randomization distribution. As Γ increases, the confidence interval becomes longer reflecting the possible impact of the unknown $\mathbf{u}$.

Let T be the rank sum statistic under the null hypothesis of no treatment effect. From Proposition 19 in the appendix, §4.7.3, $\mathrm{prob}\{T \geq a_1|m\}$ is maximized at some $\mathbf{u} \in U^+$ and $\mathrm{prob}\{T \leq a_0|m\}$ is maximized at some $\mathbf{u} \in U^-$. Suppose that there are numbers a_0 and a_1 so that $\mathrm{prob}\{T \geq a_1|m\} \leq \alpha/2$ for all $\mathbf{u} \in U^+$ with equality for some $\mathbf{u} \in U^+$ and $\mathrm{prob}\{T \leq a_0|m\} \leq \alpha/2$ for all $\mathbf{u} \in U^-$ with equality for some $\mathbf{u} \in U^-$. In other words, if $T \leq a_0$ or $T \geq a_1$, then the null hypothesis of no effect is rejected at level α for all $\mathbf{u} \in U$.

In large samples, in a one-sided 0.025 level test, a value τ is rejected as too small for all for all $\mathbf{u} \in U$ if

$$\min_{\mathbf{u} \in U^+} \frac{t(\mathbf{Z}, \mathbf{R} - \mathbf{Z}\tau) - \mu_{\mathbf{u}}}{\sigma_{\mathbf{u}}} \geq 1.96, \tag{4.38}$$

and τ is rejected as too large for $\mathbf{u} \in U$ if

$$\max_{\mathbf{u} \in U^-} \frac{t(\mathbf{Z}, \mathbf{R} - \mathbf{Z}\tau) - \mu_{\mathbf{u}}}{\sigma_{\mathbf{u}}} \leq -1.96, \tag{4.39}$$

where $\mu_{\mathbf{u}}$ and $\sigma_{\mathbf{u}}$ are approximated by (4.34) and (4.35). Since $t(\cdot, \cdot)$ is effect increasing, $t(\mathbf{Z}, \mathbf{R} - \tau\mathbf{R})$ declines as τ increases, and the set of points τ satisfying neither (4.38) nor (4.39) is therefore an interval.

4.6.10   An Example: Sensitivity Interval for the Increase in Chromosome Aberrations

In the example of §4.6.6 the Hodges–Lehmann point estimate of the increase in chromosome aberrations was $\hat{\tau} = 1.7\%$ in the absence of hidden bias, $\Gamma = 1$, and the point estimates ranged from 0.7 to 2.2% for $\Gamma = 3$. Consider now the 95% confidence interval for τ.

In the absence of hidden bias, $\Gamma = 1$, the largest value of τ that is rejected as too small is 0.2% and the smallest value that is rejected as too large is 2.8%, so the 95% confidence interval is $[0.2, 2.8]$. In other words, if 0.2 were subtracted from the responses of each subject who ate contaminated fish, and the rank sum test were applied to the resulting adjusted responses, the standardized deviate would just equal 1.96, whereas if 2.8 were subtracted instead, the deviate would just pass -1.96. Notice that, even in the absence of hidden bias, the confidence interval $[0.2, 2.8]$ is fairly long.

If there were hidden bias of magnitude $\Gamma = 2$, the largest value τ barely rejected as too small in (4.38) is 0.0 and the smallest value τ rejected as too large in (4.39) is 3.5%, so the 95% sensitivity interval is $[0.0, 3.5]$. Keep in mind that this is a two-sided interval, while the test in §4.6.6 was a one-sided test. The sensitivity interval is the union of all the confidence intervals that might have been calculated had the unobserved covariate $\mathbf{u}$ been observed. For $\Gamma = 3$, the 95% sensitivity interval is $[-0.4, 4.0]$.

4.7 Appendix: Technical Results and Proofs

4.7.1 Outline and Summary

This appendix contains proofs and general results. Section 4.7.2 proves Proposition 14 concerning the values $\mathbf{u}^+$ and $\mathbf{u}^-$ providing the bounds for sign-score statistics. Section 4.7.3 discusses sensitivity analysis for arrangement increasing statistics. Exact and approximate moment formulas for sum statistics are derived in §4.7.4. Finally, §4.7.5 discusses the effect of certain transformations of $\mathbf{u}$, most importantly, the equivalence of $(\Gamma, \mathbf{u})$ and $(1/\Gamma, \mathbf{1} - \mathbf{u})$.

4.7.2 Bounds for Sign-Score Statistics

The proof of Proposition 14 in §4.4.1 is now given.

Proof. The proof uses Holley's inequality, Theorem 9 in the appendix to Chapter 2. Fix a $\mathbf{u} \in U$. Set $\mathbf{v} = \mathbf{u}^+ - \mathbf{u}$. Renumber units in each subclass so $c_{s1} \geq c_{s2} \geq \cdots \geq c_{s,ns}$ and $v_{s1} \geq v_{s2} \geq \cdots \geq v_{s,n_s}$, which is possible by the definition of $\mathbf{u}^+$ and the fact that $\mathbf{u} \in U$. With this ordering of the units, $t(\mathbf{z}, \mathbf{r})$ is an isotonic function of $\mathbf{z} \in \Omega$ in the sense of §2.10.4, since interchanging $z_{si} = 0$ and $z_{s,i+1} = 1$ leaves $\sum_i c_{si} z_{si}$ unchanged if $c_{si} = c_{s,i+1}$ or increases $\sum_i c_{si} z_{si}$ by one if $1 = c_{si} > c_{s,i+1} = 0$. To prove the first inequality in Proposition 14, it therefore suffices to show that the premise of Holley's inequality holds, that is, to show

$$\frac{\exp\left\{\gamma (\mathbf{z} \vee \mathbf{z}^*)^T \mathbf{u}^+\right\}}{\sum_{\mathbf{b} \in \Omega} \exp(\gamma \mathbf{b}^T \mathbf{u}^+)} \cdot \frac{\exp\left\{\gamma (\mathbf{z} \wedge \mathbf{z}^*)^T \mathbf{u}\right\}}{\sum_{\mathbf{b} \in \Omega} \exp(\gamma \mathbf{b}^T \mathbf{u})}$$
$$\geq \frac{\exp\left\{\gamma (\mathbf{z}^*)^T \mathbf{u}^+\right\}}{\sum_{\mathbf{b} \in \Omega} (\gamma \mathbf{b}^T \mathbf{u}^+)} \cdot \frac{\exp(\gamma \mathbf{z}^T \mathbf{u})}{\sum_{\mathbf{b} \in \Omega} \exp(\gamma \mathbf{b}^T \mathbf{u})}$$

for all $\mathbf{z}, \mathbf{z}^* \in \Omega$. To show this, it suffices to show that

$$(\mathbf{z} \vee \mathbf{z}^*)^T \mathbf{u}^+ + (\mathbf{z} \wedge \mathbf{z}^*)^T \mathbf{u} \geq \mathbf{z}^T \mathbf{u} + (\mathbf{z}^*)^T \mathbf{u}^+ \quad \text{for all } \mathbf{z}, \mathbf{z}^* \in \Omega. \quad (4.40)$$

Notice how convenient Holley's inequality is: To prove that one random variable is stochastically larger than another, all that needs to be shown is that a certain linear inequality (4.40) holds.

First add and subtract $(\mathbf{z} \vee \mathbf{z}^*)^T \mathbf{u}$ on the left-hand side of (4.40), and then apply Lemma 2.9 to get

$$(\mathbf{z} \vee \mathbf{z}^*)^T \mathbf{u}^+ + (\mathbf{z} \wedge \mathbf{z}^*)^T \mathbf{u}$$
$$= \{(\mathbf{z} \vee \mathbf{z}^*) + (\mathbf{z} \wedge \mathbf{z}^*)\}^T \mathbf{u} + (\mathbf{z} \vee \mathbf{z}^*)^T (\mathbf{u}^+ - \mathbf{u}) \qquad (4.41)$$
$$= (\mathbf{z} + \mathbf{z}^*)^T \mathbf{u} + (\mathbf{z} \vee \mathbf{z}^*)^T (\mathbf{u}^+ - \mathbf{u}). \qquad (4.42)$$

Write the right-hand side of (4.40) as $(\mathbf{z} + \mathbf{z}^*)^T \mathbf{u} + (\mathbf{z}^*)^T (\mathbf{u}^+ - \mathbf{u})$ and compare this with (4.42). From this comparison, to prove (4.40) it suffices to show that $(\mathbf{z} \vee \mathbf{z}^*)^T (\mathbf{u}^+ - \mathbf{u}) \geq (\mathbf{z}^*)^T (\mathbf{u}^+ - \mathbf{u})$, or equivalently that $\mathbf{v}^T (\mathbf{z} \vee \mathbf{z}^*) \geq \mathbf{v}^T \mathbf{z}^*$; however, this follows from $v_{s1} \geq v_{s2} \geq \cdots \geq v_{s,n_s}$, since the 1s in $\mathbf{z} \vee \mathbf{z}^*$ are to the left of the 1s in $\mathbf{z}^*$. This proves the first inequality in the proposition. The second inequality is proved in the same way after replacing $\mathbf{u}^+$ by $\mathbf{u}$ and $\mathbf{u}$ by $\mathbf{u}^-$, so $\mathbf{v}$ becomes $\mathbf{u} - \mathbf{u}^-$. ∎

4.7.3 Some Properties of Arrangement-Increasing Statistics

Let $\mathbf{v}$ be a fixed N-tuple, and let $h(\cdot, \cdot)$ be an arrangement-increasing function as defined in §2.4.4. This section discusses properties of the expectation of $h(\mathbf{Z}, \mathbf{v})$ under the model (4.6), that is, properties of

$$\omega(\mathbf{v}, \mathbf{u}) = E\{h(\mathbf{Z}, \mathbf{v})\} = \frac{\sum_{\mathbf{z} \in \Omega} h(\mathbf{z}, \mathbf{v}) \exp(\gamma \mathbf{z}^T \mathbf{u})}{\sum_{\mathbf{b} \in \Omega} \exp(\gamma \mathbf{b}^T \mathbf{u})}. \qquad (4.43)$$

These properties justify the sensitivity analysis for sum statistics in §4.7, but they also have other uses later.

Before discussing the properties of $\omega(\mathbf{v}, \mathbf{u})$, consider a few cases. Let $T = \mathbf{Z}^T \mathbf{q}$ be a sum statistic and set $\mathbf{v} = \mathbf{q}$. If $h(\mathbf{Z}, \mathbf{q}) = \mathbf{Z}^T \mathbf{q}$, then $h(\cdot, \cdot)$ is arrangement-increasing, and $\omega(\mathbf{q}, \mathbf{u})$ is the expectation of $\mathbf{Z}^T \mathbf{q}$. Alternatively, if $h(\mathbf{Z}, \mathbf{q}) = 1$ if $\mathbf{Z}^T \mathbf{q} \geq a$ and $h(\mathbf{Z}, \mathbf{q}) = 0$ otherwise, then $\omega(\mathbf{q}, \mathbf{u})$ is the probability that $\mathbf{Z}^T \mathbf{q} \geq a$, that is, the tail probability used to determine a significance level. Similar considerations apply with $\mathbf{v} = \mathbf{r}$ if $T = t(\mathbf{Z}, \mathbf{r})$ is any arrangement-increasing statistic.

The following proposition is an immediate consequence of the composition theorem for arrangement-increasing functions. The composition theorem is due to Hollander, Proschan, and Sethuraman (1977); see also Marshall and Olkin (1979, §6.F.12) or Eaton (1987, §3.4). The composition theorem concerns expressions such as $\sum_{\mathbf{z} \in \Omega} h(\mathbf{z}, \mathbf{v}) \exp(\gamma \mathbf{z}^T \mathbf{u})$ in (4.43); it asserts that this expression is an arrangement-increasing function of $\mathbf{v}$ and $\mathbf{u}$ because $h(\mathbf{z}, \mathbf{v})$ and $\exp(\gamma \mathbf{z}^T \mathbf{u})$ are each arrangement-increasing functions and Ω is a symmetrical set. The following proposition says that $\omega(\mathbf{v}, \mathbf{u})$ increases—or more precisely, $\omega(\mathbf{v}, \mathbf{u})$ does not decrease—as the coordinates of $\mathbf{u}$ are permuted within strata into the same order as the coordinates of $\mathbf{v}$. In particular, the expectation $\omega(\mathbf{v}, \mathbf{u})$ is largest when $\mathbf{u}$ and $\mathbf{v}$ are ordered in the same way within each stratum and $\omega(\mathbf{v}, \mathbf{u})$ is smallest when $\mathbf{u}$ and $\mathbf{v}$ are ordered in opposite ways.

Proposition 17 $\omega(\mathbf{v}, \mathbf{u})$ *is arrangement-increasing.*

Pick any coordinate (s, i) and let $\boldsymbol{\varepsilon}_{si}$ be the N-tuple with a one in coordinate (s, i) and zeros in the other $N - 1$ coordinates. The following proposition from Rosenbaum and Krieger (1990) says that $\omega(\mathbf{v}, \mathbf{u} + \delta\boldsymbol{\varepsilon}_{si})$ is monotone in δ for each fixed $\mathbf{u}$ and $\mathbf{v}$. Note carefully that while this is true for each fixed $\mathbf{u}$ and $\mathbf{v}$ and for each (s, i), the direction of the monotonicity—increasing or decreasing—may change as $\mathbf{u}$, $\mathbf{v}$, and (s, i) are varied.

Proposition 18 *For each fixed* $\mathbf{u}$, $\mathbf{v}$, *and* (s, i), *the expectation* $\omega(\mathbf{v}, \mathbf{u} + \delta\boldsymbol{\varepsilon}_{si})$ *is monotone in* δ.

Proof. Let $\Omega_0 = \{\mathbf{z} \in \Omega : z_{si} = 0\}$ and $\Omega_1 = \{\mathbf{z} \in \Omega : z_{si} = 1\}$ so $\Omega = \Omega_0 \cup \Omega_1$, and a sum over Ω may be broken up into a sum over Ω_0 plus a sum over Ω_1. Write

$$
\begin{aligned}
&\omega(\mathbf{v}, \mathbf{u} + \delta\boldsymbol{\varepsilon}_{si}) \\
&= \frac{\sum_{\mathbf{z} \in \Omega_0} h(\mathbf{z}, \mathbf{v}) \exp(\gamma \mathbf{z}^T \mathbf{u}) + \exp(\gamma\delta) \sum_{\mathbf{z} \in \Omega_1} h(\mathbf{z}, \mathbf{v}) \exp(\gamma \mathbf{z}^T \mathbf{u})}{\sum_{\mathbf{b} \in \Omega_0} \exp(\gamma \mathbf{b}^T \mathbf{u}) + \exp(\gamma\delta) \sum_{\mathbf{b} \in \Omega_1} \exp(\gamma \mathbf{b}^T \mathbf{u})} \\
&= \frac{A_0 + \exp(\gamma\delta) A_1}{D_0 + \exp(\gamma\delta) D_1}, \qquad \text{say,}
\end{aligned}
$$

where A_0, A_1, D_0, and D_1 are constants not varying with δ; moreover, D_0 and D_1 are strictly positive. Differentiating with respect to δ gives

$$
\frac{\partial \omega(\mathbf{v}, \mathbf{u} + \delta\boldsymbol{\varepsilon}_{si})}{\partial \delta} = \frac{(D_0 A_1 - A_0 D_1)\gamma \exp(\gamma\delta)}{\{D_0 + D_1 \exp(\gamma\delta)\}^2}, \tag{4.44}
$$

so the sign of this derivative does not change as δ changes; hence $\omega(\mathbf{v}, \mathbf{u} + \delta\boldsymbol{\varepsilon}_{si})$ is monotone in δ. ∎

Let $\vec{\mathbf{v}}$ be the N-tuple $\mathbf{v}$ after its coordinates have been arranged in decreasing order within each stratum, $v_{si} \geq v_{s,i+1}$ for $i = 1, \ldots, n_s - 1$, and $s = 1, \ldots, S$. Let U^+ be the set of all N-tuples $\mathbf{b}$ of zeros and ones such that $b_{si} \geq b_{s,i+1}, i = 1, \ldots, n_s - 1$, and $s = 1, \ldots, S$, and let U^- be the set of all N-tuples $\mathbf{b}$ of zeros and ones such that $b_{si} \leq b_{s,i+1}, i = 1, \ldots, n_s - 1$, and $s = 1, \ldots, S$.

The following proposition bounds the unknown $\omega(\mathbf{v}, \mathbf{u})$ by two quantities that can be directly calculated. The bounds are sharp in that they are attained for some $\mathbf{u} \in U$. For instance, if $h(\mathbf{Z}, \mathbf{q}) = 1$ if $\mathbf{Z}^T \mathbf{q} \geq a$ and $h(\mathbf{Z}, \mathbf{q}) = 0$ otherwise, then Proposition 19 gives bounds on the one-sided significance level obtained using the sum statistic $\mathbf{Z}^T \mathbf{q}$.

Proposition 19 *For all* $\mathbf{u} \in U$,

$$\min_{\mathbf{b} \in U^-} \omega(\overrightarrow{\mathbf{v}}, \mathbf{b}) \leq \omega(\mathbf{v}, \mathbf{u}) \leq \max_{\mathbf{b} \in U^+} \omega(\overrightarrow{\mathbf{v}}, \mathbf{b}). \tag{4.45}$$

Proof. Suppose that, contrary to the upper bound in (4.45), there is a $\mathbf{u} \in U$ such that

$$\omega(\mathbf{v}, \mathbf{u}) > \max_{\mathbf{b} \in U^+} \omega(\vec{\mathbf{v}}, \mathbf{b}).$$

A sequence $\mathbf{u}_0, \mathbf{u}_1, \dots, \mathbf{u}_J$, will be constructed such that $\mathbf{u} = \mathbf{u}_0, \mathbf{u}_J \in U^+$, and $\omega(\mathbf{v}, \mathbf{u}_j) \leq \omega(\mathbf{v}, \mathbf{u}_{j+1})$, thereby establishing a contradiction. The construction is as follows. Suppose $\mathbf{u}_j$ has at least one coordinate that is not equal to zero or one. Then form $\mathbf{u}_{j+1}$ from $\mathbf{u}_j$ by picking any one such coordinate of $\mathbf{u}_j$, and setting it to either zero or one so that $\omega(\mathbf{v}, \mathbf{u}_j) \leq \omega(\mathbf{v}, \mathbf{u}_{j+1})$; this is possible by Proposition 18. If every coordinate of $\mathbf{u}_j$ is either zero or one, then set $J = j+1$ and let $\mathbf{u}_J$ be obtained by sorting the coordinates of $\mathbf{u}_j$ into decreasing order within each stratum. By Proposition 17, $\omega(\mathbf{v}, \mathbf{u}_j) \leq \omega(\vec{\mathbf{v}}, \mathbf{u}_J)$. Also, $\mathbf{u}_J \in U^+$, so there is a contradiction, and the upper bound in (4.45) is proved. The lower bound is proved similarly. ∎

4.7.4 Moments of Sum Statistics at Extreme Values of $\mathbf{u}$

This section determines a simplified expression for the expectation and variance of a sum statistic $\mathbf{Z}^T \mathbf{q}$ under the model (4.6) when $\mathbf{u}$ has binary coordinates, $u_{si} = 0$ or $u_{si} = 1$ for each (s, i). Also, large sample approximations to the expectation and variance are given. These results were used in §4.4.6 and 4.4.7 for the $2 \times 2 \times S$ contingency table and in §4.6 for the rank sum statistic.

Under model (4.6), a sum statistic is the sum of S independent contributions, one from each stratum. The expectation and variance of a sum statistic is, therefore, the sum of the S separate expectations and variances of the contributions from each of the S strata. As a result, it suffices to consider the case of a single stratum, $S = 1$, so in this section the subscript s is omitted. Let $u_i = 1$ or $u_i = 0$ for $i = 1, \dots, n$; let $k = \sum u_i$, so $\mathbf{u}$ has k coordinates equal to 1 and $n - k$ equal to zero.

An important role is played by the random variable $C = \mathbf{Z}^T \mathbf{u}$, which is the number of treated units for which $u_i = 1$. The variable C has the extended hypergeometric distribution, which arises in other contexts as the conditional distribution of one binomial random variable given the sum of this variable and a second independent binomial variable. For detailed discussion of the extended hypergeometric distribution, see Johnson, Kotz, and Kemp (1992, §6.11) or Plackett (1981, §4.2). Write the expectation and variance of C as $\overline{E} = E(C)$ and $\overline{V} = \text{var}(C)$. The expectation and variance of $\mathbf{Z}^T \mathbf{q}$ have easily computed formulas defined in terms of $\overline{E}$ and $\overline{V}$. (For reasons that are apparent shortly, the symbols $\overline{E}$ and $\overline{V}$ used here are the same as the symbols used in §4.4.6.)

Find the mean and variance of the q_i separately for units with $u_i = 1$ and $u_i = 0$, as follows:

$$\bar{q}_1 = \tfrac{1}{k}\sum_{i:u_i=1} q_i, \qquad\qquad \bar{q}_0 = \tfrac{1}{n-k}\sum_{i:u_i=0} q_i,$$
$$w_1 = \tfrac{1}{k-1}\sum_{i:u_i=1}(q_i - \bar{q}_1)^2, \qquad w_0 = \tfrac{1}{n-k-1}\sum_{i:u_i=0}(q_i - \bar{q}_0)^2.$$

Proposition 20 *Under model (4.6) in which $\mathbf{u}$ has binary coordinates, the expectation and variance of $\mathbf{Z}^T\mathbf{q}$ are*

$$E(\mathbf{Z}^T\mathbf{q}) = \overline{E}\bar{q}_1 + (m - \overline{E})\bar{q}_0, \tag{4.46}$$

$$\mathrm{var}(\mathbf{Z}^T\mathbf{q}) = (w_1 - w_0)\overline{E} - (\overline{E}^2 + \overline{V})\left(\frac{w_1}{k} + \frac{w_0}{n-k}\right)$$
$$+ \frac{m(n - k - m + 2\overline{E})w_0}{n - k} + \overline{V}(\bar{q}_1 - \bar{q}_0)^2. \tag{4.47}$$

Proof. Notice that in (4.6), $C = \mathbf{Z}^T\mathbf{u}$ is the sufficient statistic, so the conditional distribution of $\mathbf{Z}$ given $C = Z^T\mathbf{u}$ does not depend on γ. More than this, $\mathrm{prob}(\mathbf{Z} = \mathbf{z}|C = c)$ is uniform on the subset of Ω such that $\mathbf{z}^T\mathbf{u} = C$, that is, uniform on the set containing the $\binom{k}{c}\binom{n-k}{m-c}$ vectors with c coordinates equal to 1 among the k coordinates with $u_i = 1$, $m - c$ coordinates equal to 1 among the $n - k$ coordinates with $u_i = 0$, and all other coordinates equal to 0. In other words, the distribution $\mathrm{prob}(\mathbf{Z} = \mathbf{z}|C = c)$ picks at random c of the k coordinates with $u_i = 1$ and independently picks at random $m - c$ of the $n - k$ coordinates with $u_i = 0$. It follows that $\mathrm{prob}(\mathbf{Z}^T\mathbf{q}|C = c)$ is the distribution of the sum of c scores selected at random from among the k scores q_i such that $u_i = 1$ plus the sum of $m - c$ scores independently selected at random from among the $n - k$ scores q_i such that $u_i = 0$. Therefore,

$$E(\mathbf{Z}^T\mathbf{q}) = E\{E(\mathbf{Z}^T\mathbf{q}|C)\} = E\{C\bar{q}_1 + (m - C)\bar{q}_0\} = \overline{E}\bar{q}_1 + (m - \overline{E})\bar{q}_0,$$

proving the first part of (4.46). Similarly,

$$\mathrm{var}(\mathbf{Z}^T\mathbf{q}) = E\{\mathrm{var}(\mathbf{Z}^T\mathbf{q}|C)\} + \mathrm{var}\{E(\mathbf{Z}^T\mathbf{q}|C)\}$$
$$= E\left\{\frac{C(k - C)w_1}{k} + \frac{(m - C)(n - k - m + C)w_0}{n - k}\right\}$$
$$\quad + \mathrm{var}\{C\bar{q}_1 + (m - C)\bar{q}_0\}$$
$$= E\left\{(w_1 - w_0)C - C^2\left(\frac{w_1}{k} + \frac{w_0}{n-k}\right) + \frac{m(n - k - m + 2C)w_0}{n - k}\right\}$$
$$\quad + \mathrm{var}\{C(\bar{q}_1 - \bar{q}_0)\}$$
$$= (w_1 - w_0)\overline{E} - (\overline{E}^2 + \overline{V})\left(\frac{w_1}{k} + \frac{w_0}{n-k}\right) + \frac{m(n - k - m + 2\overline{E})w_0}{n - k}$$
$$\quad + \overline{V}(\bar{q}_1 - \bar{q}_0)^2$$

proving the second part of (4.46). ∎

There are $\binom{k}{c}\binom{n-k}{m-c}$ elements of Ω such that $\mathbf{z}^T\mathbf{u} = c$ for each integer $c, \max(0, m + k - n) \leq c \leq \min(m, k)$. Using (4.6) gives:

$$\overline{E} = \frac{\sum_{c=\max(0,m+k-n)}^{\min(m,k)} c\binom{k}{c}\binom{n-k}{m-c}\exp(\gamma c)}{\sum_{c=\max(0,m+k-n)}^{\min(m,k)} \binom{k}{c}\binom{n-k}{m-c}\exp(\gamma c)} \tag{4.48}$$

and

$$\overline{V} = \frac{\sum_{c=\max(0,m+k-n)}^{\min(m,k)} c^2\binom{k}{c}\binom{n-k}{m-c}\exp(\gamma c)}{\sum_{c=\max(0,m+k-n)}^{\min(m,k)} \binom{k}{c}\binom{n-k}{m-c}\exp(\gamma c)} - \overline{E}^2. \tag{4.49}$$

An equivalent but simpler expression for $\overline{V}$ is given by Johnson, Kotz, and Kemp (1992, p. 280, expression (6.160)); namely,

$$\overline{V} = \frac{\Gamma mk - \{n - (m + k)(1 - \Gamma)\}\overline{E}}{(1 - \Gamma)} - \overline{E}^2, \tag{4.50}$$

where $\Gamma = \exp(\gamma)$.

When $m, k, n - m, n - k$ are large, $\overline{E}$ and $\overline{V}$ can be cumbersome to compute, but in this case a large sample approximation is available. The approximations to $\overline{E}$ and $\overline{V}$ are $\tilde{E}$ and $\tilde{V}$ where $\tilde{E}$ is the root of the quadratic equation

$$\tilde{E}^2(\Gamma - 1) - \tilde{E}\{(\Gamma - 1)(m + k) + n\} + \Gamma km = 0 \tag{4.51}$$

with $\max(0, k + m - n) \leq \tilde{E} \leq \min(k, m)$, and

$$\tilde{V} = \left(\frac{1}{\tilde{E}} + \frac{1}{k - \tilde{E}} + \frac{1}{m - \tilde{E}} + \frac{1}{n - k - m + \tilde{E}}\right)^{-1}. \tag{4.52}$$

These expressions are due to Stevens (1951). Hannan and Harkness (1963) give mild conditions such that C has a limiting normal distribution with expectation $\tilde{E}$ and variance $\tilde{V}$. If instead, m is fixed as $n \to \infty$ and $k \to \infty$ with $k/n \to \zeta$, then Harkness (1965) shows that C has a limiting binomial distribution with sample size m and probability $\zeta\Gamma/\{(1 - \zeta) + \zeta\Gamma\}$.

These facts about the moments of $\mathbf{Z}^T\mathbf{q}$ were used several times in this chapter. Consider the upper bound for the significance level in a 2×2 contingency table in §4.4.6 and 4.4.7. Here, r_i is binary, $q_i = r_i$ and $u_i = r_i$, so $k = r_+, \bar{q}_1 = 1, \bar{q}_0 = 0, w_1 = w_0 = 0$; therefore, from (4.46), $E(\mathbf{Z}^T\mathbf{q}) = \overline{E}$ and $\text{var}(\mathbf{Z}^T\mathbf{q}) = \overline{V}$. The Normal approximation was used in §4.4.6 and the binomial approximation in §4.4.7.

Now consider the upper bound for the sum statistic in §4.6. Here, the q_i are sorted into decreasing order, and $u_1 = 1, u_2 = 1, \dots, u_k = 1, u_{k+1} = 0, \dots, u_n = 0$ for some k and (4.46) is used with the approximations in place of the exact moments.

Under mild conditions on the behavior of the scores q_{si}, the limiting Normal distribution for $\mathbf{Z}^T\mathbf{q}$ under model (4.6) as min $n_s \to \infty$ follows from, for example, Theorem 2.1 of Bickel and van Zwet (1978). They also provide an asymptotic expansion of the tail area and a uniform bound on the error.

4.7.5   The Effects of Certain Transformations of $\mathbf{u}$

It is useful to observe that, in the model (4.6), certain transformations of $\mathbf{u}$ have simple consequences. For instance, consider the linear transformation $\mathbf{u}^* = (1/\beta)(\mathbf{u} - \alpha\mathbf{1})$ for $\beta > 0$. Write $\gamma^* = \beta\gamma$. Then

$$
\begin{aligned}
\mathrm{prob}(\mathbf{Z} = \mathbf{z}|\mathbf{m}) &= \frac{\exp(\gamma\mathbf{z}^T\mathbf{u})}{\sum_{\mathbf{b}\in\Omega}\exp(\gamma\mathbf{b}^T\mathbf{u})} = \frac{\exp\{\gamma\mathbf{z}^T(\alpha\mathbf{1} + \beta\mathbf{u}^*)\}}{\sum_{\mathbf{b}\in\Omega}\{\gamma\mathbf{b}^T(\alpha\mathbf{1} + \beta\mathbf{u}^*)\}} \\
&= \frac{\exp(\gamma^*\mathbf{z}^T\mathbf{u}^*)}{\sum_{\mathbf{b}\in\Omega}\exp(\gamma^*\mathbf{b}^T\mathbf{u}^*)},
\end{aligned}
$$

where the last equality follows from the observation that $\mathbf{b}^T\mathbf{1} = \sum m_s$ for all $\mathbf{b} \in \Omega$. In other words, a change in the location of $\mathbf{u}$ does not change the model, while a change in the scale of $\mathbf{u}$ changes the sensitivity parameter γ by a corresponding multiple. The practical consequence is that there is no need to consider other locations and scales for $\mathbf{u}$, since they are implicitly covered by (4.6).

A more important transformation is $\mathbf{u}^{**} = \mathbf{1} - \mathbf{u}$. Now, $\mathbf{u} \in [0,1]^N$ if and only if $\mathbf{u}^{**} \in [0,1]^N$. Then

$$
\begin{aligned}
\mathrm{prob}(\mathbf{Z} = \mathbf{z}|\mathbf{m}) &= \frac{\exp(\gamma\mathbf{z}^T\mathbf{u})}{\sum_{\mathbf{b}\in\Omega}\exp(\gamma\mathbf{b}^T\mathbf{u})} = \frac{\exp\{\gamma\mathbf{z}^T(\mathbf{1} - \mathbf{u}^{**})\}}{\sum_{\mathbf{b}\in\Omega}\exp\{\gamma\mathbf{b}^T(\mathbf{1} - \mathbf{u}^{**})\}} \\
&= \frac{\exp(-\gamma\mathbf{z}^T\mathbf{u}^{**})}{\sum_{\mathbf{b}\in\Omega}\exp(-\gamma\mathbf{b}^T\mathbf{u}^{**})},
\end{aligned}
$$

where the last equality again follows from $\mathbf{b}^T\mathbf{1} = \sum m_s$ for all $\mathbf{b} \in \Omega$. In other words, replacing $\mathbf{u}$ by $\mathbf{1} - \mathbf{u}$ has the effect of changing the sign of γ, or equivalently of replacing Γ by $1/\Gamma$. This has several practical consequences. First, it suffices to consider $\gamma \geq 0$, since any distribution with $\gamma < 0$ at a point $\mathbf{u}$ is the same distribution as one with $\gamma > 0$ at $\mathbf{1} - \mathbf{u}$. Second, in sensitivity analyses for sum statistics, the sets U^+ and U^- appear; see §4.6.2, and Proposition 19 in §4.7.3. Now, $\mathbf{u} \in U^-$ if and only if $\mathbf{1} - \mathbf{u}$ is in U^+. It follows that the minimum of a quantity, say $\mu_{\mathbf{u}}$ in (4.27), as $\mathbf{u}$ ranges over U^- with $\gamma > 0$ equals a minimum of the same quantity as $\mathbf{u}$ ranges over U^+ with γ replaced by $-\gamma$.

4.8   Bibliographic Notes

Cornfield, Haenszel, Hammond, Lilienfeld, Shimkin, and Wynder (1959) proposed a sensitivity analysis for risk ratios; see also Greenhouse (1982) and Gastwirth, Krieger, and Rosenbaum (1998a). Independently, Bross (1966, 1967) proposed a sensitivity analysis for 2×2 tables. Gastwirth (1988, 1992a,b) extended the method of Cornfield et al. (1959) in several directions. Related methods are discussed by Schlesselmann (1978) and Greenland (1996). Rosenbaum and Rubin (1983) discussed sensitivity analysis for a point estimate of a difference in proportions in a $2 \times 2 \times S$ table using $3S$ sensitivity parameters. The methods discussed in this chapter are largely taken from Rosenbaum (1987, 1988, 1991a, 1993, 1995, 1999a,b, 2001), Rosenbaum and Krieger (1990), and Gastwirth, Krieger, and Rosenbaum (2000). For several applications, see Normand, Landrum, Guadagnoli, et al. (2001), Aakvik (2001), and Davanzo, Thomas, and Yue, et al. (2001). Gastwirth, Krieger and Rosenbaum (1998b) propose a method of sensitivity analysis for permutation tests with two sensitivity parameters, one linking treatment assignment with u and the other linking response with u. Several methods of sensitivity analysis for various types of regression models are discussed by Rosenbaum (1986), Copas and Li (1997), Marcus (1997), and Lin, Psaty, and Kronmal (1998). Manski (1990, 1995) proposes "worst-case bounds" which are somewhat analogous to letting $\Gamma \to \infty$ in a sensitivity analysis; see also Balke and Pearl (1997) and Pearl (2000, §8.2).

4.9   Problems

1. **McNemar's test.** Starting with 10,872 death certificates with the diagnosis of sporadic motor neuron disease (MND), Graham, Macdonald, and Hawkes (1997, Table 3) examined their birth certificates and found that 70 of these cases with MND had a living twin free of MND. These 70 twin pairs formed the basis for a case-referent study. Because little is known about the causes of MND, they examined "many variables" as potential causes in their exploratory study. Indeed, many case-referent studies are exploratory: they run a substantial risk of false positives by performing many significance tests on many potential causes. Nonetheless, exploratory studies have played an important role as the first of a series of increasingly focused studies to determine the causes of a disease. The strongest association Graham, Macdonald, and Hawkes found was with "carrying out car or vehicle maintenance." There were 16 twin pairs discordant for this variable, and 14/16 had an exposed case, while 2/16 had an exposed referent, yielding an estimated odds ratio of $14/2 = 7$ in the absence

of hidden bias. Do a sensitivity analysis for the significance level from McNemar's test.

2. Wilcoxon's signed rank test. Many drugs used to treat cancer are quite harsh, and there is the possibility that these drugs can harm hospital workers who are exposed by accident. Kevekordes, Gebel, Hellwig, Dames, Dunkelberg (1998) studied this possibility when a "malfunction of a safety hood result[ed] in air flowing from the hood along the arms of the person preparing infusions of antineoplastic drugs" (p. 145). They studied 10 nurses who may have experienced substantial exposures, matching each nurse to a control based on gender, age, and intensity of smoking. They measured genetic damage using the cytokinesis block micronucleus test, reporting mean micronuclei/10^3 binucleate lymphocytes (mm/10^3), as follows:

Pair	Ages	Smoking	Exposed	Control
1	37/37	ns	20	11
2	24/25	s	10	9
3	33/32	s	22	19
4	29/30	ns	13	9
5	23/23	ns	13	7
6	28/29	ns	14	11
7	25/24	ns	12	6
8	38/40	s	21	23
9	32/32	ns	9	4
10	33/34	s	21	14

Kevekordes et. al. used Wilcoxon's signed rank test. Do a sensitivity analysis for this test, the Hodges–Lehmann point estimate and the 95% confidence interval. Compare the sensitivity in this example to the examples in the chapter. Do you notice any interesting relationships involving the matched covariates, age and smoking?

3. Sensitivity to finite biases despite an infinite odds ratio. Using data from the Los Angeles County Cancer Surveillance Program, Peters, Preston-Martin and Yu (1981) matched 92 cases of brain tumors in children less than 10-years old to referent children without brain tumors. They used friends or neighbors as referents, matching for age, race, and year of birth. Attention focused on occupational exposures of parents, including specific industries and chemical exposures. Several of many comparisons gave significant results by McNemar's test. In particular, there were 10 pairs discordant for whether the father worked in the aircraft industry, and in all 10 pairs, it was the case whose father was so employed. The point estimate of the odds ratio in the absence of bias is, therefore, infinite: $10/0 = \infty$. Do a sensitivity analysis for the significance level from McNemar's test.

4. **Mantel–Haenszel test.** In a case-referent study of eosinophilia-myalgia syndrome, Eidson, Philen, Sewell, Voorhees and Kilbourne (1990) matched 11 cases each to 2 noncases or referents, matching for residence, age, and gender. In all 11 matched triples, the case had consumed L-tryptophan, which is an amino acid health food supplement. Reported reasons for taking the supplement included insomnia, stress, and premenstrual syndrome. In 2 matched triples, one referent had used L-tryptophan. In 9 matched triples, neither referent had used L-tryptophan. Perform a sensitivity analysis with the Mantel–Haenszel test.

5. **A simple derivation.** Consider the bounding probabilities (4.17) for sign-score statistics in matching with multiple controls. These were justified by reference to the general Proposition 3, but a much simpler proof is possible in this special case. Give a brief, direct derivation of (4.17) from (4.15). In this special case, how many treatment assignments $\mathbf{z}_s$ are in Ω_s in (4.6)? (Rosenbaum 1988)

6. **A simple derivation.** Obtain the bounds (4.19) for a case-referent study by evaluating the model (4.6) at $\mathbf{u} = \mathbf{u}^+$ and $\mathbf{u} = \mathbf{u}^-$. (Hint: Corollary 15 may be helpful.)

7. **A simple derivation.** In the absence of ties and zero differences, obtain the simple formulas (4.11) for the expectation and variance of the upper bounding distribution T^+ for Wilcoxon's signed rank test. (Hints: First show that $\sum_{i=1}^{S} i = \frac{S(S+1)}{2}$ and $\sum_{i=1}^{S} i^2 = \frac{S(S+1)(2S+1)}{6}$ or see Lehmann 1975 who derives these expressions.)

8. **A simple derivation.** Derive (4.6) from (4.2). If the general case (4.6) appears difficult, first try the simple special case of matched pairs (4.7), then the slightly more general case of matching with multiple controls (4.15), and finally return to the general case (4.6).

9. **Unbounded covariates.** Consider S matched pairs, but replace the assumption that $1 \geq u_{si} \geq 0$ by the weaker assumption that $1 \geq |u_{s1} - u_{s2}|$ for at least $S^* < S$ pairs. In other words, a few pairs, specifically $S - S^*$ pairs, may have very large differences in u_{si}. For Wilcoxon's signed rank statistic T, find two new random variables replacing T^+ and T^- in Proposition 13 that bound the distribution of T. (See Rosenbaum 1987.)

10. **Quade's test in matching with multiple controls.** With continuous responses and matching with multiple controls in §4.5, Quade's test is an alternative to the stratified rank sum test which gives more weight to matched sets with more dispersion in their responses. However, like the stratified rank sum test, and unlike the aligned rank test, it is possible to tabulate the extreme moments for Quade's test in a

sensitivity analysis, so only simple arithmetic is required. For each matched set s, find the range of responses, $\max_i R_{si} - \min_i R_{si}$, and let d_s be the rank of the range for set s when the S ranges are ranked from 1 to S with average ranks for ties. If q_{si} is the rank of R_{si} when the responses in matched set s are ranked within set s from 1 to n_s, then Quade's test statistic is $\sum_{s=1}^{S} d_s \sum_{i=1}^{n_s} Z_{si} q_{si}$. The ranks of treated subjects within matched sets are weighted by the ranks of the dispersions within sets. With matched pairs, $n_s = 2$, $m_s = 1$, the stratified rank sum test reduces to the sign test, but Quade's test reduces to the signed rank test. Show that the table of exact moments in §4.5.3 for the stratified rank sum test may be adapted for use with Quade's test. (Hints: Write $\tilde{q}_{si} = d_s q_{si}$. How is the maximum expectation of $d_s \sum_{i=1}^{n_s} Z_{si} q_{si}$ related to the maximum expectation of $\sum_{i=1}^{n_s} Z_{si} q_{si}$?) Developing an idea of Tukey (1957), Quade (1979) proposed this test for unreplicated, randomized complete blocks. Tardif (1987) discussed Quade's test with replicated treatments, so Tardif's discussion includes the form of the test described in this problem.

11. **Lower bounds in matching with multiple controls.** In the discussion of matching with multiple controls with continuous responses, the minimum expectation of $\sum_{i=1}^{n_s} Z_{si} q_{si}$ and the maximum variance at this minimum expectation are needed to obtain the approximate upper bound on the lower tail probability, $\text{prob}\,(T \le t | \mathbf{m})$. Show that the expectation and variance of $\sum_{i=1}^{n_s} Z_{si} q_{si}$ when $u_{s1} = 1, \ldots, u_{sa} = 1$, $u_{s,a+1} = 0, \ldots, u_{s,n_s} = 0$ are:

$$\tilde{\mu}_{sa} = \frac{\Gamma \sum_{i=1}^{a} q_{si} + \sum_{i=a+1}^{n_s} q_{si}}{\Gamma a + (n_s - a)}.$$

and

$$\tilde{\nu}_{sa}^2 = \frac{\Gamma \sum_{i=1}^{a} q_{si}^2 + \sum_{i=a+1}^{n_s} q_{si}^2}{\Gamma a + (n_s - a)} - \tilde{\mu}_{sa}^2.$$

With $q_{s,n_s} \ge \ldots \ge q_{s1}$, use Proposition 19 to show that the minimum possible expectation of $\sum_{i=1}^{n_s} Z_{si} q_{si}$ is $\min_{0 < a < n_s} \tilde{\mu}_{sa}$. In providing an upper bound on $\text{prob}\,(T \le t | \mathbf{m})$, why is the *minimum* expectation used with the *maximum* variance?

12. **Full matching.** In a full matching, each matched set contains either one treated subject and at least one control, or one control and at least one treated subject, so $\min\{m_s, (n_s - m_s)\} = 1$ for each s. Full matching is the optimal form of stratification: it makes subjects in the same stratum as comparable as possible; see Rosenbaum (1991c) for proofs and Gu and Rosenbaum (1993) for simulation results. Are new methods of sensitivity analysis needed for full matching? Or, alternatively, can the methods in §4.5 for matching with multiple

controls be easily adapted for use in full matching? (Hint: Consider a matched set with one control and several treated subjects, so $m_s = n_s - 1$. In this set, express $\sum_{i=1}^{n_s} Z_{si} q_{si}$ in terms of $\tilde{Z}_{si} = 1 - Z_{si}$ and $\tilde{q}_{si} = \left(\sum_{j=1}^{n_s} q_{sj}\right) - q_{si}.$)

13. Write S-Plus code to perform a sensitivity analysis for the significance level from Wilcoxon's signed rank statistic using (4.9). Hint:

```
> signedrank
function(dif, gamma)
{
# Performs a sensitivity analysis for the significance
# level for Wilcoxon's signed rank statistic.
# Uses the large sample Normal approximation with
# expectation and variance formulas that allow for
# ties and zero differences.
#
# dif is a vector of matched pair,
# treated-minus-control, differences
#
# gamma is Upper Case Gamma, the sensitivity parameter.
#
# Output is the minimum and maximum significance level for a
# one-sided, upper tailed test.
#
  rk <- rank(abs(dif))
  s1 <- 1 * (dif > 0)
  s2 <- 1 * (dif < 0)
  W <- sum(s1 * rk)
  Eplus <- sum((s1 + s2) * rk * gamma)/(1 + gamma)
  Eminus <- sum((s1 + s2) * rk)/(1 + gamma)
  V <- sum((s1 + s2) * rk * rk * gamma)/((1 + gamma)^2)
  Dplus <- (W - Eplus)/sqrt(V)
  Dminus <- (W - Eminus)/sqrt(V)
  c(1 - pnorm(Dminus), 1 - pnorm(Dplus))
}
```

For the lead data in Table 3.1, the differences are:
```
> lead
 [1] 22 5 23 -6 18 25 13 47 15 16 6 1 2 7 0 4 -9 -3 36 25
 1 16 42 30 25 23 32 17 9 -3 60 14 14
```
For $\Gamma = 3$ and $\Gamma = 4$ in Table 4.2, the bounds for the significance level are:
```
> signedrank(lead,3)
[1] 6.661338e-016 1.361531e-002
> signedrank(lead,4)
```

[1] 0.00000000 0.03878978

4.10 References

Aakvik, A. (2001) Bounding a matching estimator: The case of a Norwegian training program. *Oxford Bulletin of Economics and Statistics*, **63**, 115–143.

Aiken, L. H., Smith, H. L., and Lake, E. T. (1994) Lower Medicare mortality among a set of hospitals known for good nursing care. *Medical Care*, **32**, 771–787.

Armitage, P. (1977) *Statistical Methods in Medical Research* (fourth printing), Oxford, UK: Blackwell.

Balke, A. and Pearl, J. (1997) Bounds on treatment effects from studies with imperfect compliance. *Journal of the American Statistical Association*, **92**, 1172–1176.

Bickel, P. and van Zwet, (1978) Asymptotic expansions for the power of distribution free tests in the two-sample problem. *Annals of Statistics*, **6**, 937–1004.

Boston Collaborative Drug Project (1972) Excess of ampicillin rashes associated with allopurinol or hyperuricemia. *New England Journal of Medicine*, **286**, 505–507.

Breslow, N. and Day, N. (1980) *Statistical Methods in Cancer Research, I: The Analysis of Case-Control Studies.* Lyon, France: International Agency for Research on Cancer.

Bross, I. D. J. (1966) Spurious effects from an extraneous variable. *Journal of Chronic Diseases*, **19**, 637–647.

Bross, I. D. J. (1967) Pertinency of an extraneous variable. *Journal of Chronic Diseases*, **20**, 487–495.

Cameron, E. and Pauling, L. (1976) Supplemental ascorbate in the supportive treatment of cancer: Prolongation of survival times in terminal human cancer. *Proceedings of the National Academy of Sciences (USA)*, **73**, 3685–3689.

Copas, J. B. and Li, H. G. (1997) Inference for non-random samples (with discussion). *Journal of the Royal Statistical Society*, **B**, **59**, 55–96.

Cornfield, J., Haenszel, W., Hammond, E., Lilienfeld, A., Shimkin, M., and Wynder, E. (1959). Smoking and lung cancer: Recent evidence and a discussion of some questions. *Journal of the National Cancer Institute*, **22**, 173–203.

Davanzo, P., Thomas, M. A., Yue, K., Oshiro, T., Belin, T., Strober, M., and McCracken, J. (2001) Decreased anterior cingulate myo-inositol/creatine spectoscopy resonance with lithium treatment in children with bipolar disorder. *Neuropsychopharmacology*, **24**, 359–369.

Eaton, M. (1987) *Lectures on Topics in Probability Inequalities*. Amsterdam: Centrum voor Wiskunde en Informatica.

Eidson, M., Philen, R. M., Sewell, C. M., Voorhees, R., and Kilbourne, E. M. (1990) L-tryptophan and eosinophilia-myalgia syndrome in New Mexico. *Lancet*, **335**, 645–648.

Fisher, R.A. (1958) Lung cancer and cigarettes? *Nature*, **182**, July 12, 108.

Freidlin, B. and Gastwirth, J. L. (2000) Should the median test be retired from general use? *American Statistician*, **54**, 161–164.

Gastwirth, J. L. (1988) *Statistical Reasoning in Law and Public Policy*. New York: Academic.

Gastwirth, J. L. (1992a) Employment discrimination: A statistician's look at analysis of disparate impact claims. *Law and Inequality*, **11**, 151–179.

Gastwirth, J. L. (1992b) Methods for assessing the sensitivity of statistical comparisons used in Title VII cases to omitted variables. *Jurimetrics Journal*, 19–34.

Gastwirth, J. L. and Greenhouse, S. (1987) Estimating a common relative risk: Application in equal employment. *Journal of the American Statistical Association*, **82**, 38–45.

Gastwirth, J. L., Krieger, A. M. and Rosenbaum, P. R. (1998a) Cornfield's inequality. In: *Encyclopedia of Biostatistics*, P. Armitage and T. Colton, eds., New York: Wiley, pp. 952–955.

Gastwirth, J. L., Krieger, A. M., and Rosenbaum, P. R. (1998b) Dual and simultaneous sensitivity analysis for matched pairs. *Biometrika*, 85, 907–920.

Gastwirth, J. L., Krieger, A. M., and Rosenbaum, P. R. (2000) Asymptotic separability in sensitivity analysis. *Journal of the Royal Statistical Society*, Series **B**, **62**, 545–555.

Gibbons, J. D. (1982) Brown-Mood median test. In: *Encyclopedia of Statistical Sciences*, Volume 1, S. Kotz and N. Johnson, eds., New York: Wiley, pp. 322–324.

Graham, A. J., Macdonald, A. M., and Hawkes, C. H. (1997) British motor neuron disease twin study. *Journal of Neurology, Neurosurgery and Psychiatry*, **6**, 562=569.

Greenhouse, S. (1982) Jerome Cornfield's contributions to epidemiology. *Biometrics*, **38S**, 33–46.

Greenland, S. (1996) Basic methods of sensitivity analysis of biases. *International Journal of Epidemiology*, 25, 1107–1116.

Gu, X. S. and Rosenbaum, P. R. (1993) Comparison of multivariate matching methods: Structures, distances and algorithms. *Journal of Computational and Grapical Statistics*, **2**, 405–420.

Hájek, J., Sidák, Z. and Sen, P. K. (1999) *Theory of Rank Tests* (Second Edition). New York: Academic.

Hakulinen, T. (1981) A Mantel–Haenszel statistic for testing the association between a polychotomous exposure and a rare outcome. *American Journal of Epidemiology*, **113** ,192–197.

Hammond, E. C. (1964) Smoking in relation to mortality and morbidity: Findings in first thirty-four months of follow-up in a prospective study started in 1959. *Journal of the National Cancer Institute*, **32**, 1161–1188.

Hannan, J. and Harkness, W. (1963) Normal approximation to the distribution of two independent binomials, conditional on a fixed sum. *Annals of Mathematical Statistics*, **34**, 1593–1595.

Harkness, W. (1965) Properties of the extended hypergeometric distribution. *Annals of Mathematical Statistics*, **36**, 938–945.

Herbst, A., Ulfelder, H., and Poskanzer, D. (1971) Adenocarcinoma of the vagina: Association of maternal stilbestrol therapy with tumor appearance in young women. *New England Journal of Medicine*, **284**, 878–881.

Hettmansperger, T. (1984) *Statistical Inference Based on Ranks*. New York: Wiley.

Hodges, J. and Lehmann, E. (1962) Rank methods for combination of individual experiments in the analysis of variance. *Annals of Mathematical Statistics*, **33**, 482–497.

Hodges, J. and Lehmann, E. (1963) Estimates of location based on rank tests. *Annals of Mathematical Statistics*, **34**, 598–611.

Hollander, M., Proschan, F., and Sethuraman, J. (1977) Functions decreasing in transposition and their applications in ranking problems. *Annals of Statistics*, **5**, 722–733.

Hollander, M. and Wolfe, D. (1973, 1999) *Nonparametric Statistical Methods*. New York: Wiley.

Holley, R. (1974) Remarks on the FKG inequalities. *Communications in Mathematical Physics*, **36**, 227–231.

Jick, H., Miettinen, O., Neff, R., et al. (1973) Coffee and myocardial infarction. *New England Journal of Medicine*, **289**, 63–77.

Johnson, N., Kotz, S., and Kemp, A. (1992) *Univariate Discrete Distributions*. New York: Wiley.

Kelsey, J. and Hardy, R. (1975) Driving of motor vehicles as a risk factor for acute herniated lumbar intervertebral disc. *American Journal of Epidemiology*, **102**, 63–73.

Kevekordes, S., Gebel, T. W., Hellwig, M., Dames, W., and Dunkelberg, H. (1998) Human effect monitoring in cases of occupational exposure to antineoplastic drugs: a method comparison. *Occupational and Environmental Medicine*, **55**, 145–149.

Krieger, A. M. and Rosenbaum, P. R. (1994) A stochastic comparison for arrangement increasing functions. *Combinatorics, Probability and Computing*, **3**, 345–348.

Lehmann, E. (1975) *Nonparametrics: Statistical Methods Based on Ranks*. San Francisco: Holden-Day.

Lin, D. Y., Psaty, B. M., and Kronmal, R. A. (1998) Assessing the sensitivity of regression results to unmeasured confounders in observational studies. *Biometrics*, **54**, 948–963.

Mack, T., Pike, M., Henderson, B., Pfeffer, R., Gerkins, V., Arthur, B., and Brown, S. (1976) Estrogens and endometrial cancer in a retirement community. *New England Journal of Medicine*, **294**, 1262–1267.

Mann, H. and Whitney, D. (1947) On a test of whether one of two random variables is stochastically larger than the other. *Annals of Mathematical Statistics*, **18**, 50–60.

Manski, C. (1990) Nonparametric bounds on treatment effects. *American Economic Review*, 319–323.

Manski, C. (1995) *Identification Problems in the Social Sciences.* Cambridge, MA: Harvard University Press.

Mantel, N. and Haenszel, W. (1959) Statistical aspects of retrospective studies of disease. *Journal of the National Cancer Institute,* **22,** 719–748.

Marcus, S. (1997) Using omitted variable bias to assess uncertainty in the estimation of an AIDS education treatment effect. *Journal of Educational and Behavioral Statistics,* **22,** 193–202.

Marshall, A. and Olkin, I. (1979) *Inequalities: Theory of Majorization and Its Applications.* New York: Academic.

McNemar, Q. (1947) Note on the sampling error of the differences between correlated proportions or percentages. *Psychometrika,* **12,** 153–157.

Miettinen, O. (1969) Individual matching with multiple controls in the case of all or none responses. *Biometrics,* **22,** 339–355.

Ming, K. and Rosenbaum, P. R. (2000) Substantial gains in bias reduction from matching with a variable number of controls. *Biometrics,* **56,** 118–124.

Moertel, C., Fleming, T., Creagan, E., Rubin, J., O'Connell, M., and Ames, M. (1985) High-dose vitamin C vs placebo in the treatment of patients with advanced cancer who have had no prior chemotherapy: A randomized double-blind comparison. *New England Journal of Medicine,* **312,** 137–141.

Morton, D., Saah, A., Silberg, S., Owens, W., Roberts, M., and Saah, M. (1982) Lead absorption in children of employees in a lead related industry. *American Journal of Epidemiology,* **115,** 549–555.

Mosteller, F. and Tukey, J. (1977) *Data Analysis and Regression.* Reading, MA: Addison-Wesley.

Nicholson, W., Selikoff, I., Seidman, H., Lilis, R., and Formby, P. (1979) Long-term mortality experience of chrysotile miners and millers in Thetford Mines, Quebec. *Annals of the New York Academy of Sciences,* **330,** 11–21.

Normand, S. T., Landrum, M. B., Guadagnoli, E., Ayanian, J. Z., Ryan, T. J., Cleary, P. D., and McNeil, B. J. (2001) Validating recommendations for coronary angiography following acute myocardial infarction in the elderly: A matched analysis using propensity scores. *Journal of Clinical Epidemiology,* **54,** 387–398.

O'Brien, P. C. and Fleming, T. R. (1987) A paired Prentice–Wilcoxon test for censored paired data. *Biometrics*, **43**, 169–180.

Pearl, J. (2000) *Causality*. New York: Cambridge University Press.

Peters, J. M., Preston-Martin, S., and Yu, M. C. (1981) Brain tumors in children and occupational exposure of parents. *Science*, 213, 235–236.

Plackett, R. L. (1981) *The Analysis of Categorical Data* (second edition). New York: Macmillan.

Psaty, B. M., Koepsell, T. D., Lin, D., Weiss, N., Siscovick, D. S., Rosendaal, F. R., Pahor, M., and Furberg, C. D. (1999) Assessment and control for confounding by indication in observational studies. *Journal of the American Geriatrics Society*, **47**, 749–754.

Quade, D. (1979) Using weighted rankings in the analysis of complete blocks with additive block effects. *Journal of the American Statistical Association*, **74**, 680–683.

Rosenbaum, P. R. (1986) Dropping out of high school in the United States: An observational study. *Journal of Educational Statistics*, **11**, 207–224.

Rosenbaum, P. R. (1987) Sensitivity analysis for certain permutation inferences in matched observational studies. *Biometrika*, **74**, 13–26.

Rosenbaum, P. R. (1988) Sensitivity analysis for matching with multiple controls. *Biometrika*, **75**, 577–581.

Rosenbaum, P. R. (1989) On permutation tests for hidden biases in observational studies: An application of Holley's inequality to the Savage lattice. *Annals of Statistics*, **17**, 643–653.

Rosenbaum, P. R. (1991a) Sensitivity analysis for matched case-control studies. *Biometrics*, **47**, 87–100.

Rosenbaum, P. R. (1991b) Discussing hidden bias in observational studies. *Annals of Internal Medicine*, **115**, 901–905.

Rosenbaum, P. R. (1991c) A characterization of optimal designs for observational studies. *Journal of the Royal Statistical Society*, Series **B**, **53**, 597–610.

Rosenbaum, P. R. (1993) Hodges–Lehmann point estimates of treatment effect in observational studies. *Journal of the American Statistical Association*, **88**, 1250–1253.

Rosenbaum, P. R. (1995) Quantiles in nonrandom samples and observational studies. *Journal of the American Statistical Association*, **90**, 11424–1431.

Rosenbaum, P. R. (1999a) Using combined quantile averages in matched observational studies. *Applied Statistics*, **48**, 63–78.

Rosenbaum, P. R. (1999b) Reduced sensitivity to hidden bias at upper quantiles in observational studies with dilated effects. *Biometrics*, **55**, 560–564.

Rosenbaum, P. R. (1999) Holley's inequality. In: *Encyclopedia of Statistical Sciences*, Update Volume 3, S. Kotz, C. B. Read, and D. L. Banks, eds., New York: Wiley, pp. 329–331.

Rosenbaum, P. R. (2001) Effects attributable to treatment: Inference in experiments and observational studies with a discrete pivot. *Biometrika*, **88**, 219–232.

Rosenbaum, P. R. and Krieger, A. M. (1990) Sensitivity analysis for two-sample permutation inferences in observational studies. *Journal of the American Statistical Association*, **85**, 493–498.

Rosenbaum, P. R. and Rubin, D. B. (1983) Assessing sensitivity to an unobserved binary covariate in an observational study with binary outcome. *Journal of the Royal Statistical Society*, Series **B**, **45**, 212–218.

Rubin, D. B. (1978) Bayesian inference for causal effects: The role of randomization. *Annals of Statistics*, **6**, 34–58.

Savage, I. R. (1964) Contributions to the theory of rank order statistics: Applications of lattice theory. *Review of the International Statistical Institute*, **32**, 52–63.

Schlesselmann, J. J. (1978) Assessing the effects of confounding variables. *American Journal of Epidemiology*, **108**, 3–8.

Skerfving, S., Hansson, K., Mangs, C., Lindsten, J., and Ryman, N. (1974) Methylmercury-induced chromosome damage in man. *Environmental Research*, **7**, 83–98.

Smith, H. L. (1997) Matching with multiple controls to estimate treatment effects in observational studies. *Sociological Methodology*, **27**, 325–353.

Stevens, W. L. (1951) Mean and variance of an entry in a contingency table. *Biometrika*, **38**, 468–470.

Tardif, S. (1987) Efficiency and optimality results for tests based on weighted rankings. *Journal of the American Statistical Association*, **82**, 637–644.

Tukey, J. W. (1957) Sums of random partitions of ranks. *Annals of Mathematical Statistics*, 28, 987–992.

Wilcoxon, F. (1945) Individual comparisons by ranking methods. *Biometrics*, **1**, 80–83.

5

Models for Treatment Effects

5.1 Effects That Vary from Person to Person

The effect of a treatment may vary from one person to the next. One person may benefit or suffer greatly from treatment, while another person may experience little or no effect. In other words, the effect of the treatment on the ith person in stratum s, namely $r_{Tsi} - r_{Csi}$, may not be constant, but may change with i and s.

With only brief exceptions, in Chapters 2 to 4, the focus has been on testing the null hypothesis of no treatment effect, $H_0 : r_{Tsi} = r_{Csi}$ for $i = 1, \dots, n_s$, $s = 1, \dots, S$, and on inference about an additive treatment effect, $r_{Tsi} = r_{Csi} + \tau$ for $i = 1, \dots, n_s$, $s = 1, \dots, S$. In applications, these are two common tasks. Moreover, these two tasks serve to illustrate the conceptual differences between inference in a randomized experiment, in an observational study free of hidden bias, and in an observational study that addresses hidden bias using sensitivity analysis. Nonetheless, treatment effects that vary from person to person are quite common, and associated methods of inference are needed.

In this chapter, several situations are discussed.

1. *Binary responses.* When the responses under treatment and under control, r_{Tsi} and r_{Csi}, are each binary, 1 or 0, say dead or alive, it is often useful to test the null hypothesis of no treatment effect, but the model of an additive treatment effect is not useful. With binary responses, if $r_{Tsi} = r_{Csi} + \tau$, $i = 1, \dots, n_s$, $s = 1, \dots, S$, then τ must be 1, 0, or -1, and if $\tau = 1$, then every one of the N subjects

must die under treatment and survive under control. A treatment may possibly have no effect on a binary response, but if it has some effect, then it is very likely to affect some individuals and not others. Methods for binary responses are discussed in §5.5.

2. *Continuous responses that are not shifted.* In a large randomized experiment, the model of an additive effect has clear implications for the distribution of observed responses in treated and control groups. Examination of the data from such an experiment may clearly indicate that the effect is not additive. Consider as a simple example the case of a randomized experiment that is *balanced* in the specific sense that the number m_s of treated subjects and the number of controls $n_s - m_s$ do not vary from one stratum s to another. For instance, a completely randomized experiment, $S = 1$, and a matched pairs randomized experiment, $m_s = n_s - m_s = 1$ for all s, are both balanced in the sense that the term is used here. In such balanced randomized experiments, with a large sample size N, the model of an additive effect, $r_{Tsi} = r_{Csi} + \tau$, $i = 1, \ldots, n_s$, $s = 1, \ldots, S$, for individuals implies that the distribution of observed responses, R_{si}, among treated subjects, $Z_{si} = 1$, has the same shape and dispersion as the distribution of observed responses, R_{si}, among controls, $Z_{si} = 0$, but the treated distribution is shifted by τ. For example, boxplots or histograms of the distributions of responses in treated and control groups would look the same, except one would be shifted upwards by τ. If, in a large, balanced, randomized experiment, the distributions of observed responses are clearly not shifted, then the model of an additive effect is not applicable, and other methods are needed. In this case, several of the models in this chapter may be used instead.

3. *Active ingredients and instrumental variables.* In some contexts, the treatment may contain an active ingredient, and it is only the active ingredient that has an effect. Assignment to treatment is associated with a higher level of the active ingredient, but doses of this ingredient vary. In this case, it may be natural to think that the treatment effect $r_{Tsi} - r_{Csi}$ varies with the level of the active ingredient, and is not constant for all subjects who received the treatment. The assigned treatment is an instrument for manipulating the active ingredient; see §5.4.

The material in this chapter is of practical importance for inference in experiments and observational studies, but it does not add new concepts beyond those in Chapters 2 to 4 and the material in this chapter is not used extensively in later chapters. Moreover, after §5.2, the remaining sections may be read in any order, except that §5.5 should be read before §5.6.

5.2   Order Statistics

Nonadditive models are often described using order statistics. The N subjects present N potential responses to the control, r_{Csi}, $i = 1, \ldots, n_s$, $s = 1, \ldots, S$. Sort these N values into order and label them $r_{C(N)} \geq r_{C(N-1)} \geq \cdots \geq r_{C(1)}$; these are the N order statistics of the potential responses to control. In randomization inference, the potential responses, r_{Csi}, are fixed features of the finite population of N subjects; hence, the order statistics of potential responses to control, $r_{C(k)}$, are also fixed. Notice carefully that we observe the response to control, r_{Csi}, from each subject who received the control, $Z_{si} = 0$, but we do not observe r_{Csi} for treated subjects, $Z_{si} = 1$, so we do not observe any of the order statistics $r_{C(k)}$. Nonetheless, it is possible to use the observed data to draw inferences about $r_{C(k)}$. Similar considerations apply to the N order statistics $r_{T(N)} \geq r_{T(N-1)} \geq \cdots \geq r_{T(1)}$ formed by sorting the N potential responses to treatment r_{Tsi}, $i = 1, \ldots, n_s$, $s = 1, \ldots, S$. If the treatment effect were additive, $r_{Tsi} = r_{Csi} + \tau$, $i = 1, \ldots, n_s$, $s = 1, \ldots, S$, then the effect on the order statistics would also be additive, $r_{T(k)} = r_{C(k)} + \tau$, for $k = 1, \ldots, N$.

The N observed responses, $R_{si} = Z_{si} r_{Tsi} + (1 - Z_{si}) r_{Csi}$, may also be sorted into order, yielding the N observed order statistics, $R_{(N)} \geq R_{(N-1)} \geq \cdots \geq R_{(1)}$. These differ in two important ways from the order statistics of the potential responses. First, the $R_{(k)}$ are observed while the $r_{C(k)}$ and $r_{T(k)}$ are not. Second, since the observed responses R_{si} vary with the treatment assignment, Z_{si}, their order statistics $R_{(k)}$ are random variables, whereas the $r_{C(k)}$ and $r_{T(k)}$ are fixed.

5.3   Dilated Effects

5.3.1   Definition of Dilated Effects

In a large balanced randomized experiment, the model of an additive treatment effect cannot produce treated responses that are both higher and more dispersed than control responses, although this pattern is not uncommon in applications. In contrast, the model of dilated treatment effects can produce this pattern.

In a dilated effect, the effect is larger when the response under control is larger. The treatment has a *dilated effect* if $r_{Tsi} = r_{Csi} + \Delta(r_{Csi})$, $i = 1, \ldots, n_s$, $s = 1, \ldots, S$, for some nonnegative, nondecreasing function, $\Delta(\cdot)$, so $\Delta(r) \geq 0$ for all r and $\Delta(r) \geq \Delta(r^*)$ for all $r \geq r^*$. Notice that, with a dilated effect, the effect of the treatment, $r_{Tsi} - r_{Csi}$, is nonnegative and is larger, or at least no smaller, when the response that would have been observed under control, r_{Csi}, is higher.

Consider several examples of dilated effects. The null hypothesis of no treatment effect, $r_{Tsi} = r_{Csi}$, $i = 1, \ldots, n_s$, $s = 1, \ldots, S$, and the model of an additive treatment effect, $r_{Tsi} = r_{Csi} + \tau$, $i = 1, \ldots, n_s$, $s = 1, \ldots, S$, are both trivial examples of dilated effects. A multiplicative effect, $r_{Tsi} = \beta r_{Csi}$ with $\beta \geq 1$ is a dilated effect with $\Delta(r_{Csi}) = (\beta - 1) r_{Csi}$. A linear effect, $r_{Tsi} = \alpha + \beta r_{Csi}$ with $\alpha \geq 0$ and $\beta \geq 1$, is a dilated effect with $\Delta(r_{Csi}) = \alpha + (\beta - 1) r_{Csi}$. As a final example, suppose $\Delta(r) = 0$ for $\tilde{r} \geq r$ and $\Delta(r) > 0$ for $r > \tilde{r}$ with $\Delta(\cdot)$ nondecreasing; then individuals who would exhibit low responses under control, $\tilde{r} \geq r_{Csi}$, are not susceptible to the treatment, but the treatment does affect other individuals with $r_{Csi} > \tilde{r}$.

Because larger control responses r_{Csi} entail larger treated responses $r_{Tsi} = r_{Csi} + \Delta(r_{Csi})$ when the effect is dilated, it follows that the r_{Tsi}'s and r_{Csi}'s are ordered in the same way, so the person with the kth largest r_{Tsi} also has the kth largest r_{Csi}. It follows that $r_{T(k)} = r_{C(k)} + \Delta\{r_{C(k)}\}$, for $k = 1, \ldots, N$.

When the effect is dilated, the potential responses under treatment are not only higher than the potential responses under control, but also more dispersed. A common way to measure dispersion is by the difference in two order statistics, such as the range, which is the difference between the maximum and the minimum, or the interquartile range, which is the difference between the upper and lower quartiles. When the effect is dilated, the order statistics of potential responses to treatment are farther apart than the order statistics of potential responses to control; that is, for every $k > j$,

$$r_{T(k)} - r_{T(j)} = \left[r_{C(k)} + \Delta\{r_{C(k)}\}\right] - \left[r_{C(j)} + \Delta\{r_{C(j)}\}\right] \geq r_{C(k)} - r_{C(j)},$$

because $\Delta\{r_{C(k)}\} \geq \Delta\{r_{C(j)}\}$. Bickel and Lehmann (1976) and Shaked (1982) discuss this sort of dispersive ordering of distributions.

Fix a k and let $\rho = r_{C(k)}$ be the k/N quantile of the potential responses to control, and consider drawing inferences about the effect of the treatment at this quantile, $\Delta(\rho)$. For instance, if N were odd and $k = (N + 1)/2$, then $\rho = r_{C(k)}$ is the median response that would have been observed had all N subjects received control, and $\Delta(\rho)$ is the effect of the treatment at this median response. In this case, because $\Delta(\cdot)$ is nondecreasing, $\Delta(\rho)$ is also the median of the N treatment effects, $r_{Tsi} - r_{Csi}$.

5.3.2 Example: Kidney Function in Cadmium Workers

Thun et al. (1989) compared cadmium workers to controls in an effort to estimate the effects of cadmium exposure on kidney function; see also Thun (1993). Specifically, they compared male workers at a cadmium recovery plant in Colorado to unexposed male workers at a Colorado hospital, after "frequency matching" for an important covariate, age. Kidney dysfunction was measured by β-2-microglobulin in $\mu g/g$ of creatinine. In Table 5.1, frequency matching is replaced by pair matching for age, yielding 23 pairs,

TABLE 5.1. Kidney Function of Cadmium Workers and Hospital Controls.

Pair	Cadmium Worker	Hospital Worker
1	107,143	311
2	33,679	338
3	18,836	159
4	173	110
5	389	226
6	1,144	305
7	513	222
8	211	242
9	24,288	250
10	67,632	256
11	488	135
12	700	96
13	328	142
14	98	120
15	122	376
16	2,302	173
17	10,208	178
18	892	213
19	2,803	257
20	201	81
21	148	199
22	522	114
23	941	247

one cadmium worker, one hospital control; see Rosenbaum (1996a, §4.3) for details. For some cadmium workers, the β-2-microglobulin levels are much higher and much more dispersed than among hospital controls, and the model of an additive effect is not appropriate. This pattern is not uncommon in studies of the effects of occupational hazards, where many exposed subjects appear comparatively normal, while many others have extreme values of a measure of biological impact.

The model of a dilated treatment effect can generate the pattern in Table 5.1. There are $46 = 2 \times 23$ potential response pairs, (r_{Tsi}, r_{Csi}). Consider first $\rho = r_{C(23)}$ and $\Delta(\rho)$, essentially the median potential response to control and the median treatment effect.

5.3.3 Adjusted Responses with Dilated Effects

The logic of inference about an additive effect requires some changes before it can be applied to a dilated effect. When the treatment has an additive effect $r_{Tsi} = r_{Csi} + \tau$ for $i = 1, \ldots, n_s$, $s = 1, \ldots, S$, the observed response

$R_{si} = Z_{si}r_{Tsi} + (1 - Z_{si})\, r_{Csi}$ is simply $R_{si} = r_{Csi} + Z_{si}\tau$, and this fact permits inference about τ to be derived from a test of the null hypothesis of no treatment effect. For instance, the hypothesis $H_0 : \tau = \tau_0$ is tested by calculating adjusted responses, $R_{si} - Z_{si}\tau_0$, which equal r_{Csi} if the null hypothesis is true, and then testing whether the adjusted responses are consistent with the null hypothesis of no treatment effect. This logic was applied many times in earlier chapters, but it does not quite work with dilated effects, $r_{Tsi} = r_{Csi} + \Delta(r_{Csi})$. The reason is that, under the model of a dilated effect, the adjusted response, $R_{si} - Z_{si}\Delta(\rho)$, does not equal r_{Csi}, does not satisfy the null hypothesis of no treatment effect, and continues to depend on the treatment assignment Z_{si} through $R_{si} - Z_{si}\Delta(\rho) = r_{Csi} + Z_{si}\{\Delta(r_{Csi}) - \Delta(\rho)\}$.

The following proposition says that although the magnitudes of the adjusted responses are not equal to the magnitudes of the responses under control, there is a sense in which they have the correct sign. More precisely, the adjusted response, $R_{si} - Z_{si}\Delta(\rho)$ is above ρ just when r_{Csi} is above ρ. This will provide a basis for exact randomization inference and sensitivity analysis for $\Delta(\rho)$. Write $sign(a) = 1$ if $a > 0$, $sign(a) = 0$ if $a = 0$, and $sign(a) = -1$ if $a < 0$.

Proposition 21 *If the treatment has a dilated effect, for $i = 1, \ldots, n_s$, $s = 1, \ldots, S$,*

$$sign\{R_{si} - Z_{si}\Delta(\rho) - \rho\} = sign(r_{Csi} - \rho).$$

Proof. Recall that $\Delta(\cdot)$ is nonnegative and nondecreasing. It follows that if $r_{Csi} > \rho$ then $r_{Tsi} = r_{Csi} + \Delta(r_{Csi}) > \rho + \Delta(\rho)$, so that $R_{si} - Z_{si}\Delta(\rho) > \rho$. Similarly, if $r_{Csi} < \rho$ then $r_{Tsi} = r_{Csi} + \Delta(r_{Csi}) < \rho + \Delta(\rho)$, so that $R_{si} - Z_{si}\Delta(\rho) < \rho$. Finally, if $r_{Csi} = \rho$ then $r_{Tsi} = r_{Csi} + \Delta(r_{Csi}) = \rho + \Delta(\rho)$, so that $R_{si} - Z_{si}\Delta(\rho) = \rho$. ■

5.3.4   Testing Hypotheses About $\Delta(\rho)$

Under the model of a dilated effect, for fixed k, consider testing the null hypothesis $H_0 : \Delta(\rho) = \Delta_0$. Calculate the adjusted responses, $A_{si} = R_{si} - Z_{si}\Delta_0$, and let $A_{(N)} \geq A_{(N-1)} \geq \ldots \geq A_{(1)}$ be their order statistics. If the null hypothesis is true, then the proposition implies $\rho = A_{(k)}$. Let $q_{si} = 1$ if $A_{si} \geq A_{(k)}$ and $q_{si} = 0$ if $A_{si} < A_{(k)}$. Again, by the proposition, if the null hypothesis is true, $q_{si} = 1$ if $r_{Csi} \geq \rho$ and $q_{si} = 0$ if $r_{Csi} < \rho$, so the test statistic, $T = \sum_{s=1}^{S}\sum_{i=1}^{n_s} Z_{si}q_{si}$ is the number of treated subjects whose responses under control, r_{Csi}, would have exceeded $\rho = r_{C(k)}$.

A moment's thought reveals that the distribution of T under the null hypothesis has already been determined in earlier chapters. Since the events $r_{Csi} \geq \rho$ and $r_{Csi} < \rho$ involve only fixed quantities — that is, they do not involve the treatment assignment Z_{si} — it follows that the scores q_{si}

TABLE 5.2. Testing $H_0 : \Delta\left(r_{C(23)}\right) = 100$ in the Cadmium Data.

		Hospital Control	
		$A_{si} \geq 247$	$A_{si} < 247$
Cadmium Worker	$A_{si} \geq 247$	7	9
	$A_{si} < 247$	1	6

are fixed when the null hypothesis is true, and T is the sum of fixed binary scores for the treated subjects. From Chapters 2 and 3 it follows that, in a uniform randomized experiment or an observational study free of hidden bias, the distribution of T under the null hypothesis is simply the sum of independent hypergeometric random variables. Under the sensitivity analysis model of Chapter 4, the distribution of T is bounded by two known distributions for each fixed $\Gamma \geq 1$, namely two sums of independent extended hypergeometric random variables; see Corollary 16.

In the cadmium data in Table 5.1, consider testing the hypothesis that the effect at the median $\rho = r_{C(23)}$ is $H_0 : \Delta\left(\rho\right) = 100\ \mu g/g$. To do this, first subtract 100 from the response of each cadmium worker, yielding 46 adjusted responses, A_{si}. Sort the 46 adjusted responses into order, and find that $A_{(23)} = 247$. If the hypothesis $H_0 : \Delta\left(\rho\right) = 100$ were true, then $\rho = r_{C(23)} = A_{(23)} = 247$. Then determine which adjusted responses A_{si} are greater than or equal to 247, setting $q_{si} = 1$ if $A_{si} \geq 247$ and $q_{si} = 0$ if $A_{si} < 247$. Then summarize the results in Table 5.2. Because the cadmium data are matched pairs, if $(q_{s1}, q_{s2}) = (1, 1)$ then $1 = \sum_{i=1}^{n_s} Z_{si}q_{si}$, and if $(q_{s1}, q_{s2}) = (0, 0)$ then $0 = \sum_{i=1}^{n_s} Z_{si}q_{si}$, so under the null hypothesis, T receives a stochastic contribution from pair s only if the pair is discordant, $q_{s1} + q_{s2} = 1$.

In a study free of hidden bias, the hypothesis $H_0 : \Delta\left\{r_{C(23)}\right\} = 100$ would be tested by comparing the $9 + 1 = 10$ discordant pairs in Table 5.2 to a binomial distribution with 10 trials and probability of success $1/2$. This is analogous to McNemar's test for paired binary data. The chance of 9 or more heads in 10 independent trials with probability of success $1/2$ is 0.011, and this is the one-sided significance level. If there were no hidden bias, the data would strongly suggest the effect at the median response under control is greater than 100.

The sensitivity analysis for $\Gamma \geq 1$ uses Table 5.2 and parallels the sensitivity analysis for McNemar's test in Chapter 4. Specifically, the upper bound on the one-sided significance level is obtained by comparing 9 heads in 10 trials to a binomial with 10 trials and probability of success $\Gamma/\left(1 + \Gamma\right)$. For $\Gamma = 1$, 1.5, and 2, these upper bounds are, respectively, 0.011, 0.046, and 0.104, so an unobserved covariate that doubled the odds of exposure to cadmium within pairs matched for age could render plausible the hypothesis $H_0 : \Delta\left(r_{C(23)}\right) = 100$.

5.3.5   Confidence Intervals for $\Delta(\rho)$

Confidence intervals are obtained by inverting the test, that is, by testing each hypothesis $H_0 : \Delta(\rho) = \Delta_0$ and retaining in the interval the values not rejected. For instance, in the absence of hidden bias, the one-sided 95% confidence interval for $\Delta\{r_{C(23)}\}$ in the cadmium data is $\Delta \geq 147$, because the hypothesis $H_0 : \Delta\{r_{C(23)}\} = 147$ yields 9 successes in 11 discordant pairs with binomial tail probability 0.033, while $H_0 : \Delta\{r_{C(23)}\} = 147.0001$ yields 8 successes in 10 discordant pairs with binomial tail probability 0.055. If there were no hidden bias, we would be 95% confident that the effect at the median was at least 147 $\mu g/g$.

   The sensitivity analysis finds the minimum endpoint of the 95% confidence interval for several Γ. For $\Gamma = 1$, $1\frac{2}{3}$, and 2, the 95% intervals are, respectively, $\Delta \geq 147$, $\Delta \geq 81$, and $\Delta \geq -39$. A bias of magnitude $\Gamma = 2$ could explain away the ostensible effect of the treatment, $\Delta\{r_{C(23)}\}$, at the median response under control, $r_{C(23)}$.

5.3.6   Point Estimates of $\Delta(\rho)$

Each hypothesis $H_0 : \Delta(\rho) = \Delta_0$ yields a value of the test statistic $T = \sum_{s=1}^{S} \sum_{i=1}^{n_s} Z_{si} q_{si}$ which, in the absence of hidden bias, $\Gamma = 1$, has expectation $\sum_{s=1}^{S}(m_s\, q_{s+})/n_s$, where $m_s = \sum_{i=1}^{n_s} Z_{si}$, $q_{s+} = \sum_{i=1}^{n_s} q_{si}$, and q_{si} varies with Δ_0. Using SOLVE($\cdot$), find $\widehat{\Delta}$ by equating T to its expectation.

   In the absence of hidden bias in the cadmium data, both T and its expectation equal 12 for all Δ_0 on the open interval $(309, 501)$, so SOLVE($\cdot$) returns the point estimate $\widehat{\Delta} = (309 + 501)/2 = 405$.

   The sensitivity analysis finds the maximum expectation of T as $\sum_{s=1}^{S} \overline{E}_s$, where $\overline{E}_s$ is given by the extended hypergeometric distribution,

$$\overline{E}_s = \frac{\sum_{t=\max(0,m_s+q_{s+}-n_s)}^{\min(m_s,q_{s+})} t\binom{q_{s+}}{t}\binom{n_s-q_{s+}}{m_s-t}\Gamma^t}{\sum_{t=\max(0,m_s+q_{s+}-n_s)}^{\min(m_s,q_{s+})} \binom{q_{s+}}{t}\binom{n_s-q_{s+}}{m_s-t}\Gamma^t},$$

and the minimum point estimate is obtained by equating T to its expectation and solving for $\widehat{\Delta}_{\min}$ using SOLVE($\cdot$). The minimum expectation and maximum point estimate are obtained in the same way, but with $1/\Gamma$ in place of Γ.

   In the cadmium data, for $\Gamma = 1$, 2, and 3, the minimum point estimates of $\Delta\{r_{C(23)}\}$ are, respectively, $\widehat{\Delta}_{\min} = 405$, $\widehat{\Delta}_{\min} = 291$, and $\widehat{\Delta}_{\min} = 262$. For instance, with $\Gamma = 2$, the hypothesis $H_0 : \Delta\{r_{C(23)}\} = 291$ yields $A_{(23)} = 222$ and the counts in Table 5.3. Here, $T = 8 + 6 = 14$; that is, in 14/23 pairs, including 8 concordant pairs, the cadmium worker had an adjusted response of 222 or more. In the 8 concordant pairs with $q_{s+} = 2$, the expected contribution to T is $\overline{E}_s = 1$, while in the 6 concordant pairs

TABLE 5.3. Testing $H_0 : \Delta \left\{ r_{C(23)} \right\} = 291$ in the Cadmium Data.

		Hospital Control	
		$A_{si} \geq 222$	$A_{si} < 222$
Cadmium Worker	$A_{si} \geq 222$	8	3
	$A_{si} < 222$	6	6

with $q_{s+} = 0$, the expected contribution to T is $\overline{E}_s = 0$. In the $9 = 6 + 3$ discordant pairs with $q_{s+} = 1$, the maximum expected contribution to T is $\Gamma / (\Gamma + 1)$ and the minimum contribution is $1 / (\Gamma + 1)$. It follows that, under the hypothesis $H_0 : \Delta \left\{ r_{C(23)} \right\} = 291$, the expectation of T is bounded by $8 + 9\Gamma / (\Gamma + 1)$ and $8 + 9 / (\Gamma + 1)$. For $\Gamma = 2$, the maximum expectation is $8 + (9 \times 2) / (2 + 1) = 8 + 18/3 = 14$ which equals T. Moreover, for the hypothesis $H_0 : \Delta \left\{ r_{C(23)} \right\} = 291.0001$, the statistic is $T = 13$ with null expectation $13\frac{2}{3}$, but for the hypothesis $H_0 : \Delta \left\{ r_{C(23)} \right\} = 290.9999$, the statistic is $T = 14$ with null expectation $13\frac{2}{3}$; therefore, $SOLVE\,(\cdot)$ returns $\widehat{\Delta}_{\min} = 291$.

5.3.7 Reduced Sensitivity at Upper Quantiles

With a dilated treatment effect, the effect is larger when responses under control are higher. Larger effects are often less sensitive to hidden bias than smaller effects. One reason dilated effects are important is that there may be less sensitivity to hidden bias for upper quantile effects than for lower quantile effects.

Consider again the cadmium workers in Table 5.1. As $32/43 = .744$, the upper quartile of the 43 potential responses to control may be taken to be $r_{C(32)}$ and the effect at the upper quartile is $\Delta \left\{ r_{C(32)} \right\}$. Consider testing two hypotheses, namely the hypothesis of no effect at the median, $H_0 : \Delta \left\{ r_{C(23)} \right\} = 0$, and the hypothesis of no effect at the upper quartile, $H_0 : \Delta \left\{ r_{C(32)} \right\} = 0$. Under the model of an additive treatment effect, these two hypotheses would imply each other, but under the model of a dilated effect, $\Delta \left\{ r_{C(32)} \right\}$ may be larger than $\Delta \left\{ r_{C(23)} \right\}$. Table 5.4 compares the sensitivity of these two tests, reporting upper bounds on one-sided significance levels. Notice that the test of no effect at the median becomes sensitive to hidden bias at $\Gamma = 2$, whereas the test of no effect at the upper quartile becomes sensitive at $\Gamma = 5$. A moderate hidden bias could explain the somewhat poorer typical kidney function found among cadmium workers, but only a very large hidden bias could explain the extremely poor kidney function of some cadmium workers.

The conclusion is similar if significance tests are replaced by confidence intervals. For instance, for $\Gamma = 2$, the minimum one-sided 95% confidence intervals are $\Delta \left\{ r_{C(23)} \right\} \geq -39$ and $\Delta \left\{ r_{C(32)} \right\} \geq 211$. With a moderate bias of $\Gamma = 2$, no effect at the median is plausible, but the smallest plausible effect at the upper quartile is substantial.

TABLE 5.4. Sensitivity of Two Hypothesis Tests in the Cadmium Data.

Γ	$H_0 : \Delta \left\{ r_{C(23)} \right\} = 0$	$H_0 : \Delta \left\{ r_{C(32)} \right\} = 0$
1	.0032	.000031
2	.054	.0023
3	.16	.013
4	.27	.035
5	.38	.065

5.4 Instrumental Variables

5.4.1 The Effects of an Active Ingredient: Efficacy and Effectiveness

Some treatments work because of an active ingredient, and sometimes the effects of the ingredient are more interesting than the effects of the treatment. More precisely, a treatment may contain a single active ingredient, and may have no effect but through this one active ingredient. In this case, the effect of the treatment may vary from one person to the next because the dose of the active ingredient varies. For instance, in a randomized clinical trial comparing a drug to no treatment, some patients may not consume the full dose of their assigned drug, with the possible result that a negligible dose of a potent drug had negligible effects. When subjects are randomly assigned to treatment or control, but the doses are set by a process that is outside experimental control, one may want to use the randomization as the basis for inference, and yet model the effect in terms of the actual dose of the active ingredient that was received. This is possible, and it is the simplest example of an instrumental variable: randomization acts as an instrument for the dose. The approach taken in this section builds upon the work of Angrist, Imbens and Rubin (1996) along the specific lines developed in Rosenbaum (1996b, 1999b).

Suppose that the ith subject in stratum s receives the treatment at dose d_{Tsi} if assigned to treatment, $Z_{si} = 1$, and receives the treatment at dose d_{Csi} if assigned to control, $Z_{si} = 0$, so the dose actually received by this subject is $D_{si} = Z_{si}d_{Tsi} + (1 - Z_{si})\, d_{Csi}$. If the treatment effect were $r_{Tsi} - r_{Csi} = \beta\,(d_{Tsi} - d_{Csi})$, then the treatment effect would depend solely on the doses. For instance, if controls are assigned to zero dose, $d_{Csi} = 0$, and treated subjects are assigned to a 10 mg dose, but the ith subject in stratum s takes none of her assigned drug, then $d_{Tsi} = 0$, and the effect of the treatment for her is $r_{Tsi} - r_{Csi} = \beta\,(d_{Tsi} - d_{Csi}) = 0$. In contrast, a patient who would take the full dose would have effect $\beta \times 10$, and a patient who would take half of the assigned dose would have effect $\beta \times 5$.

Sommer and Zeger (1991) distinguish the "biologic efficacy" of a treatment from its "programmatic effectiveness." The efficacy is, essentially, the effect the drug would have if used in the way it was intended to be used, whereas the programmatic effectiveness reflects also the way the drug is actually used. In the model $r_{Tsi} - r_{Csi} = \beta (d_{Tsi} - d_{Csi})$, a 10 mg dose has efficacy $\beta \times 10$, but if the ith patient in stratum s refuses to take the 10 mg dose and takes 5 mg instead, the effectiveness for this patient is $\beta \times 5$. The effectiveness for this patient is only half the biologic efficacy of the 10 mg dose because the patient took only half the dose. Efficacy and effectiveness are both of interest, but different methods of inference are required.

If the model $r_{Tsi} - r_{Csi} = \beta (d_{Tsi} - d_{Csi})$ were true, then one might be interested in estimating β, rather than estimating a constant or average treatment effect. Notice, however, that $r_{Tsi} - r_{Csi} = \beta (d_{Tsi} - d_{Csi})$ is a model, and it can be false. For instance, some treatments have programmatic effectiveness but no biologic efficacy—for instance, this is true of placebos. The effect of a placebo is: (i) fully realized by assigning a patient to the treatment, $Z_{si} = 1$, and convincing him that he received it, (ii) the effect is absent if assigned to control, $Z_{si} = 0$; and (iii) the magnitude of the effect $r_{Tsi} - r_{Csi}$ is undiminished by covertly setting the dose to zero, $d_{Tsi} = d_{Csi} = 0$. The model that says treatment effect is proportional to dose increase, $r_{Tsi} - r_{Csi} = \beta (d_{Tsi} - d_{Csi})$, is false for placebos.

5.4.2 Exclusion Restriction

The model that says treatment effect is proportional to dose, $r_{Tsi} - r_{Csi} = \beta (d_{Tsi} - d_{Csi})$, satisfies what is known as an exclusion restriction. Informally, the exclusion restriction says that treatment assignment, Z_{si}, is related to observed response, $R_{si} = Z_{si} r_{Tsi} + (1 - Z_{si}) r_{Csi}$, only through the realized dose, $D_{si} = Z_{si} d_{Tsi} + (1 - Z_{si}) d_{Csi}$. To see this, notice that if $r_{Tsi} - r_{Csi} = \beta (d_{Tsi} - d_{Csi})$, then $r_{Tsi} - \beta d_{Tsi} = r_{Csi} - \beta d_{Csi} = R_{si} - \beta D_{si}$, so that $R_{si} - \beta D_{si}$ is a constant, not varying with the treatment assignment Z_{si}.

5.4.3 A Numerical Illustration: Noncompliance in a Randomized Trial

Instrumental variable methods are illustrated using a contrived, artificial example in this section, and using a practical example in §5.4.6. The artificial example is simple and clear because, unlike a practical example, the correct answer is known and the responses under both treatment and control are available. As a numerical illustration, consider the simple, artificial data in Table 5.5 concerning a randomized trial of encouragement to exercise for patients with chronic obstructive pulmonary disease. The trial is not stratified, $S = 1$, so the s subscript is dropped. The treatment, Z_i, is encouragement to exercise, $Z_i = 1$ signifying encouragement, $Z_i = 0$ signifying no encouragement. The encouraged exercises are walking, jogging, or

TABLE 5.5. Noncompliance with Encouragment (Z_i) to Exercise (D_i).

i	d_{Ti}	d_{Ci}	r_{Ti}	r_{Ci}	Z_i	D_i	R_i
1	1	1	71	71	1	1	71
2	1	1	68	68	0	1	68
3	1	0	64	59	1	1	64
4	1	0	62	57	0	0	57
5	1	0	59	54	0	0	54
6	1	0	58	53	1	1	58
7	1	0	56	51	1	1	56
8	1	0	56	51	0	0	51
9	0	0	42	42	0	0	42
10	0	0	39	39	1	0	39

cycling. There are $n = 10$ patients, $i = 1, \ldots, 10$, of whom $m = 5 = \sum Z_i$ were randomly selected to be encouraged to exercise, the remainder being controls. For example, $i = 1$ was encouraged but $i = 2$ was not. The pair (d_{Ti}, d_{Ci}) indicates whether person i would exercise, with or without encouragement, where 1 signifies exercise and 0 indicates no exercise. For example, $i = 1$ would exercise whether encouraged or not, $(d_{Ti}, d_{Ci}) = (1, 1)$, whereas $i = 10$ would not exercise in either case, $(d_{Ti}, d_{Ci}) = (0, 0)$, but $i = 3$ exercises only if encouraged, $(d_{Ti}, d_{Ci}) = (1, 0)$.

The response, (r_{Ti}, r_{Ci}) is a measure of lung function, FEV or forced expiratory volume, on a convenient scale, with higher numbers signifying better lung function. Notice that $r_{Tsi} - r_{Csi} = \beta (d_{Tsi} - d_{Csi})$ with $\beta = 5$, so switching from no exercise to exercise, $d_{Tsi} - d_{Csi} = 1$, raises lung function by 5, but encouragement that is ignored, $d_{Tsi} - d_{Csi} = 0$, does nothing. For instance, encouragement did nothing for subject $i = 1$, whose response was unchanged, $r_{Tsi} - r_{Csi} = 71 - 71 = 0$, because this subject would have exercised anyway, $\beta (d_{Tsi} - d_{Csi}) = 5 (1 - 1) = 0$. However, $i = 3$ exercises only if encouraged, $d_{Tsi} - d_{Csi} = 1$, so would improve with encouragement, $r_{Tsi} - r_{Csi} = 64 - 59 = 5 = 5 (1 - 0)$.

Notice who responds to encouragement. Subjects $i = 1$ and $i = 2$ would have the best lung function without encouragement, and they will exercise with or without encouragement. Subjects $i = 9$ and $i = 10$ would have the poorest lung function without encouragement, and they will not exercise even if encouraged. Subjects $i = 3, 4, \ldots, 8$ have intermediate lung function without exercise, and they will exercise only if encouraged. Although encouragement, Z_i, is randomized, compliance with assigned treatment, (d_{Ti}, d_{Ci}), is strongly confounded with the health of the patient. In this context, how can we estimate the efficacy β?

The two obvious estimates are obviously wrong. The difference between the mean response of patients who exercised and those who did not is

$$\frac{71 + 68 + 64 + 58 + 56}{5} - \frac{57 + 54 + 51 + 42 + 39}{5} = 14.8,$$

which is nearly three times $\beta = 5$. The problem with this estimate is that the people who exercised were in better health than the people who did not exercise; that is the main reason the people who exercised had better lung function. Similarly, the mean difference between those who were encouraged and those who were not is:

$$\frac{71 + 64 + 58 + 56 + 39}{5} - \frac{68 + 57 + 54 + 51 + 42}{5} = 3.2,$$

which is much less than $\beta = 5$. The estimate, 3.2, is called the *intent-to-treat estimate*, and it is a useful estimate of "program effectiveness" but not of "efficacy". Exercise is more effective, in this artificial example, than a program to encourage exercise, because some patients exercise without encouragement and others will not exercise even if encouraged. In practice, one must ask, with different answers in different contexts, whether one is interested in the effects of encouragement or of exercise, of effectiveness or of efficacy, and choose an estimator accordingly.

How can we estimate the efficacy β?

5.4.4   Exact Randomization Inference in Randomized Trials with Noncompliance

Exact randomization inference about β is straightforward and involves no new principles. Under the model that says response is proportional to dose, $r_{Tsi} - r_{Csi} = \beta(d_{Tsi} - d_{Csi})$, consider first testing the null hypothesis $H_0 : \beta = \beta_0$. First, compute the adjusted responses, $R_{si} - \beta_0 D_{si}$, which equal the constant $r_{Tsi} - \beta d_{Tsi} = r_{Csi} - \beta d_{Csi} = a_{si}$, say, when the null hypothesis is true. For testing the true hypothesis, $H_0 : \beta = 5$, the adjusted responses are given in column five of Table 5.6. Now perform a randomization test of the null hypothesis of no effect of encouragement on the adjusted responses. For instance, the Wilcoxon rank sum test might be used, ranking the adjusted responses, as in the last column of Table 5.6, and summing the ranks for encouraged patients, $T = 10 + 8 + 5 + 3.5 + 1 = 27.5$. In this randomized experiment, there are $\binom{10}{5} = 252$ ways to pick $m = 5$ of $n = 10$ subjects at random to receive encouragement, and each assignment has probability $1/252$. A random choice of 5 of the 10 slightly tied ranks in Table 5.6 produces a rank sum of 35.5 or more with probability greater than 0.05, so $T = 27.5$ does not lead to rejection of the hypothesis $H_0 : \beta = 5$.

Notice carefully that both actual exercise, D_i, and encouragement, Z_i, play a role in this test. The effect of the treatment is expressed in terms of actual exercise, D_i, whereas the test relies on the random assignment of encouragement, Z_i. Actual exercise, D_i, is strongly confounded with initial health; it cannot be the basis for a randomization test. Encouragement, Z_i, has an effect only if exercise behavior changes; encouragement alone does not describe the effect of exercise.

TABLE 5.6. Testing $H_0 : \beta = 5$ Using the Rank Sum Statistic.

i	FEV R_i	Exercise D_i	Encouragement Z_i	$R_i - 5D_i$	Rank
1	71	1	1	66	10
2	68	1	0	63	9
3	64	1	1	59	8
4	57	0	0	57	7
5	54	0	0	54	6
6	58	1	1	53	5
7	56	1	1	51	3.5
8	51	0	0	51	3.5
9	42	0	0	42	2
10	39	0	1	39	1

The test just described is an exact test about efficacy derived from the random assignment of encouragement, Z_{si}, and the model which asserts that programmatic effectiveness is proportional to the change in exercise, $r_{Tsi} - r_{Csi} = \beta(d_{Tsi} - d_{Csi})$, and it is valid in giving correct significance levels. Specifically, if the null hypothesis $H_0 : \beta = \beta_0$ is true, then the adjusted responses $R_{si} - \beta_0 D_{si}$ equal $r_{Tsi} - \beta d_{Tsi} = r_{Csi} - \beta d_{Csi} = a_{si}$, which satisfy the null hypothesis of no treatment effect, and $t(\mathbf{Z}, \mathbf{R} - \beta_0 \mathbf{D}) = t(\mathbf{Z}, \mathbf{a})$ has its usual null randomization distribution, where $\mathbf{D} = (D_{11}, D_{12}, \dots, D_{S,n_S})^T$ and $\mathbf{a} = (a_{11}, a_{12}, \dots, a_{S,n_S})^T$.

When does this test have power? Suppose that we test $H_0 : \beta = \beta_0$ when in fact $\beta \neq \beta_0$. This test will have no power — more precisely, it will have power equal to the level of the test — if no one responds to encouragement by starting to exercise, that is, if $d_{Tsi} = d_{Csi}$ for all s, i. To see this, notice that if $d_{Tsi} = d_{Csi}$ then $r_{Tsi} - r_{Csi} = \beta(d_{Tsi} - d_{Csi}) = 0$, and therefore $r_{Tsi} - \beta_0 d_{Tsi} = r_{Csi} - \beta_0 d_{Csi}$ and $R_{si} - \beta_0 D_{si}$ is constant, not varying with Z_{si}, for every for every β_0. Encouragement must produce some increase in exercise — that is, $(d_{Tsi}, d_{Csi}) = (1, 0)$ must occur frequently — if this test is to have much power.

Notice that the randomization test of no efficacy, $H_0 : \beta = 0$, is identical to the randomization test of no treatment effect, $H_0 : r_{Tsi} = r_{Csi}$, for all s, i in Chapter 2. The two tests give identical significance levels in every application. The test of $H_0 : \beta = 0$ can discover efficacy only if the treatment has some effectiveness, as judged by the randomization test for the intent-to-treat analysis that ignores the dose received, D_{si}.

As discussed in Chapter 2, a Hodge-Lehmann estimate equates a statistic to its null expectation and solves approximately for the parameter using $SOLVE(\cdot)$. In a randomized experiment with $S = 1$, the rank sum statistic has null expectation $m(n+1)/2$. To estimate β, compute the rank sum statistic from the adjusted responses, $R_i - \beta_0 D_i$, and use $SOLVE(\cdot)$ to find the $\widehat{\beta}$ that gives approximately $m(n+1)/2$ as the value of the statistic.

In the example, $m(n+1)/2 = 5(10+1)/2 = 27.5$, and the rank sums computed with $\beta_0 = 4.9999$, 5, and 5.0001 are, respectively, $T = 28$, 27.5, and 27, so the estimate, $\widehat{\beta} = 5$, is exactly the correct value, $\beta = 5$, in this artificial example.

A $100(1-\alpha)\%$ confidence set for β is obtained by testing each hypothesis $H_0 : \beta = \beta_0$ at level α and retaining the values of β_0 not rejected by the test. The confidence set need not be an interval, but if an interval is desired, the confidence interval may be defined as the shortest closed interval that contains the confidence set—adding points to a confidence set cannot decrease its chance of covering the true β.

The confidence set behaves in a manner that is both sensible and unusual. If the model $r_{Tsi} - r_{Csi} = \beta(d_{Tsi} - d_{Csi})$ is severely wrong, the test may reject every value of β, returning an empty confidence set. This might happen, for example, with a treatment whose effects are dramatic in size but are essentially placebo effects. In this case, the assigned treatment, Z_{si}, may strongly predict the response R_{si}, but the dose D_{si} may not. If the model is clearly wrong, then it is reasonable for a confidence interval to reject the model. If the noncompliance with encouragement is very extensive, so what people are encouraged to do, Z_{si}, is weakly related to what they actually do, D_{si}, then Z_{si} is said to be a "weak instrument" for D_{si}, and in this case, the data may contain little information about β. A weak instrument will be correctly reflected in the confidence interval, which may maintain its $100(1-\alpha)\%$ exact coverage by becoming longer, perhaps infinite in length. Of course, when the data contain little information, a confidence interval should say so.

The example above illustrates this. With $n = 10$, $m = 5$, the null distribution of the rank sum statistic has $\Pr(T \geq 36) = .048$ and $\Pr(T \geq 30) = .345$. In testing $H_0 : \beta = \beta_0$, no matter how small β_0 is set, it is not possible in the example to produce a rank sum of $T = 36$, so the one-sided 95% confidence interval is $\beta \geq -\infty$. However, the rank sum is $T = 30$ for $H_0 : \beta = 1.9999$ and $T = 29$ for $H_0 : \beta = 2.0001$, so the $(2/3) - confidence$ interval is $\beta \geq 2$. Due to noncompliance and a small sample size, there is not much information about β in these data, but there is a little. (The minimum wage example in §5.4.6 is larger, with a stronger instrument, and it produces informative 95% confidence intervals.)

Traditional discussions of instrumental variables distinguish between the assumptions that make the problem either "identified" or "unidentified." Roughly speaking, if the problem is identified, then the confidence interval for β gradually shrinks to the single point β as the sample size increases. The idea of identification draws a bright red line between two types of problems. Is this red line useful? In some problems that are formally identified, the instrument is very weak, and with practical sample sizes, there is little information about β. In principle, in a problem that is formally not identified, there may be quite a bit of information about β, perhaps enough for some particular practical decision; see Rosenbaum (1997) for discussion. Arguably, a bright red line relating assumptions to asymptotics is less interesting than an exact confidence interval describing what has been learned from the evidence actually at hand.

5.4.5   Sensitivity Analysis in Observational Studies with Instrumental Variables

Instrumental variables are used in observational studies as well as experiments, but in an observational study, the treatment or instrument, Z_{si}, is not randomly assigned. This section describes sensitivity analysis with an instrumental variable in parallel with the randomization inference with instrumental variables in the previous section. The analysis makes two assumptions: first, treatments, Z_{si}, are governed by the model used in Chapter 4; second, that effects are proportional to dose, $r_{Tsi} - r_{Csi} = \beta (d_{Tsi} - d_{Csi})$.

Consider testing the hypothesis $H_0 : \beta = \beta_0$ using one of the test statistics $T = t(\cdot, \cdot)$ discussed in Chapter 4, perhaps the signed rank statistic with matched pairs. If the null hypothesis were true, then the adjusted responses, $R_{si} - \beta_0 D_{si}$, would equal $r_{Tsi} - \beta_0 d_{Tsi} = r_{Csi} - \beta d_{Csi} = a_{si}$, which are constant, not varying with the treatment assignment, Z_{si}, so the a_{si} would satisfy the null hypothesis of no treatment effect. In other words, if $H_0 : \beta = \beta_0$ is true, the adjusted responses $R_{si} - \beta_0 D_{si}$ satisfy the null hypothesis of no treatment effect, and $t(\mathbf{Z}, \mathbf{R} - \beta_0 \mathbf{D}) = t(\mathbf{Z}, \mathbf{a})$ has the null distribution described in Chapter 4. This leads to tests, confidence intervals by inverting tests, and Hodges–Lehmann point estimates. In other words, sensitivity analysis with an instrumental variable entails a different model for treatment effect, namely $r_{Ti} - r_{Ci} = \beta (d_{Ti} - d_{Ci})$, and hence a different definition of adjusted responses $R_{si} - \beta_0 D_{si}$, but beyond this no new principles or techniques are involved. An example is given in the next section.

5.4.6   Example: Minimum Wages and Employment

In an interesting, thoughtful, careful, but nonetheless controversial study, Card and Krueger (1994, 1995, 1998, 2000) examined the possible effects of increases in the minimum wage on employment. Economic theory is usually understood as predicting that a forced increase in the price of labor will reduce demand for it, that is, put people out of work. Economists are often critical of the minimum wage arguing that it tends to hurt the very people it is intended to help. Card and Krueger compared New Jersey to adjacent eastern Pennsylvania before and after New Jersey increased its minimum wage from \$4.25 to \$5.05 per hour, or about 19%, on 1 April 1992. They looked at changes in employment in the fast-food industry, specifically

Burger King, Kentucky Fried Chicken, Wendy's, and Roy Rogers. Here, their data will be used to illustrate sensitivity analysis with an instrumental variable. Card and Krueger have made their data publicly available using FTP; see Card and Krueger (1995, p18) for specifics. Card and Krueger found "no evidence that the rise in New Jersey's minimum wage reduced employment at fast-food restaurants in the state."

The appendix, §5.10, contains a portion of Card and Krueger's data, intended solely to illustrate methodology. The data describe New Jersey and Pennsylvania, before and after New Jersey's wage increase, for 66 matched pairs of restaurants. Two variables are described: full-time equivalent (FTE) employment, and starting wages. For example, the New Jersey restaurant in pair #5 reported 23 full-time equivalent employees before the increase and 19 afterward, while its paired Pennsylvania restaurant reported 22.75 FTEs before and 21 afterwards. In this same pair, the starting wage rose from $4.25 to $5.05 in the New Jersey restaurant, but stayed steady at $4.25 in the Pennsylvania restaurant. Pair #5 exhibits the pattern anticipated by economic theory: wages rose in New Jersey from the old minimum to the new one, and employment declined, whereas the Pennsylvania restaurant changed little. Obviously, different patterns are seen in different restaurants. The restaurants were paired for chain and for starting wages before the increase. Pairs 1 through 30 are Burger Kings, pairs 31 through 40 are Kentucky Fried Chicken restaurants, pairs 41 through 55 are Roy Rogers, and pairs 56 through 66 are Wendy's. A few of the employment numbers indicate that the interviewers and respondents may now and then have misunderstood one another, as happens in most if not all surveys. (For instance, in pair #29, the Pennsylvania Burger King reported a decline from 70.5 FTE before to 29 after.)

Notice that some New Jersey restaurants were paying close to or more than $5.05 as a starting wage before the increase in the minimum wage, and many of these restaurants did not substantially increase their starting wages. For example, in pair #29, the New Jersey Burger King was paying $5.12 before the increase and $5.05 afterwards. One might reasonably expect that changes in minimum wage laws will have effects only to the extent that they change wages, so that the New Jersey Burger King in pair #29 should experience little or no effect.

Let the response be the increase in the log of employment, so $R_{11} = \log(26.50) - \log(15.50)$, and let the dose be the increase in the log of starting wages, so $D_{11} = \log(5.05) - \log(4.25)$. The model $r_{Tsi} - r_{Csi} = \beta(d_{Tsi} - d_{Csi})$ predicts a larger effect of the minimum wage, $r_{Tsi} - r_{Csi}$, on a restaurant that was forced to make a larger change in wages, $d_{Tsi} - d_{Csi}$. Because of the log scale, the parameter β is the elasticity. Notice that β relates wage changes in either state to employment, as distinct from relating a change in the minimum wage law to employment.

The hypothesis $H_0 : \beta = \beta_0$ is tested by computing the adjusted responses, $R_{si} - \beta_0 D_{si}$, and applying Wilcoxon's signed rank test to them, as in §4.3.3. Indeed, the test of $H_0 : \beta = 0$ is identical to the test in §4.3.3. Even in the absence of hidden bias, $\Gamma = 1$, the null hypothesis $H_0 : \beta = 0$ is plausible when tested against the predicted decline in employment, $H_0 : \beta < 0$, with approximate one-sided significance level from the randomization distribution of the signed rank test of 0.87. In the absence of bias, the one-sided, approximate 95% confidence interval is $\beta \geq -0.29$. For a restaurant that increased its starting wage from the old minimum to the new minimum, $D_{si} = \log(5.05) - \log(4.25) = 0.172$, a $\beta = -0.29$ yields $-0.29 \times 0.172 = -0.050$ or about a 5% decline in employment. In other words, in the absence of hidden bias, the confidence interval rejects hypotheses that say the decline in employment is more than 5% for restaurants that faced the full impact of the wage increase. In the absence of hidden bias, the Hodges–Lehmann point estimate of β is actually positive, specifically 0.64, suggesting an increase rather than a decrease in employment.

Although the data give no sign of a decline in employment associated with wage increases, the results are sensitive to moderate biases. For example, the hypothesis $H_0 : \beta = -1$ entails roughly a $17.2\% = -1 \times 0.172$ decline in employment for a restaurant raising wages from the old minimum of \$4.25 to the new minimum of \$5.05, a substantial decline in employment. In the absence of hidden bias, $\Gamma = 1$, the hypothesis $H_0 : \beta = -1$ is clearly rejected with one-sided significance level of 0.004, but the hypothesis becomes barely plausible when $\Gamma = 1.34$, with upper bound on the significance level of 0.0506, and plausible when $\Gamma = 1.5$ with upper bound on the significance level of 0.10. The minimum Hodges–Lehmann point estimate equals -1 at about $\Gamma = 2.2$. The 66 matched pairs provide no sign of a decline in employment with wage increases, and they are consistent with increases in employment, but they are also sensitive to moderate biases, so a substantial decline in employment could be hidden by biases of moderate size.

5.5   Attributable Effects: Binary Responses

5.5.1   What Are Attributable Effects?

The attributable effect is the effect of the treatment on the treated subjects. Attributable effects permit randomization inferences in contexts where the treatment effect cannot be additive.

In this section, there is a single stratum, $S = 1$, so the s subscript is dropped, and the responses, r_{Ti} and r_{Ci}, are binary, 1 or 0, with a nonnegative effect, $r_{Ti} \geq r_{Ci}$. Then the effect of the treatment on subject i is $\delta_i = r_{Ti} - r_{Ci}$, where $\delta_i = 0$ or $\delta_i = 1$. Notice that $\delta_i = 1$ if $(r_{Ti}, r_{Ci}) = (1, 0)$ so that exposure to the treatment causes a response in subject i who would not have responded in the absence of treatment. Also, $\delta_i = 0$ if either $(r_{Ti}, r_{Ci}) = (0, 0)$ or $(r_{Ti}, r_{Ci}) = (1, 1)$ so exposure to the

TABLE 5.7. The Observed 2×2 Contingency Table.

Response	Treated	Control
1	$\sum Z_i\, r_{Ti}$	$\sum (1 - Z_i)\, r_{Ci}$
0	$\sum Z_i\, (1 - r_{Ti})$	$\sum (1 - Z_i)\, (1 - r_{Ci})$
Total	m	$n - m$

treatment does not alter the response of subject i, but that response may be 1 or 0. Write $\boldsymbol{\delta} = (\delta_1, \dots, \delta_n)^T$, which is an n–dimensional parameter with binary coordinates. Hamilton (1979) uses $\boldsymbol{\delta}$ to express the common measures of association in a 2×2 contingency table.

The attributable effect is $A = \sum_{i=1}^{n} Z_i\, (r_{Ti} - r_{Ci}) = \sum_{i=1}^{n} Z_i \delta_i$; it is the number of treated responses actually caused by exposure to the treatment. If there are $m = 5$ treated subjects, and $2 = \sum_{i=1}^{n} Z_i r_{Ti}$ exhibit responses, but exactly one of these, for subject j with $Z_j = 1$, would have occurred anyway under control, $r_{Tj} = r_{Cj} = 1$, then the attributable effect is $A = 1$ response caused by exposure to treatment. Under the null hypothesis of no treatment effect, $H_0 : r_{Ti} = r_{Ci}$, $i = 1, \dots, n$, the attributable effect A is 0. The attributable effect A depends not only on the parameter $\boldsymbol{\delta}$, but also on who received treatment, $\mathbf{Z}$. In a randomized experiment, A would vary with the random assignment of treatments $\mathbf{Z}$. That is, the attributable effect A is an unobserved random variable, not a fixed parameter. Can one draw inferences about the value of A?

5.5.2 2×2 Tables and Attributable Effects

Table 5.7 is the observed 2×2 contingency table. Notice carefully that in this table, treated subjects reveal their response to treatment, r_{Ti}, and control subjects reveal their response to control, r_{Ci}.

In contrast, Table 5.8 is the 2×2 contingency table that would have been observed had all subjects received the control, so this table involves only potential responses to control, r_{Ci}. Notice that the marginal row and column totals in Table 5.8 are fixed, not varying with the treatment assignment $\mathbf{Z}$, but the row totals of Table 5.7 change as $\mathbf{Z}$ changes when the treatment has an effect. Under the null hypothesis of no treatment effect, $H_0 : r_{Ti} = r_{Ci}$, $i = 1, \dots, n$, Tables 5.7 and 5.8 are equal. In a randomized experiment, the upper corner cell, $\sum Z_i\, r_{Ci}$, in Table 5.8 has the hypergeometric distribution, and this is the basis for Fisher's exact test of the hypothesis of no effect. As noted in Chapter 4, for this test of the hypothesis of no effect, the sensitivity to hidden bias in an observational study is examined by comparing the observed Table 5.7 to two extended hypergeometric distributions; see Corollary 15. The question at hand is whether we can use similar reasoning to draw inferences about the attributable effect A when the null hypothesis of no effect does not hold.

TABLE 5.8. Responses Had Everyone Received Control.

Response	Treated	Control	Total
1	$\sum Z_i r_{Ci}$	$\sum (1 - Z_i) r_{Ci}$	$\sum r_{Ci}$
0	$\sum Z_i (1 - r_{Ci})$	$\sum (1 - Z_i)(1 - r_{Ci})$	$\sum (1 - r_{Ci})$
Total	m	$n - m$	n

TABLE 5.9. The Observed Table Adjusted for the Attributable Effect.

Response	Treated	Control
1	$\sum Z_i r_{Ti} - \sum Z_i \delta_i$	$\sum (1 - Z_i) r_{Ci}$
0	$\{\sum Z_i (1 - r_{Ti})\} + \sum Z_i \delta_i$	$\sum (1 - Z_i)(1 - r_{Ci})$
Total	m	$n - m$

In Table 5.9, the observed Table 5.7 is adjusted by subtracting the attributable effect $A = \sum Z_i \delta_i$ from the upper left corner cell and adding it to the lower left corner cell. Inference about A is based on the observation that Table 5.9 equals Table 5.8. Informally, one asks: After what adjustments $A = \sum Z_i \delta_i$ does Table 5.9 appear to satisfy the null hypothesis of no treatment effect?

5.5.3   Testing Hypotheses About δ

Consider testing the hypothesis $H_0 : \delta = \delta_0$ where δ_0 is an n–dimensional vector with 1 or 0 coordinates. For some hypotheses δ_0, inspection of the data immediately shows the hypothesis is false. Since the treatment has a nonnegative effect, $r_{Ti} \geq r_{Ci}$, a treated subject without a response, $Z_i = 1$ and $R_i = 0$, must have $\delta_i = 0$, and a control subject with a response, $Z_i = 0$ and $R_i = 1$, must have $\delta_i = 0$. Call the hypothesis $H_0 : \delta = \delta_0$ compatible if it satisfies these conditions and incompatible otherwise. The one true hypothesis is compatible for every $\mathbf{Z}$.

If the hypothesis $H_0 : \delta = \delta_0$ is incompatible, reject it with certainty, that is, with type 1 error rate of zero. Otherwise, if the hypothesis is compatible, compute $A_0 = \sum Z_i \delta_{0i}$. If the hypothesis $H_0 : \delta = \delta_0$ is true, then $A_0 = \sum Z_i \delta_{0i} = A = \sum Z_i \delta_i$, so Table 5.9 may be computed, and it equals Table 5.8. In a randomized experiment, the hypothesis $H_0 : \delta = \delta_0$ is tested by comparing this table to the hypergeometric distribution, whereas in an observational study, the sensitivity analysis compares the table to two extended hypergeometric distributions to bound the significance level.

5.5.4   Confidence Sets for δ and Inference About A

A confidence set for δ may be obtained by testing every hypothesis $H_0 : \delta = \delta_0$ and retaining the compatible hypotheses not rejected by the test. For instance, if the test in the previous section rejects at the 0.05 level

TABLE 5.10. Station Design and Mortality.

	No Pit	Pit
Dead	16	14
Alive	5	18

hypotheses $H_0 : \delta = \delta_0$ if $A_0 < a$ and accepts if $A_0 \geq a$, then a one-sided confidence set for δ is the set of all compatible δ_0 with $A_0 = \sum Z_i \, \delta_{0i} \geq a$.

The confidence set for δ is awkward to examine because it is a set of n–dimensional vectors with binary coordinates. However, it is easy to describe the confidence set in a useful manner. Consider, for instance, the one-sided confidence set for δ consisting of all compatible δ_0 with $A_0 = \sum Z_i \, \delta_{0i} \geq a$. It is the set of all treatment effects δ with at least a responses among treated subjects actually caused by the treatment, that is, with at least a responses attributable to treatment. An n–dimensional confidence set for δ becomes a one–dimensional set of plausible values for A.

5.5.5   Example: Death in the London Underground

In the London Underground, some train stations have a drainage pit below the tracks. The pit is about a meter in depth and runs the length of the station platform. When a passenger falls or jumps or is pushed from the station platform under a train — an "incident" in the language of the British Railway Regulations Act of 1893—such a pit is a place to escape contact with the wheels of the train. Using legally required records of 53 such incidents, Coats and Walter (1999) compared mortality in stations with and without such a pit. Their data are in Table 5.10. In stations without a pit, 16 of 21 incidents resulted in death, whereas in stations with a pit, 14 of 32 incidents resulted in death.

If Fisher's exact randomization test for a 2×2 table is applied, the one-sided significance level is 0.0193, so if there were no hidden bias, $\Gamma = 1$, then it would not be plausible that $A = 0$ deaths were caused by the absence of a pit. In the absence of hidden bias, $\Gamma = 1$, all hypothesis $H_0 : \delta = \delta_0$ with $A_0 = \sum Z_i \, \delta_{0i} = 1$ are rejected with one-sided significance level 0.0439, so neither zero nor one attributable death is plausible. However, in the absence of hidden bias, $\Gamma = 1$, all hypotheses $H_0 : \delta = \delta_0$ with $A_0 = \sum Z_i \, \delta_{0i} = 2$ attributable deaths are accepted with one-sided significance level 0.0875, so with 95% confidence, at least $6.7\% = 2/30$ of the deaths following incidents would be attributable to the absence of a pit. The test for $A = 1$ is performed by applying Fisher's exact test to Table 5.10 after subtracting $A = 1$ from 16 and adding it to 5, altering the row totals.

In conducting an analysis of sensitivity to hidden bias, as discussed in Chapter 4, exact, sharp bounds on the significance levels from Fisher's exact test are obtained by comparing a 2×2 table to two extended hypergeometric distributions, one with parameter Γ and the other with parameter $1/\Gamma$.

This comparison is made after adjusting the table for the hypothesized value of A. For instance, with $\Gamma = 1.5$, the upper bound on the one sided significance level with $A = 0$ is 0.0885 and for $A = 1$ it is 0.1634, so the study is sensitive to biases of moderate size.

5.6 Attributable Effects: Displacements

5.6.1 What Are Displacements?

When a treatment has a nonnegative effect, displacements are a fairly general way of measuring the magnitude of the treatment effect. Displacements compare the observed treated responses to the unobserved distribution of responses that would have been seen had all subjects received the control.

This section considers a study without strata, in which the treatment has a nonnegative effect, $r_{Ti} \geq r_{Ci}$. Recall that the n subjects offer n potential responses to treatment and control, (r_{Ti}, r_{Ci}), $i = 1, \ldots, n$, and these give rise to two sets of order statistics, $r_{T(n)} \geq r_{T(n-1)} \geq \cdots \geq r_{T(1)}$ and $r_{C(n)} \geq r_{C(n-1)} \geq \cdots \geq r_{C(1)}$. Since some subjects receive treatment and others receive control, these order statistics are not observed. The n observed responses, $R_i = Z_i r_{Ti} + (1 - Z_i) r_{Ci}$, yield n observed order statistics, $R_{(n)} \geq R_{(n-1)} \geq \cdots \geq R_{(1)}$.

Fix an integer k so that the k/n quantile of potential responses to control is $r_{C(k)}$. Assume that $r_{C(k+1)} > r_{C(k)}$, and let θ be any value strictly between $r_{C(k)}$ and $r_{C(k+1)}$, so $r_{C(k+1)} > \theta > r_{C(k)}$. For instance, if n is even and $k = n/2$, then θ is between the two middle order statistics of responses to control, and is a median. As seen below, the observed data may reveal that no such θ exists under certain hypotheses, so such hypotheses about θ may be rejected with certainty.

The ith subject has a displacement around θ if $r_{Ti} > \theta > r_{Ci}$. In words, the ith subject has a displacement if the subject would have a response below θ under control, but would have a response above θ under treatment. For instance, if n is even and $k = n/2$, then the ith subject has a displacement if the response under control is below the unobserved median of the n potential responses under control, but the response under treatment is above the control median. If $k/n = 95\%$, then a displacement signifies a fairly typical response r_{Ci} under control together with a response r_{Ti} under treatment that would be unusually high under control.

Write $\delta_i = 1$ if subject i has a displacement, that is, if $r_{Ti} > \theta > r_{Ci}$. The number of treated subjects who experienced displacements because they were exposed to treatment is $A = \sum Z_i \delta_i$.

Write $\tilde{r}_{Ti} = 1$ if $r_{Ti} > \theta$, and $\tilde{r}_{Ti} = 0$ otherwise. In parallel, write $\tilde{r}_{Ci} = 1$ if $r_{Ci} > \theta$, and $\tilde{r}_{Ci} = 0$ otherwise. Then $\delta_i = \tilde{r}_{Ti} - \tilde{r}_{Ci}$. If θ were known, then inference about A could be based on the method in the previous section for inference about binary responses. However, θ is not known, because it depends on the n unobserved order statistics of responses to control, $r_{C(n)} \geq r_{C(n-1)} \geq \cdots \geq r_{C(1)}$. Given that θ is not known, how can inferences be drawn?

5.6.2 Inference About Displacements

Inference about displacements is based on the following proposition from Rosenbaum (2001). Although θ is not known, the proposition says that the observed order statistics, $R_{(n)} \geq R_{(n-1)} \geq \cdots \geq R_{(1)}$, together with a hypothesis about $A = \sum Z_i \delta_i$ pin down θ narrowly enough to permit inferences.

Proposition 22 *If $a = \sum_i Z_i \delta_i$, then $R_{(k+1-a)} > \theta > R_{(k-a)}$.*

Proof. By assumption $r_{C(k+1)} > r_{C(k)}$, so there are exactly $n - k$ subjects with $r_{Ci} > \theta$, and since $r_{Ti} \geq r_{Ci}$ it follows that these $n - k$ subjects all have $R_i > \theta$. Since $a = \sum_i Z_i \delta_i$ there are exactly a other subjects, not included among the $n - k$ subjects, with $R_i = r_{Ti} > \theta > r_{Ci}$. For the remaining $k - a$ subjects, $\theta > R_i$. So there are exactly $n - k + a$ subjects with $R_i > \theta$ and exactly $k - a$ subjects with $\theta > R_i$. This means that $R_{(n)} \geq R_{(n-1)} \geq \cdots \geq R_{(k+1-a)} > \theta > R_{(k-a)} \geq \cdots \geq R_{(1)}$. ∎

Notice carefully that if $R_{(k+1-a)} = R_{(k-a)}$ then there is no θ such that $a = \sum_i Z_i \delta_i$ displacements around θ have occurred among treated subjects.

Inference proceeds as follows. To test a hypothesis, $H_0 : \delta = \delta_0$ with $A_0 = \sum Z_i \delta_{0i}$, find $R_{(k+1-A_0)}$ and $R_{(k-A_0)}$. If $R_{(k+1-A_0)} = R_{(k-A_0)}$, reject the hypothesis with certainty, that is, with type one error rate of zero. Otherwise, if $R_{(k+1-A_0)} > R_{(k-A_0)}$, then under the null hypothesis, a treated subject, $Z_i = 1$, has $\tilde{r}_{Ti} = 1$ if $R_i > R_{(k-A_0)}$ and $\tilde{r}_{Ti} = 0$ otherwise, while a control subject, $Z_i = 0$, has $\tilde{r}_{Ci} = 1$ if $R_i > R_{(k-A_0)}$ and $\tilde{r}_{Ci} = 0$ otherwise. The hypothesis $H_0 : \delta = \delta_0$ may now be tested using the methods in the previous section for attributable effects with binary responses.

5.6.3 Example: Cytogenetic Effects of Benzene Among Shoe Workers

In Bursa, Turkey, Tunca and Egeli (1996) identified $m = 58$ shoe workers who used glues containing substantial quantities of benzene. The workers were exposed for periods ranging from 5 to 50 years, in conditions with insufficient ventilation, and with benzene exposure levels 10 to 30 times the maximum allowable concentration. In an effort to estimate the cytogenetic changes caused by long-term occupational exposure to benzene, Tunca and Egeli compared the shoe workers to $n - m = 20$ controls who were also residents of Bursa, Turkey, but who were believed to not have exposures to benzene.

For each of the $n = 78$ subjects, roughly 20 metaphases were analyzed for chromosomal aberrations, but the number of metaphases analyzed varied from person to person. Here, attention focuses on the percentage of gaps found for each subject. The data are fairly coarse exhibiting ties. Sorted into order, the percentages of gaps for the 58 shoe workers are 0.00, 0.00, 5.55, 6.66, 7.69, 8.33, 8.33, 8.69, 9.09, 9.09, 9.52, 10.00, 10.00, 10.00, 10.00, 10.52, 11.11, 11.11, 11.76, 11.76, 11.76, 11.76, 11.76, 12.00, 12.50, 13.04, 13.33, 13.33, 13.33, 13.33, 13.51, 14.28, 15.38, 15.38, 15.38, 16.66, 17.64, 17.64, 18.18, 18.60, 19.04, 19.23, 19.23, 20.00, 20.00, 20.00, 20.00, 20.00, 20.00, 21.05, 23.07, 23.80, 24.13, 25.00, 27.27, 29.41, 30.00, and 33.33. For the 20 controls, the sorted percentages of gaps are: 0, 0, 0, 0, 0, 0, 0, 0, 0, 0, 0, 4, 5, 5, 5, 5, 5, 5, 8, and 10. Many of the shoe workers had many more gaps than did most of the controls.

The unobserved median of $n = 78$ potential responses to control is the average of the 39th and 40th order statistics, $\{r_{C(39)} + r_{C(40)}\}/2$. Therefore, let $k = 39$, so that subject i is defined to have a displacement above the control median if there is a θ such that $r_{C(40)} > \theta > r_{C(39)}$ and $r_{Ti} > \theta > r_{Ci}$. Roughly speaking, subject i has a displacement above the median if he would have had fewer than the median number of gaps that would have been seen had all $n = 78$ subjects escaped exposure to benzene, but instead would have had strictly more than that median had he been exposed to benzene.

In Table 5.11, three groups of hypotheses are tested assuming no hidden bias, $\Gamma = 1$, namely each compatible hypothesis $H_0 : \delta = \delta_0$ with $A_0 = 1$, with $A_0 = 19$, and with $A_0 = 25$. Since $R_{(39)} = 11.76 > 11.11 = R_{(38)}$, if $A = 1$, then θ is between 11.11 and 11.76. Since none of the observed responses for controls is above 11.11, the first 2×2 table in Table 5.11 is obtained, with significance level $P < 0.00001$, so it is not plausible that only $A = 1$ treated subject experienced a displacement from 11.11 or below to 11.76 or above. This significance level is based on the hypergeometric distribution. Similarly, it is not plausible that only $A = 19$ treated subjects had displacements from $R_{(20)} = 5$ or below to $R_{(21)} = 5.55$ or above. Now, $R_{(15)} = \ldots = R_{(20)} = 5$. It is plausible that $A = 25$ or fewer treated subjects had displacements from $R_{(14)} = 4$ or below to $R_{(15)} = 5$ or above.

The sensitivity analysis for $\Gamma \geq 1$ is similar except that the resulting 2×2 tables in Table 5.11 are compared to extended hypergeometric distributions to obtain bounds on the significance level. For instance, for testing hypotheses $H_0 : \delta = \delta_0$ yielding $A_0 = 19$ displacements attributable to treatment, the upper bounds on the significance level for $\Gamma = 1, 2, 3$, and 4 are 0.000025, 0.0027, 0.020, and 0.058. For $A_0 = 19$ to be even marginally plausible, failure to control for an unobserved covariate would need to alter the odds of treatment by at least a factor of $\Gamma = 4$.

TABLE 5.11. Testing Three Hypotheses in the Benzene Data.

$A = 1, \quad R_{(39-1)} = 11.11, \quad P < 0.00001$			
	Treated	Control	Total
$r_{Ci} > 11.11$	39	0	39
$11.11 \geq r_{Ci}$	19	20	39
Total	58	20	78

$A = 19, \quad R_{(39-19)} = 5, \quad P < 0.000025$			
	Treated	Control	Total
$r_{Ci} > 5$	37	2	39
$5 \geq r_{Ci}$	21	18	39
Total	58	20	78

$A = 25, \quad R_{(39-25)} = 4, \quad P = 0.219$			
	Treated	Control	Total
$r_{Ci} > 4$	31	8	39
$4 \geq r_{Ci}$	27	12	39
Total	58	20	78

5.7 Inference About a Quantile

5.7.1 A Method Based on the Extended Hypergeometric Distribution

An alternative to modeling the effect of the treatment is to draw nonparametric inferences about the distributions of responses that would be seen under treatment or control. Consider the case of a single stratum, $S = 1$, and drop the s subscript. Then the order statistics of the n potential responses to treatment are $r_{T(n)} \geq r_{T(n-1)} \geq \cdots \geq r_{T(1)}$, and these are not observed because only m subjects received treatment. This section discusses inference about the k/n quantile, $r_{T(k)}$. For instance, if n is odd and $k = (n+1)/2$, then $r_{T(k)}$ is the median response that would have been observed had all n subjects been exposed to treatment. A confidence interval for $r_{T(k)}$ and its sensitivity analysis will be constructed.

It is convenient to assume, at first, that there are no ties, so that $r_{T(n)} > r_{T(n-1)} > \cdots > r_{T(1)}$. It is not difficult to show that the method to be described works with ties, and that ties simply increase the probability that the confidence interval covers the parameter, so a 95% confidence interval has coverage slightly greater than 95%; see Rosenbaum (1995, §2.2). Write $\widetilde{R}_{T(m)} > \widetilde{R}_{T(m-1)} > \cdots > \widetilde{R}_{T(1)}$ for the m observed order statistics from the m treated subjects. The confidence interval for $r_{T(k)}$ is formed using two of the observed order statistics, $\left[\tilde{R}_{T(a)}, \tilde{R}_{T(b)}\right]$ with $a < b$. Specifically, the argument that follows uses Corollary 15 to show that, under the sensitivity analysis model, the confidence interval $\left[\tilde{R}_{T(a)}, \tilde{R}_{T(b)}\right]$ covers $r_{T(k)}$ with probability at least $\Upsilon(n, m, k, a, 1/\Gamma) - \Upsilon(n, m, k-1, b, \Gamma)$.

Let $c_i = 1$ if $r_{Ti} \leq r_{T(k)}$ and $c_i = 0$ otherwise. Then $k = \sum_{i=1}^{n} c_i = c_+$ from the definition of $r_{T(k)}$. Notice that $\sum_{i=1}^{n} Z_i c_i \geq a$ if and only if $\tilde{R}_{T(a)} \leq r_{T(k)}$. As noted in Corollary 15, under the sensitivity analysis model, the sign score statistic $\sum_{i=1}^{n} Z_i c_i$ has a distribution bounded by two extended hypergeometric distributions with parameters $\Gamma \geq 1$ and $1/\Gamma$. It follows that the one-sided confidence interval $r_{T(k)} \geq \tilde{R}_{T(a)}$ has coverage probability that is sharply bounded by:

$$\Upsilon(n, m, k, a, \Gamma) \geq \Pr\left(\sum_{i=1}^{n} Z_i c_i \geq a\right)$$

$$= \Pr\left\{r_{T(k)} \geq \tilde{R}_{T(a)}\right\} \geq \Upsilon\left(n, m, k, a, \frac{1}{\Gamma}\right),$$

In parallel, replacing k by $k-1$, yields:

$$\Upsilon(n, m, k-1, b, \Gamma) \geq \Pr\left\{r_{T(k-1)} \geq \tilde{R}_{T(b)}\right\}$$

$$\geq \Upsilon\left(n, m, k-1, b, \frac{1}{\Gamma}\right),$$

but $\Pr\left\{\tilde{R}_{T(b)} \geq r_{T(k)}\right\} = 1 - \Pr\left\{r_{T(k-1)} \geq \tilde{R}_{T(b)}\right\}$, so the one-sided confidence interval $\tilde{R}_{T(b)} \geq r_{T(k)}$ has coverage probability that is sharply bounded by:

$$1 - \Upsilon\left(n, m, k-1, b, \frac{1}{\Gamma}\right) \geq \Pr\left\{\tilde{R}_{T(b)} \geq r_{T(k)}\right\}$$

$$\geq 1 - \Upsilon(n, m, k-1, b, \Gamma).$$

Finally consider the two-sided interval $\left[\tilde{R}_{T(a)}, \tilde{R}_{T(b)}\right]$, where $a < b$. This interval covers $r_{T(k)}$ if the event $r_{T(k)} \geq \tilde{R}_{T(a)}$ occurs but the event $r_{T(k-1)} \geq \tilde{R}_{T(b)}$ does not occur, so $\left[\tilde{R}_{T(a)}, \tilde{R}_{T(b)}\right]$ covers $r_{T(k)}$ with probability

$$\Pr\left\{r_{T(k)} \geq \tilde{R}_{T(a)}\right\} - \Pr\left\{r_{T(k-1)} \geq \tilde{R}_{T(b)}\right\},$$

which is at least $\Upsilon(n, m, k, a, 1/\Gamma) - \Upsilon(n, m, k-1, b, \Gamma)$.

In parallel, confidence intervals for $r_{C(k)}$ are constructed using the $n-m$ observed order statistics $\tilde{R}_{C(n-m)} > \tilde{R}_{C(n-m-1)} > \cdots > \tilde{R}_{C(1)}$ from the $n-m$ controls. Of course, one must replace m by $n-m$, so that, for example, $\left[\tilde{R}_{C(a)}, \tilde{R}_{C(b)}\right]$ covers $r_{C(k)}$ with probability at least $\Upsilon(n, n-m, k, a, 1/\Gamma) - \Upsilon(n, n-m, k-1, b, \Gamma)$.

TABLE 5.12. Sensitivity of Confidence Intervals for the Median Mercury Level.

Γ	$r_{T(20)}$	$r_{C(20)}$
1	$[100, 196]$	$[6, 11]$
2	$[70, 236]$	$[6, 12]$
5	$[69, 270]$	$[4, 15]$

5.7.2 Example: Methylmercury in Fish, Continued

In the study by Skerfving, Hansson, Mangs, Lindsten, and Ryman (1974), discussed in Chapter 4, $m = 23$ subjects who ate large quantities of fish contaminated with methylmercury were compared to $n - m = 16$ controls who ate little or no contaminated fish. There are $n = 39$ subjects in total, so the median for these $n = 39$ subjects is the $(39 + 1)/2 = 20^{th}$ order statistic. Consider, here, the level of mercury found in the blood as the response. Had all $n = 39$ subjects eaten contaminated fish, the median response would have been $r_{T(20)}$, whereas had all $n = 39$ subjects avoided contaminated fish, the median would have been $r_{C(20)}$, but neither quantity is observed, because $m = 23$ ate contaminated fish and $n - m = 16$ did not.

The $m = 23$ treated subjects yield $m = 23$ observed order statistics, $\widetilde{R}_{T(1)} = 12.8$, $\widetilde{R}_{T(2)} = 40, \ldots, \widetilde{R}_{T(23)} = 1100$. The chance that the random interval $\left[\widetilde{R}_{T(a)}, \widetilde{R}_{T(b)}\right]$ covers the fixed parameter $r_{T(20)}$ is at least $\Upsilon(39, 23, 20, a, 1/\Gamma) - \Upsilon(39, 23, 20 - 1, b, \Gamma)$. For a 95% interval one selects a and b so that $\Upsilon(39, 23, 20, a, 1/\Gamma) - \Upsilon(39, 23, 20 - 1, b, \Gamma) \geq 0.95$, perhaps by insisting that $\Upsilon(39, 23, 20, a, 1/\Gamma) \geq 0.975$ and $0.25 \geq \Upsilon(39, 23, 20 - 1, b, \Gamma)$. Similarly, the $n - m = 16$ controls yield $n - m = 16$ order statistics, $\widetilde{R}_{C(1)} = 3$, $\widetilde{R}_{C(2)} = 4, \ldots, \widetilde{R}_{C(16)} = 17$. Then $\left[\widetilde{R}_{C(a)}, \widetilde{R}_{C(b)}\right]$ covers $r_{C(20)}$ with probability at least

$$\Upsilon(39, 16, 20, a, 1/\Gamma) - \Upsilon(39, 16, 20 - 1, b, \Gamma).$$

Table 5.12 gives the sensitivity analysis for the 95% confidence intervals for the two median mercury levels, $r_{T(20)}$ and $r_{C(20)}$. The coverage is actually somewhat greater than 95% both because of ties and because of the discreteness of the extended hypergeometric distribution, which does not always permit exact equality, $\Upsilon\left(n, m, k, a, \frac{1}{\Gamma}\right) - \Upsilon(n, m, k - 1, b, \Gamma) = 0.95$. As always, the confidence intervals become somewhat longer as Γ increases. However, even if substantial biases are present, $\Gamma = 5$, it would be clear that eating contaminated fish caused a substantial increase in typical blood mercury levels. Even with $\Gamma = 5$, the median under control, $r_{C(20)}$, is much lower and more precisely determined than the median under treatment, $r_{T(20)}$.

5.8   Bibliographic Notes

Doksum (1974), Doksum and Sievers (1976), Lehmann (1975), and Switzer (1976) express treatment effects in terms of a function $\Delta(\cdot)$ such that $r_{Tsi} = r_{Csi} + \Delta(r_{Csi})$, and dilated effects, with $\Delta(\cdot)$ required to be nonnegative and nondecreasing, are discussed in Rosenbaum (1999a). The fine paper by Angrist, Imbens, and Rubin (1996) shaped my presentation of randomization as an instrumental variable for studying the effects of an active ingredient; see also Angrist and Imbens (1994) and Angrist (1998). Instrumental variables have long been a standard topic in the fitting of econometric models, and links to randomized experiments have been discussed by Holland (1988) and Sommer and Zeger (1991), among others. Rosenbaum (1996b, 1999b) discusses exact randomization inference and sensitivity analysis with an instrumental variable, as presented in this chapter.

Attributable effects are related to attributable risks, as discussed by MacMahon and Pugh (1970), Walter (1975), and Hamilton (1979). Displacement effects are related to the control median test, as discussed by Gart (1963) and Gastwirth (1968), quantile comparison functions, as discussed by Li, Tiwari, and Wells (1996), and nonparametric inference based on placements, as discussed by Orban and Wolfe (1982). The discussion of attributable effects is based on Rosenbaum (2001).

The confidence interval for a finite population quantile in §5.7 is discussed by Wilks (1962, §11.4) and by Sedransk and Meyer (1978) when there is no hidden bias, $\Gamma = 1$, and Rosenbaum (1995) discusses sensitivity analysis for $\Gamma \geq 1$.

5.9   Problems

1. Quantiles in large finite populations. Suppose the treated group is a small fraction of the n subjects in the study. Specifically, in §5.7, suppose the number m of treated subjects is fixed, and the number of controls increases, so $n \to \infty$, with $k/n \to \theta$ where $0 < \theta < 1$. Show how to use the binomial distribution with sample size m and success probability $\theta\Gamma/(\theta\Gamma + 1 - \theta)$ in place of the extended hypergeometric distribution to approximate the confidence interval. (Hint: The paper by Harkness (1965) discusses the limiting behavior of the extended hypergeometric distribution and might be helpful.) Notice that the limiting binomial distribution does not require knowledge of n, so it may be used to compare a treated group with a population distribution. (Solution: See Rosenbaum (1995, §2.3).)

2. Worst case bounds. Some authors, such as Manski (1990, 1995), do not introduce a sensitivity parameter, Γ, and instead report a worst

case bound. What happens to the confidence interval in §5.7 for a quantile, $r_{T(k)}$, when $\Gamma \to \infty$? What information is there about $r_{T(k)}$ when the bias may be infinitely large? Consider a simple case, with n is odd, $k = (n+1)/2$ so $r_{T(k)}$ is the median, and set $m = (n+1)/2$, so just slightly more than half of the subjects received treatment. (Solution: See Rosenbaum (1995, §2.4).)

3. Dilated effects without strata. The method for dilated effects in §5.3 was illustrated with paired data, $n_s = 2$, $m_s = 1$, for each s. Apply the method to the shoe worker data in §5.6 where there is a single stratum $S = 1$. (Rosenbaum (1999a) discusses an unstratified example.)

4. Confidence intervals for an additive effect using differences of order statistics. Without strata, assume the treatment has an additive effect, $r_{Ti} = r_{Ci} + \tau$. Obtain sensitivity bounds for

$$\Pr\left\{\tau \geq \widetilde{R}_{T(a)} - \widetilde{R}_{C(k-a+1)}\right\},$$

in the notation of §5.7. (Solution: Rosenbaum (1995, §4.2).)

5.10 Appendix: Minimum Wage Data

This appendix contains 66 matched pairs of restaurants from Card and Krueger's (1994, 1995) data; see also Rosenbaum (1999b). Two variables are described: full-time equivalent (FTE) employment, and starting wages. The restaurants were paired for chain and for starting wages before the increase. Pairs 1 through 30 are Burger Kings, pairs 31 through 40 are Kentucky Fried Chicken restaurants, pairs 41 through 55 are Roy Rogers, and pairs 56 through 66 are Wendy's. See §5.4.6 for detailed discussion.

Matched Pairs of Restaurants,
Before and After the Minimum Wage Increase

| | FTE Employment | | | | Starting Wages | | | |
| | New Jersey | | Pennsylvania | | New Jersey | | Pennsylvania | |
#	Before	After	Before	After	Before	After	Before	After
1	15.50	26.50	15.50	14.00	4.25	5.05	4.25	4.40
2	18.00	32.00	18.00	19.00	4.25	5.05	4.25	4.25
3	19.00	20.00	19.00	25.00	4.25	5.25	4.25	4.38
4	19.50	47.50	19.50	27.00	4.25	5.05	4.25	4.25
5	23.00	19.00	22.75	21.00	4.25	5.05	4.25	4.25
6	24.00	26.50	24.00	23.00	4.25	5.05	4.25	4.75
7	24.00	25.50	24.50	43.50	4.25	5.05	4.25	4.75
8	26.00	25.00	26.00	20.00	4.25	5.05	4.25	6.25
9	27.00	35.00	27.00	25.50	4.25	5.05	4.25	4.25

10	30.00	26.00	32.50	25.50	4.25	5.05	4.25	4.25
11	33.00	29.00	36.50	27.00	4.25	5.05	4.25	4.25
12	38.00	24.50	39.00	26.50	4.25	5.05	4.25	4.25
13	41.00	22.50	41.50	23.50	4.25	5.05	4.25	4.90
14	41.50	21.50	45.00	37.50	4.25	5.05	4.25	4.25
15	13.00	19.00	21.00	17.50	4.37	5.05	4.35	4.25
16	24.00	32.00	24.00	16.00	4.50	5.05	4.50	4.25
17	30.50	22.50	32.00	21.00	4.50	5.05	4.50	4.50
18	50.00	30.00	48.50	27.00	4.50	5.05	4.50	4.35
19	18.50	24.75	52.50	34.00	4.55	5.05	4.50	5.05
20	16.00	15.50	15.50	34.50	4.75	5.25	4.75	4.75
21	17.00	21.50	18.50	41.25	4.75	5.05	4.75	4.50
22	20.50	39.00	21.00	24.50	4.75	5.05	4.75	4.50
23	22.50	44.00	22.50	35.50	4.75	5.25	4.75	4.50
24	18.50	13.00	15.50	27.50	4.87	5.05	4.87	5.00
25	16.00	24.50	13.50	18.00	5.00	5.50	5.00	4.50
26	23.00	23.25	23.50	36.50	5.00	5.05	5.00	5.00
27	27.50	27.00	26.50	30.50	5.00	5.25	5.00	4.50
28	35.00	10.50	58.00	29.00	5.00	5.05	5.00	5.00
29	46.50	23.75	70.50	29.00	5.12	5.05	5.00	4.75
30	20.00	20.50	28.50	26.00	5.50	5.05	5.50	4.75
31	8.00	8.50	7.50	9.50	4.25	5.05	4.25	4.25
32	10.50	14.00	10.50	16.50	4.25	5.05	4.25	4.25
33	9.50	8.00	8.50	14.00	4.50	5.05	4.50	5.00
34	8.50	9.00	8.50	13.00	4.50	5.05	4.50	4.50
35	14.00	16.50	14.00	17.00	4.50	5.25	4.50	4.75
36	11.50	14.00	11.00	11.00	4.75	5.05	4.75	4.25
37	9.00	14.00	9.00	8.50	5.00	5.05	5.00	5.00
38	10.00	7.50	9.50	13.50	5.00	5.05	5.00	4.75
39	15.00	18.50	11.00	11.00	5.25	5.05	5.25	5.00
40	9.50	9.50	16.75	20.00	5.50	5.05	5.25	5.00
41	18.75	24.00	18.00	15.00	4.25	5.05	4.25	4.35
42	13.50	14.50	13.00	13.50	4.50	5.05	4.50	4.50
43	15.50	11.50	18.00	34.00	4.50	5.05	4.50	4.25
44	15.00	19.00	15.00	12.50	4.75	5.05	4.75	4.75
45	19.50	28.50	20.25	10.00	4.75	5.05	4.75	4.91
46	5.00	8.00	14.00	13.50	5.00	5.05	5.00	4.25
47	15.50	17.50	15.00	19.00	4.95	5.05	5.00	4.75
48	21.00	18.00	15.50	17.50	5.00	5.05	5.00	4.75
49	21.00	14.00	18.50	14.50	5.00	5.05	5.00	4.25
50	23.00	15.50	19.00	17.50	5.00	5.05	5.00	4.75
51	24.00	21.00	20.25	14.00	5.00	5.05	5.00	5.00
52	24.50	36.00	20.50	18.00	5.00	5.05	5.00	4.75
53	26.00	16.50	21.00	21.00	5.00	5.05	5.00	5.00
54	30.00	26.00	29.50	15.50	5.00	5.05	5.00	4.75

55	35.00	37.00	36.00	19.00	5.00	5.05	5.00	4.90
56	13.00	15.00	10.50	12.00	4.25	5.30	4.25	4.25
57	18.00	23.50	17.50	19.00	4.25	5.05	4.25	4.35
58	18.00	17.50	25.00	25.00	4.25	5.05	4.25	4.25
59	24.00	24.00	25.00	31.50	4.50	5.05	4.25	4.50
60	36.50	60.50	38.00	18.00	4.25	5.05	4.25	4.25
61	53.00	19.00	38.00	20.00	4.62	5.05	4.67	4.50
62	22.00	23.00	23.50	18.00	4.75	5.05	4.75	4.50
63	29.00	21.50	28.00	26.75	4.75	5.05	4.75	5.00
64	18.25	33.00	19.00	20.50	5.00	5.05	5.00	5.00
65	33.50	32.00	34.00	20.00	5.00	5.50	5.00	5.25
66	18.00	33.50	24.00	35.50	5.50	5.05	5.50	4.75

5.11 References

Angrist, J. D. (1998) Estimating the labor market impact of voluntary military service using social security data on military applicants. *Econometrica*, **66**, 249–288.

Angrist, J. D. and Imbens, G. W. (1994) Identification and estimation of local average treatment effects. *Econometrica*, **62**, 467–475.

Angrist, J. D., Imbens, G. W., and Rubin, D. B. (1996) Identification of causal effects using instrumental variables (with discussion). *Journal of the American Statistical Association*, **91**, 444–455.

Bickel, P. and Lehmann, E. (1976) Descriptive statistics for nonparametric problems, IV: Spread. In *Contributions to Statistics*, J. Juneckova, ed., Dordrecht, Holland: Reidel, pp. 33–40.

Card, D. and Krueger, A. (1994) Minimum wages and employment: A case study of the fast-food industry in New Jersey and Pennsylvania. *American Economic Review*, **84**, 772–793.

Card, D. and Krueger, A. (1995) *Myth and Measurement: The New Economics of the Minimum Wage*. Princeton, NJ: Princeton University Press.

Card, D. and Krueger, A. (1998) A reanalysis of the effect of the New Jersey minimum wage increase on the fast-food industry with representative payroll data. National Bureau of Economic Research, Working Paper 6386.

Card, D. and Krueger, A. (2000) Minimum wages and employment: A case study of the fast-food industry in New Jersey and Pennsylvania: Reply. *American Economic Review*, **90**, 1397–1420.

Coats, T. J. and Walter, D. P. (1999) Effect of station design on death in the London Underground: Observational study. *British Medical Journal*, **319**, 957.

Doksum, K. (1974) Empirical probability plots and statistical inference for nonlinear models in the two-sample case. *Annals of Statistics*, **2**, 267–277.

Doksum, K. and Sievers, G. (1976) Plotting with confidence: Graphical comparisons of two populations. *Biometrika*, **63**, 421–434.

Gart, J. J. (1963) A median test with sequential application. *Biometrika*, **50**, 55–62.

Gastwirth, J. L. (1968) The first-median test: A two-sided version of the control median test. *Journal of the American Statistical Association*, **63**, 692–706.

Hamilton, M. A. (1979). Choosing the parameter for 2×2 and $2 \times 2 \times 2$ table analysis. *American Journal of Epidemiology*, **109**, 362–375.

Harkness, W. (1965) Properties of the extended hypergeometric distribution. *Annals of Mathematical Statistics*, **36**, 938–945.

Holland, P. W. (1988) Causal inference, path analysis, and recursive structural equation models (with discussion). *Sociological Methodology*, 449–476.

Lehmann, E. (1975) *Nonparametrics: Statistical Methods Based on Ranks*. San Francisco: Holden-Day.

Li, G., Tiwari, R. C., and Wells, M. T. (1996) Quantile comparison functions in two-sample problems, with application to comparisons of diagnostic markers. *Journal of the American Statistical Association*, **91**, 689–698.

MacMahon, B. and Pugh, T. F. (1970) *Epidemiology: Principles and Methods*. Boston: Little, Brown.

Orban, J. and Wolfe, D. A. (1982) A class of distribution-free two-sample tests based on placements. *Journal of the American Statistical Association*, **77**, 666–672.

Manski, C. (1990) Nonparametric bounds on treatment effects. *American Economic Review*, 319–323.

Manski, C. (1995) *Identification Problems in the Social Sciences*. Cambridge, MA: Harvard University Press.

Rosenbaum, P. R. (1995) Quantiles in nonrandom samples and observational studies. *Journal of the American Statistical Association*, **90**, 1424–1431.

Rosenbaum, P. R. (1996a) Observational studies and nonrandomized experiments. In: *Handbook of Statistics, Volume 13, Design of Experiments*, Chapter 6, S. Ghosh and C. R. Rao, eds., New York: Elsevier, pp. 181–197.

Rosenbaum, P. R. (1996b) Comment on "Identification of causal effects using instrumental variables" by Angrist, Imbens, and Rubin. *Journal of the American Statistical Association*, **91**, 465–468.

Rosenbaum, P. R. (1997) Discussion of a paper by Copas and Li. *Journal of the Royal Statistical Society*, Series **B**, **59**, 90.

Rosenbaum, P. R. (1999a) Reduced sensitivity to hidden bias at upper quantiles in observational studies with dilated effects. *Biometrics*, **55**, 560–564.

Rosenbaum, P. R. (1999b) Using combined quantile averages in matched observational studies. *Applied Statistics*, **48**, 63–78.

Rosenbaum, P. (2001) Effects attributable to treatment: Inference in experiments and observational studies with a discrete pivot. *Biometrika*, **88**, 219–232.

Rubin, D. B. (1974) Estimating the causal effects of treatments in randomized and nonrandomized studies. *Journal of Educational Psychology*, **66**, 688–701.

Shaked, M. (1982) Dispersive ordering of distributions. *Journal of Applied Probability*, **19**, 310–320.

Sedransk, J. and Meyer, J. (1978) Confidence intervals for the quantiles of finite populations: simple random and stratified random sampling. *Journal of the Royal Statistical Society*, Series B, **40**, 239–252.

Sheiner, L. B. and Rubin, D. B. (1995) Intention-to-treat analysis and the goals of clinical trials. *Clinical Pharmacology and Therapeutics*, **57**, 6–15.

Skerfving, S., Hansson, K., Mangs, C., Lindsten, J., and Ryman, N. (1974) Methylmercury-induced chromosome damage in man. *Environmental Research*, **7**, 83–98.

Sommer, A. and Zeger, S. L. (1991) On estimating efficacy from clinical trials. *Statistics in Medicine*, **10**, 45–52.

Switzer, P. (1976) Confidence procedures for two-sample problems. *Biometrika*, **63**, 13–25.

Thun, M. (1993) Kidney dysfunction in cadmium workers. In: *Case Studies in Occupational Epidemiology*, K. Steenland, ed., New York: Oxford University Press, pp. 105–126.

Thun, M., Osorio, A., Schober, S., et al. (1989) Nephropathy in cadmium workers: Assessment of risk from airborne occupational exposure to cadmium. *British Journal of Industrial Medicine*, **46**, 689–697.

Tunca, B. T. and Egeli, U. (1996) Cytogenetic findings on shoe workers exposed long-term to benzene. *Environmental Health Perspectives*, **104**, supplement 6, 1313–1317.

Walter, S. D. (1975) The distribution of Levin's measure of attributable risk. *Biometrika*, **62**, 371–375.

Wilks, S. (1962) *Mathematical Statistics*. New York: Wiley.

6
Known Effects

6.1 Detecting Hidden Bias Using Known Effects

6.1.1 Sensitivity Analysis and Detecting Hidden Bias: Their Complementary Roles

As seen in Chapter 4, observational studies vary considerably in their sensitivity to hidden bias. The study of coffee and myocardial infarction in §4.4.5 is sensitive to small biases while the studies of smoking and lung cancer in §4.3.3 or DES and vaginal cancer in §4.4.5 are sensitive only to biases that are many times larger. If sensitive to small biases, a study is especially open to the criticism that a particular unrecorded covariate was not controlled because, in this case, small differences in an important covariate can readily explain the difference in outcomes in treated and control groups. Still, all observational studies are sensitive to sufficiently large biases, and large biases have occurred on occasion; see §4.4.3. A sensitivity analysis shows how biases of various magnitudes might alter conclusions, but it does not indicate whether biases are present or what magnitudes are plausible.

Efforts to detect hidden bias involve collecting data that have a reasonable prospect of revealing a hidden bias if it is present. If treatments are not assigned to subjects as if at random, if the distribution of $\mathbf{Z}$ is not uniform on Ω, if instead the distribution is tilted, favoring certain assignments $\mathbf{Z}$ over others, then this tilting or bias might leave visible traces in data we have or could obtain. Efforts to detect hidden bias use additional information beyond the treated and control groups, their covariates and outcomes of primary interest. Recall from §1.2 Sir Ronald Fisher's advice, "Make your theories elaborate." This additional information may consist of several control groups, or several referent groups in a case-referent study, or it may consist of additional outcomes for which the magnitude or direction of the treatment effect is known. A systematic difference between the outcomes in several control groups cannot be an effect of the treatment and must be a hidden bias. A systematic difference between treated and control groups in an outcome the treatment does not affect must be a hidden bias. Detecting hidden bias entails checking that treatment effects appear where they should, and not elsewhere.

Efforts to detect bias and sensitivity analysis complement one another in their strengths and limitations. It is difficult if not impossible to detect small hidden biases, but a sensitivity analysis might indicate that small biases would not materially alter the study's conclusions. All studies are sensitive to sufficiently dramatic biases, but dramatic biases are the ones most likely to be detected.

Statistical theory contributes in two ways to efforts to detect bias. First, it provides qualitative advice about what these efforts can and cannot do, and about the types of data and study designs that have a reasonable prospect of detecting hidden bias. Second, theory provides a quantitative link between detection and sensitivity analysis. When does failure to detect hidden bias provide evidence that biases are absent or small? Does failure to detect hidden bias make a study less sensitive to bias?

Chapters 6, through 8 discuss three methods of detecting hidden biases, namely, known effects, multiple referent groups in a case-referent study, and multiple control groups. The three methods are ordered in this way because the technical development in Chapter 6 is most similar to Chapter 4, Chapter 7 is a step away, and Chapter 8 is a larger step away. The remainder of §6.1 discusses examples of known effects and concludes by asking when an effect is known.

6.1.2   An Example: Nuclear Weapons Testing and Leukemia

Can hidden biases be detected? Is there any real hope of detecting a departure from a random assignment of treatments within strata that are homogeneous in observed covariates? Is the task ahead an impossible task? The first example is of interest because it affords several opportunities to detect hidden bias and, in each case, bias is detected. Also, there is a bit of late news produced four years after the publication of the original study.

Lyon, Klauber, Gardner, and Udall (1979) studied the possible effects of fallout from nuclear weapons testing in Nevada on the risk of childhood leukemia and other cancers in Utah. The study grouped counties in Utah into high or low exposure based on proximity to nuclear testing in Nevada. In addition, the counties were studied in three time periods, 1944 to 1950

TABLE 6.1. Leukemia Mortality Per 100,000 Person Years in Utah, Before, During and After Above Ground Nuclear Testing.

Cohort	Person-Years	Deaths	Rate
Low Exposure Counties Before: 1944-1950	1,095,997	44	4.0
High Exposure Counties Before: 1944-1950	330,177	7	2.1
Low Exposure Counties During: 1951-1958	3,898,901	152	3.9
High Exposure Counties During: 1951-1958	724,531	32	4.4
Low Exposure Counties After: 1959-1975	3,153,008	112	3.6
High Exposure Counties After: 1959-1975	451,408	10	2.2

prior to nuclear testing in Nevada, 1951 to 1958 during above-ground testing, and 1959 to 1975 during underground testing. As will be seen, this is an attractive and careful study design in that it provides several checks for hidden bias. Frequencies of childhood leukemia and other cancer were obtained from the Utah State Bureau of Vital Statistics, and person-years at risk were determined from the US Census and the Utah State Bureau of Vital Statistics; see Lyon et al. (1979) for details. This study was discussed by Land (1979), Beck and Krey (1983), Rosenbaum (1984a), and Gastwirth (1988, pp. 870–878). In particular, Gastwirth discusses the outcome of related litigation.

If above-ground nuclear testing has an effect, then how should that effect appear? Radiation could reasonably cause leukemias and other childhood cancers, but it should not prevent them, and more radiation should yield more cancers. If above-ground testing has an effect, and if there is no hidden bias, one expects the highest levels of leukemia and other cancers in the high-exposure counties during the period of above-ground testing, followed by lower but possibly still elevated levels in the low-exposure counties during the same time period, with still lower and similar levels in the high- and low-exposure counties before and after above-ground nuclear testing. This is what one would expect to see if, contrary to fact, children had been assigned at random to one of the six cohorts of children. This is an elaborate theory in Fisher's sense, one that includes several effects of known direction and several control groups.

Table 6.1 shows the leukemia mortality rates from Lyon et al. (1979) for children under the age of 15. These are mortality rates per 100,000 person-years.

Table 6.1 conforms to the anticipated effect in one respect but not in another. As anticipated, in the high-exposure counties, the leukemia mor-

TABLE 6.2. Other Childhood Cancer Mortality Per 100,000 Person Years in Utah, Before, During and After Above Ground Nuclear Testing.

Cohort	Person-Years	Deaths	Rate
Low Exposure Counties Before: 1944-1950	1,095,997	50	4.6
High Exposure Counties Before: 1944-1950	330,177	21	6.4
Low Exposure Counties During: 1951-1958	3,898,901	165	4.2
High Exposure Counties During: 1951-1958	724,531	21	2.9
Low Exposure Counties After: 1959-1975	3,153,008	106	3.4
High Exposure Counties After: 1959-1975	451,408	15	3.3

tality rises from 2.1 leukemias per 100,000 before testing to 4.4 during above-ground testing, and then returns to 2.2 during the period of underground testing. The low-exposure counties show a slight decline. On the other hand, the low-exposure counties had almost twice the leukemia of the high-exposure counties before nuclear testing (4.0 versus 2.1) and more than 50% more after nuclear testing went underground (3.6 versus 2.2). Indeed, the rise in leukemias in the high-exposure counties during above-ground testing brought the level to just slightly more than that found in the low-exposure counties at the same time (4.4 versus 3.9). It appears that the high- and low-exposure counties are not comparable in ways that matter for the leukemia rate.

Table 6.2 shows the corresponding rates for other childhood cancers. Here, the anticipated effect is not seen. For the high-exposure counties, the rate of other cancers declined during the period of above-ground testing, just as the leukemia rate was rising. This decline is not plausibly an effect of nuclear testing, and it strongly suggests that the two periods were not comparable in ways that matter for cancers other than leukemias. The low-exposure counties show a slight decline. For cancers other than leukemia, before nuclear testing, the high-exposure counties had higher rates than the low-exposure counties, but that reversed during above-ground testing. It appears that the high- and low-exposure counties are not comparable in ways that matter for cancers other than leukemias. If the differences that should not be present are subjected to a formal hypothesis test, then the differences are found to be statistically significant and not readily attributed to chance; see Rosenbaum (1984a).

In short, the mortality rates vary in ways not readily attributed to chance or to effects of above-ground nuclear testing. More than this, the variation in mortality rates that is inconsistent with an effect of nuclear testing is as

large as the variation that is consistent with an effect. This does not necessarily mean that the interesting rise in leukemia in high-exposure counties
is not an effect of nuclear testing, but it is reason for extreme caution about
any such claim.

The study by Lyon et al. (1979) is an excellent study, even though there
is strong evidence of hidden bias. Indeed, it is an excellent study *because*
there is strong evidence of hidden bias. A poor study might have looked only
at leukemia and only in high-exposure counties, and it would have found no
evidence of hidden bias—because it did not look—and therefore it would
have left its audience with an incorrect impression. The editorial policies
of major journals shape science. The *New England Journal of Medicine*
wisely published the study by Lyon et al. (1979) along with a cautious
editorial by Land (1979), despite the evidence of hidden bias, and one
hopes it would have wisely declined to publish the imagined study which
considered only leukemia in high-exposure counties, despite its absence of
evidence of hidden bias.

Four years later, Beck and Krey (1983) published a study that attempted
to estimate the doses of radiation received between 1951 and 1958 in different parts of Utah. They took soil samples in the high- and low-exposure
counties described by Lyon et al. (1979) and measured cesium-137 and plu-
tonium. They concluded: "Although the highest exposures were found in
the extreme southwest part of Utah, as expected, the residents of the pop-
ulous northern valleys around Provo, Salt Lake City, and Ogden received
higher mean dose and a significantly greater population dose (person-rads)
than did the residents of most counties closer to the test site [p. 18]." In
other words, their sampling suggests that the "low"-exposure counties may
have received higher exposures than the "high"-exposure counties. Beck
and Krey also compared the radiation they found with background radiation and global fallout and expressed their view that ". . . bone doses to
the population of southern Utah were far too low to account for the excess
childhood leukemia mortality reported by Lyon et al. (1979)."

For us, the important point is that careful study design can help to
detect hidden biases.

6.1.3 Specificity of Associations and Known Effects

In epidemiology, it is often said that a "specific" association between an
exposure and a disease provides greater evidence that the exposure causes
the disease than does a "nonspecific" association. For instance, Yerushalmy
and Palmer (1959) write:

> The demonstration of high relative frequencies in the study
> group is thus only a first step in the process of searching for
> etiologic factors. The investigation must proceed to the second
> and more crucial consideration (which, for want of a better

term, is denoted here as that of specificity of effect), i.e., to the demonstration the difference in relative frequencies reflects a specific and meaningful relationship between the characteristic under suspicion and the disease under consideration ... [p. 36].

... If the characteristic can be shown to be related only or mostly to the disease under study and not to many other disease entities, then our confidence that it is a cause-carrying vector for that disease is greatly increased.

If, on the other hand, it is found that the characteristic is also related to numerous other diseases, including those without obvious physiologic or pathologic connection with the characteristic in question, the relationship must be assumed—until further proof—to be nonspecific [p. 37].

In a comment, Lilienfeld (1959) adds several points. He suggests that specificity does not refer to the presence or absence of an association, but to its strength or degree. He argues that the association between smoking and lung diseases is specific because these associations are strong, though there are weak associations with many diseases. Moreover, he writes:

It is needless to point out that our interpretation of any relationship is limited by our biologic knowledge, and it may well be that an association which at present does not appear to be biologically plausible will turn out to be so when our knowledge has been extended. In fact, the finding of a biologically implausible association may be the first lead to this extension of knowledge.

A similar point is made by Sartwell (1960), though with a more critical tone. Still more critical is Rothman (1986, p. 18), who says that "... everyday experience teaches us repeatedly that single events may have many effects" and that the criterion of specificity "seems useless and misleading." Sir Austin Bradford Hill (1965, p. 297) regards specificity more favorably, but lists these and other difficulties.

If one agrees that the *number* of associations is not a particularly useful number, and instead emphasizes whether there are associations that are not plausibly effects of the treatment or exposure under study, then there is a strong link between specificity in this narrow sense and detecting hidden bias through known effects. If an observational study includes several outcomes, some of which may plausibly be affected by the treatment and others for which an actual effect is implausible, then the association between treatment and outcomes is specific in the narrow sense if strong associations are confined to outcomes for which effects are plausible. In this instance, the treatment is known to have no effect on certain outcomes.

Thomas Cook (1991) discusses specificity clearly in somewhat different terminology:

'Nonequivalent dependent variable' designs ... depend on
some of the indicators being theoretically responsive to a treatment but all of them being responsive to the known plausible
alternative explanations. [p. 119] ... [If a] prediction is about a
specific and complex pattern of relationships [... then ...] no
uncertainty is reduced if a theory makes a highly specific numerical prediction *that other theories also make.* Specificity (or
multivariate complexity) only facilitates causal inference when
no other theory makes the same prediction. The more specific
or complex the causal implications are, the less likely it is pre-
sumed to be that alternative theories can be found that make
the same prediction [p. 121].

See also Cook, Campbell, and Peracchio (1990, pp. 546–7 and pp. 561–
564) for discussion of nonequivalent dependent variables, see Campbell
(1966) for discussion of the principle of pattern-matching in inference, and
see Ross, Campbell, and Glass (1970) for an interesting example.

6.1.4   Example of Specificity: Acute Stress as a Cause of Sudden Death from Coronary Disease and Cancer

Is the risk of rapid death from coronary disease increased by acute stress?
Trichopoulos et al. (1983) compared coronary mortality in Athens in the
period immediately following the 1981 earthquake to time periods immediately before the earthquake and to corresponding time periods in 1980 and
1982. After the earthquake, they claim that "the psychological stress was
unquestionable, intense, and general" (p. 442). They found higher rates
of coronary mortality immediately following the earthquake than in the
comparison time periods.

It is plausible that acute stress acutely increases the risk of death from
coronary disease, but it is not particularly plausible that acute stress quickly
causes many deaths from cancer. Trichopoulos et al. conducted a parallel
analysis of cancer mortality following the earthquake, finding no increase
in risk. So the association of mortality with stress appears where an effect
caused by acute stress is plausible, but not where a causal effect is much
less plausible.

6.1.5   Example of Specificity: Supply and Demand for Cholecystectomy

If the cost of performing a medical procedure fell dramatically, would the
expenditure per person in a general population fall as well? With consumer
products, a drop in price may stimulate increased demand, so expenditure
per person may rise or fall. Can the same thing happen with medical
procedures?

Using data from US Healthcare's HMO PA in southeastern Pennsylvania, Legorreta, Silber, Costantino, Kobylinski, and Zatz (1993) asked this question of laparoscopic cholecystectomy, which was essentially unused in 1987 and comprised more than 80% of cholecystectomies in 1992. Laparoscopic or closed cholecystectomy costs less than traditional open cholecystectomy, is safer, less invasive, and requires a shorter hospital stay. They found that between 1988 and 1992, the cost per cholecystectomy fell by about 25%, but the number of cholecystectomies per 1000 enrollees nearly doubled, and total expenditures on gallbladder disease rose.

Was the increase in the number of cholecystectomies a consequence of the introduction of laparoscopic techniques? Or might it reflect some change in the patient population or in general medical practice from 1988 to 1992? In an effort to examine this question, Legorreta et al. (1993) did parallel analyses of two other surgical procedures that should not be affected by the introduction of laparoscopic cholecystectomy. One of these other procedures was appendectomy, which is not an elective procedure, and the other was inguinal herniorrhaphy, which has "a discretionary component similar to cholecystectomy" (p. 1431). The rates of appendectomy and inguinal herniorrhaphy did not increase significantly during this time period, nor did associated expenditures per 1000 enrollees. The change in the frequency and cost of cholecystectomies is specific to this procedure, and is not found in rates for two other procedures that should not be affected by the introduction of laparoscopic cholecystectomy.

6.1.6 Example of Specificity: Anger and Curiosity as Causes of Myocardial Infarction

Do brief bouts of anger increase the risk of myocardial infarction (MI)? In a case-crossover study, Mittleman et al. (1995) compared anger in the two hours before an MI to the same two hours on the previous day. In part of their study, for 881 cases of acute MI, they measured anger using the State-Trait Personality Inventory in each of the two time periods. Using Wilcoxon's signed rank test, they found anger was more often reported in the two hours before an MI than in the corresponding two hours on the previous day, the difference being highly significant, $P = 0.001$.

However, Mittleman et al. were concerned that their measures of anger were based on recall, and a patient's recall may itself be affected by the occurrence of an MI. Perhaps the events leading up to an MI appear to a patient to be more important than the events of the less dramatic, perhaps uneventful, previous day. As a partial check on this possibility, Mittleman et al. also obtained measures of curiosity from the State-Trait Personality Inventory in both time periods. Of course, they strongly doubted that curiosity causes MI. They found no sign of differences in curiosity in the two time periods using the signed rank test, $P = 0.20$. So the level of anger is associated with the timing of the MI, but the level of curiosity is not. They wrote (p. 1724): "... the specificity observed for anger ... as opposed to curiosity on the STPI subscales ... argue against recall bias."

6.1.7 Unaffected Subgroups: Abortion and Crime

Theorizing that unwanted children, perhaps children of unwed teenage mothers, might possibly be more likely to commit crimes as teens or adults, Donohue and Levitt (2001) investigated the sharp decline in crime rates in the 1990s and their relationship to the legalization of abortion in 1973 by the Supreme Court Decision in *Roe v. Wade*. Among many analyses, their study includes a test for hidden bias using a known effect.

If legalized abortion reduced crime simply by eliminating future criminals, then the effect of legalization should be confined to certain age cohorts. The availability of legal abortion in 1973 should not affect the cohort born in 1971, but might affect the cohort born in, say 1975. In one of many analyses, Donohue and Levitt relate abortion rates by state to two outcomes, namely (i) arrest rates by state for age cohorts that might be affected and (ii) arrest rates for cohorts that cannot be affected. Whether or not adjustments are made for covariates, they find that arrest rates for cohorts that cannot be affected are unrelated to abortion rates, whereas arrest rates for cohorts that can be affected are negatively associated with abortion rates. They write: "In high abortion states, only arrests of those born after abortion legalization fall relative to low abortion states."

6.1.8 When Is an Effect Known?

When is an effect known? Might we be in error in asserting that an effect is known? In discussing "specificity," Sartwell (1960) recalls the ridicule that was once attached to the idea that tiny, invisible living organisms were a cause of disease. Would anyone claim that our current knowledge is free of substantial error? There are two points.

First, as suggested at the end of §6.1.2, known effects and other attempts to detect hidden biases are not the basis for suppressing data or studies— quite the contrary. The study by Lyon et al. (1979) is a superior study because its report includes information about internal inconsistencies. If we misinterpret those inconsistencies because of some error in our current knowledge, then we are far more likely to correct the error if the report includes the inconsistencies than if they are excluded. If we attempt to detect hidden bias, we report more data, not less.

Second, there is an entirely practical, albeit circular, definition of a known effect. An effect is known if, were the study to contradict it, we would have grave doubts about the study and few doubts about the effect. It is conceivable, in principle, that our understanding of time and radiation is substantially in error, as our understanding of bacterial disease was once substantially in error, so that radiation released in the 1950s had biological effects in the 1940s; however, before entertaining this possibility, we would think that perhaps the records of the Census Bureau or the Utah registry contained errors, or that some other environmental hazard was the cause.

6.2   An Outcome Known to Be Unaffected by the Treatment

6.2.1   An Additional Outcome Intended to Detect Hidden Bias

In addition to the response $\mathbf{R}$ of primary interest, the study includes another outcome, say $\mathbf{Y}$, in an effort to detect hidden bias. In other words, $\mathbf{Y}$ was recorded in the hope of testing whether the treatment assignments $\mathbf{Z}$ given $\mathbf{m}$ are uniformly or randomly distributed over Ω, or alternatively whether there is a hidden bias.

This chapter considers two senses in which the effect of the treatment on $\mathbf{Y}$ might be known. In §6.2, $\mathbf{Y}$ is known to be unaffected by the treatment. For instance, in §6.1.3 the concern was with diseases that the treatment could not plausibly cause; see also Problem 1. In §6.3, $\mathbf{Y}$ is known either to be unaffected or to be positively affected; that is, $\mathbf{Y}$ may be affected by the treatment, but the direction of the effect is known. In §6.5, the direction of the effect on the primary outcome, $\mathbf{R}$, is known, and there is no secondary outcome, $\mathbf{Y}$. In §6.2, because $\mathbf{Y}$ is unaffected by the treatment, write $\mathbf{y}$ in place of $\mathbf{Y}$. To say $\mathbf{y}$ is unaffected by treatment is to say $\mathbf{y}$ does not vary with $\mathbf{Z}$ and so is constant.

In an observational study, suppose a test statistic $T = t(\mathbf{Z}, \mathbf{R})$ is computed from the responses $\mathbf{R}$ of primary interest, and then T is compared to the uniform or random distribution of $\mathbf{Z}$ on Ω under the null hypothesis that $\mathbf{R} = \mathbf{r}$ is unaffected by the treatment; that is, the significance level (2.5) is calculated. Recall from §3.2.2 that this significance level is the basis for testing the null hypothesis of no treatment effect *if the study is free of hidden bias*, but the test is not generally correct if hidden bias is present. If this significance level is small, it may be evidence that the null hypothesis is untrue and the treatment does affect $\mathbf{R}$, or it may be evidence that there is a hidden bias and the distribution of treatment assignments $\mathbf{Z}$ is not random.

The situation with $\mathbf{y}$ is different, because $\mathbf{y}$ is known to be unaffected by the treatment. From the unaffected outcome $\mathbf{y}$, calculate a statistic $T^* = t^*(\mathbf{Z}, \mathbf{y})$ and its significance level $|\{\mathbf{z} \in \Omega : t^*(\mathbf{z}, \mathbf{y}) \geq T^*\}|/K$. If this significance level is small, then there is evidence of hidden bias, that is, evidence that $\mathbf{Z}$ is not uniformly or randomly distributed on Ω.

6.2.2   An Example: Unrelated Health Conditions in the Study of Methylmercury in Fish

Recall from §4.6.6 the study by Skerfving, Hansson, Mangs, Lindsten, and Ryman (1974) of possible chromosome damage caused by eating fish contaminated with methylmercury. The outcomes of interest were the level of mercury in the blood and the percentage of cells exhibiting chromosome damage. In addition to these outcomes, Table 1 of their study described other health conditions of these 39 subjects, including other diseases such as hypertension and asthma, drugs taken regularly, diagnostic X-rays over the previous three years, and viral diseases such as influenza. These are outcomes since they describe the period when the exposed subjects were consuming contaminated fish. However, it is difficult to imagine that eating fish contaminated with methylmercury causes influenza or asthma, or prompts X-rays of the hip or lumbar spine. For illustration, Table 6.3 records the number of unrelated health conditions for each of the 39 subjects or $\mathbf{y}$ which will be assumed to be unaffected by eating contaminated fish.

If the rank sum statistic is calculated from $\mathbf{y}$, the sum of the ranks for exposed subjects is 483.5, with average ranks used for the many ties. The null hypothesis states that the study is free of hidden bias, so the distribution of treatment assignments $\mathbf{Z}$ given $\mathbf{m}$ is random or uniform on Ω. Under the null hypothesis, the rank sum has expectation 460 and variance 932.4 allowing for the many ties. The deviate is $(483.5 - 460)/\sqrt{932.4} = 0.77$ which is far from significance at the 0.05 level. In words, in terms of health conditions methylmercury could not reasonably cause, the treated and control subjects differ no more than we would have expected had treatments been assigned at random. The hypothesis that the study is free of hidden bias is not rejected.

Several questions arise. When does such a test have a reasonable prospect of detecting hidden bias? If no evidence of hidden bias is found, does this imply reduced sensitivity to bias in the comparisons involving the outcomes of primary interest? If evidence of bias is found, what can be said about its magnitude and its impact on the primary comparisons?

6.2.3   The Power of the Test for Hidden Bias

For a particular unobserved covariate $\mathbf{u}$, what unaffected outcome $\mathbf{y}$ would be useful in detecting hidden bias from $\mathbf{u}$? Put another way, for a given unaffected outcome $\mathbf{y}$, what unobserved covariate $\mathbf{u}$ can $\mathbf{y}$ hope to detect?

The power of the test for hidden bias based on $T^* = t^*(\mathbf{Z}, \mathbf{y})$ will now be studied in qualitative terms that indicate the relationship between $\mathbf{u}$ and $\mathbf{y}$ that leads to high power. Let "a" be a number such that

$$\alpha = \frac{1}{K} \sum_{\mathbf{z} \in \Omega} [t^*(\mathbf{z}, \mathbf{y}) \geq a] = \frac{|\{\mathbf{z} \in \Omega : t^*(\mathbf{z}, \mathbf{y}) \geq a\}|}{K},$$

TABLE 6.3. Unrelated Health Conditions and Exposure to Contaminated Fish.

Id.		Unrelated conditions
1	Control	0
2	Control	0
3	Control	0
4	Control	0
5	Control	0
6	Control	0
7	Control	0
8	Control	0
9	Control	2
10	Control	0
11	Control	0
12	Control	0
13	Control	2
14	Control	1
15	Control	4
16	Control	1
17	Exposed	0
18	Exposed	0
19	Exposed	2
20	Exposed	0
21	Exposed	0
22	Exposed	2
23	Exposed	0
24	Exposed	0
25	Exposed	1
26	Exposed	9
27	Exposed	0
29	Exposed	1
30	Exposed	0
31	Exposed	0
32	Exposed	2
33	Exposed	0
34	Exposed	0
35	Exposed	2
36	Exposed	0
37	Exposed	6
38	Exposed	1
39	Exposed	1

where $K = |\Omega|$ and $[event] = 1$ if event occurs and $[event] = 0$ otherwise, so a test for hidden bias that rejects when $t^*(\mathbf{Z}, \mathbf{y}) \geq a$ has level α. Under the model (4.6), the power of this test is

$$\beta(\mathbf{y}, \mathbf{u}) = \sum_{\mathbf{z} \in \Omega} [t^*(\mathbf{z}, \mathbf{y}) \geq a] \frac{\exp(\gamma \mathbf{z}^T \mathbf{u})}{\sum_{\mathbf{b} \in \Omega} \exp(\gamma \mathbf{b}^T \mathbf{u})}. \tag{6.1}$$

All of the statistics $t^*(\mathbf{z}, \mathbf{y})$ in §2.4.3 are arrangement-increasing, meaning that the statistics are monotone increasing as the coordinates of $\mathbf{z}$ and $\mathbf{y}$ are gradually permuted into the same order within each stratum or matched set. In other words, if in stratum s, subject i received the treatment, $z_{si} = 1$, subject j received the control $z_{sj} = 0$, but subject j had the higher value of y, that is, $y_{sj} \geq y_{si}$, then interchanging these two subjects, placing j in the treated group and i in the control group would increase—or at least not decrease—the value of the statistic $t^*(\mathbf{z}, \mathbf{y})$.

The following proposition says that the power $\beta(\mathbf{y}, \mathbf{u})$ of this test increases steadily as the coordinates of $\mathbf{y}$ and $\mathbf{u}$ are permuted into the same order within each stratum or matched set. The power is greater when $\mathbf{y}$ and $\mathbf{u}$ are strongly related.

Proposition 23 *If the test statistic $t^*(\mathbf{z}, \mathbf{y})$ is arrangement-increasing, then the power $\beta(\mathbf{y}, \mathbf{u})$ is arrangement-increasing.*

The proof follows from Proposition 17 in §4.7.3 which in turn is a consequence of the composition theorem for arrangement-increasing functions (Hollander, Proschan, and Sethuraman 1977). D'Abadie and Proschan (1984, §6) describe a test with an arrangement-increasing power function as having *isotonic power*, meaning the power increases as the departure from the null hypothesis becomes more pronounced in the direction indicated by an order. Under slightly different assumptions, it is also possible to show that a test based on an arrangement-increasing statistic $T^* = t^*(\mathbf{Z}, \mathbf{y})$ is unbiased against alternatives in which $\mathbf{y}$ and $\mathbf{u}$ are positively related; see Rosenbaum (1989a).

In short, the power $\beta(\mathbf{y}, \mathbf{u})$ of a test based on $T^* = t^*(\mathbf{Z}, \mathbf{y})$ increases with the strength of the relationship between $\mathbf{y}$ and $\mathbf{u}$. This yields two qualitative conclusions. First, if concerned about a particular unobserved covariate $\mathbf{u}$, one should search for an unaffected outcome $\mathbf{y}$ that is strongly related to $\mathbf{u}$. Second, if an unaffected outcome $\mathbf{y}$ is available, a test based on $\mathbf{y}$ provides information about unobserved covariates $\mathbf{u}$ with which $\mathbf{y}$ is strongly related.

Consider the use of these conclusions in a particular case. Sackett (1979) cataloged and illustrated with examples numerous sources of bias in observational studies. Included were situations in which the search for disease or the inclination to recall or record disease was more intense among subjects exposed to a hazardous agent than among subjects not exposed. The danger, of course, is that more disease will be found among exposed subjects not because the hazardous agent caused the disease, but rather because the agent stimulated intensive search for disease or distorted recall of disease. If **u** were an unobserved covariate measuring the inclination to search for or recall disease, then a test for bias due to **u** would have highest power if it were based on an unaffected outcome **y** strongly related to **u**. In other words, a good test would be based on records of diseases that could not reasonably be caused by the agent under study but which would likely be detected in diagnostic activities or recalled by subjects. Exposed and control groups should differ with respect to such a **y** if there are substantial distortions due to differing diagnostic efforts or recall in exposed and control groups.

As a second illustration, consider again the study of legalized abortion and crime by Donohue and Levitt (2001). In their study, the unaffected outcome **y** is the arrest rate by state for cohorts that are too old to have been affected by the *Roe v. Wade* decision in 1973. Of course, arrest rates vary by state for many reasons, some of which were not observed. If **u** is an unobserved variable describing the states, then the power of the test for hidden bias using **y** is greatest when **u** and **y** are strongly associated. If arrest rates for older cohorts **y** are related to an unobserved **u** in much the same way that arrest rates for younger cohorts are related to **u**, then a test for hidden bias based on **y** will have high power to detect bias from a variable **u** that is related to the outcome of interest.

6.2.4   Did the Test for Bias Reduce Sensitivity to Bias?

In §6.2.2, the difference between treated and control groups in unrelated health conditions appeared reasonably consistent with the absence of hidden bias. The question remains whether these same findings are consistent with hidden biases that would alter the conclusion that methylmercury causes chromosome damage.

In §4.6.6, the effect of mercury on chromosome aberrations was sensitive to hidden bias when $\Gamma = 3$ for a particular value of **u**, say $\mathbf{u}^{\circ}$. This $\mathbf{u}^{\circ}$ contained 1s and 0s and was strongly related to the outcome **r** of primary interest, namely, chromosome aberrations. In other words, if treatment assignments were governed by the model (4.6) with $(\Gamma, \mathbf{u}) = (3, \mathbf{u}^{\circ})$, then there would not be strong evidence that eating contaminated fish causes chromosome damage. For $\Gamma = 3$, this $\mathbf{u}^{\circ}$ maximized over $\mathbf{u} \in [0, 1]^{N}$ the significance level of the test for an effect on chromosome aberrations. With a different choice of **u**, a larger value of Γ would be required to change the conclusion about chromosome aberrations. The distinction between $\Gamma = 1$ and $(\Gamma, \mathbf{u}) = (3, \mathbf{u}^{\circ})$ is important because the main conclusions would be different in these two cases.

Are the observed data about unrelated health conditions **y** consistent with $(\Gamma, \mathbf{u}) = (3, \mathbf{u}^{\circ})$? The unaffected outcome **y** can test the hypothesis that $(\Gamma, \mathbf{u}) = (3, \mathbf{u}^{\circ})$ in much the same way that it tested the hypothesis that $\Gamma = 1$. If $(\Gamma, \mathbf{u}) = (3, \mathbf{u}^\circ)$ is not rejected, then $\mathbf{y}$ has done nothing to reduce our concern about a bias of the form $(\Gamma, \mathbf{u}) = (3, \mathbf{u}^\circ)$. If $(\Gamma, \mathbf{u}) = (3, \mathbf{u}^\circ)$ is rejected, a point of greatest sensitivity to bias has been rejected as implausible. Rejection is likely if the model with $(\Gamma, \mathbf{u}) = (3, \mathbf{u}^\circ)$ tends to produce large differences in values of y_{si}, in treated and control groups, but these large differences are not seen in the data. In short, the question is: Did the test for bias using $\mathbf{y}$ reduce sensitivity to bias?

In §6.2.3, the observed value $T^* = t^*(\mathbf{Z}, \mathbf{y})$ of the test statistic was compared with its distribution when there is no hidden bias, that is, the model $\text{prob}(\mathbf{Z} = \mathbf{z}|\mathbf{m}) = 1/K$ for each $\mathbf{z} \in \Omega$, leading to the exact significance level $|\{\mathbf{z} \in \Omega : t^*(\mathbf{z}, \mathbf{y}) \geq T^*\}|/K$. In fact, any specified value of $(\Gamma, \mathbf{u}^\circ)$ may be tested using $T^* = t^*(\mathbf{Z}, \mathbf{y})$ in much the same way, yielding the one-sided significance level

$$\sum_{\mathbf{z} \in \Omega} [t^*(\mathbf{z}, \mathbf{y}) \geq T^*] \frac{\exp(\gamma \mathbf{z}^T \mathbf{u}^\circ)}{\sum_{\mathbf{b} \in \Omega} \exp(\gamma \mathbf{b}^T \mathbf{u}^\circ)}, \tag{6.2}$$

where, as always, $\Gamma = \exp(\gamma)$. Expression (6.2) is the probability that the test statistic would exceed its observed value T^* if

$$\text{prob}(\mathbf{Z} = \mathbf{z}|\mathbf{m}) = \frac{\exp(\gamma \mathbf{z}^T \mathbf{u}^\circ)}{\sum_{\mathbf{b} \in \Omega} \exp(\gamma \mathbf{b}^T \mathbf{u}^\circ)}. \tag{6.3}$$

For the test statistics in §2.4.3, the exact significance level (6.3) may be approximated in large samples using the expectation $\mu_{\mathbf{u}^\circ}^*$ and variance $\sigma_{\mathbf{u}^\circ}^*$ of T^* under the model (6.3), and referring the deviate $(T^* - \mu_{\mathbf{u}^\circ}^*)/\sigma_{\mathbf{u}^\circ}^*$ to the standard normal distribution. The calculations of the moments of T^* are essentially the same as for T in Chapter 4, except that a different $\mathbf{u}^\circ$ is of interest. An example is given in §6.2.5.

If the test (6.2) rejects $(\Gamma, \mathbf{u}^\circ)$, then the test for bias based on the unaffected outcome $\mathbf{y}$ has rejected as implausible the point $(\Gamma, \mathbf{u}^\circ)$ which did most to perturb the inference about the outcome $\mathbf{R}$ of primary interest, so the sensitivity of the primary comparison is reduced. On the other hand, if $(\Gamma, \mathbf{u}^\circ)$ is not rejected, then $(\Gamma, \mathbf{u}^\circ)$ remains plausible in light of the test based on $\mathbf{y}$, so the sensitivity of the primary comparison for $\mathbf{R}$ is unchanged.

As the sample size increases, the test for hidden bias will tend to reject any value $(\Gamma^\circ, \mathbf{u}^\circ)$ for which the expectation $\mu_{\mathbf{u}^\circ}^*$ of T^* differs from the expectation $\mu_{\mathbf{u}}^*$ of T^* at the true $(\Gamma, \mathbf{u})$. Somewhat more precisely, as $N \to \infty$, if $\sigma_{\mathbf{u}^*}^*/\sqrt{N}$ tends to a constant, as is true of the statistics in §2.4.3 under mild conditions that forbid degeneracy, and if $(\mu_{\mathbf{u}^\circ}^* - \mu_{\mathbf{u}}^*)/N$ tends to a nonzero constant, then the power of the test for bias based on T^* tends to one. For details, see Rosenbaum (1992).

6.2.5 An Example: Did the Test Using Unrelated Health Conditions Reduce Sensitivity?

In §4.6.6, the association between chromosome damage and eating contaminated fish became sensitive to hidden bias when $\Gamma = 3$ for a $\mathbf{u}^\circ$ containing 21 ones and 18 zeros, where $\mathbf{u}^\circ$ was ordered in the same way as the percentages of Cu cells in Table 4.15. Is this pattern and magnitude of hidden bias, $(\Gamma, \mathbf{u}) = (3, \mathbf{u}^\circ)$, plausible in light of the test for hidden bias in §6.2.3?

The unrelated health conditions in Table 6.3 were found to be consistent with the absence of bias, $\Gamma = 1$; specifically, the rank sum was 483.5 with expectation 460, variance 932.4, and standardized deviate $(483.5 - 460)/\sqrt{932.4} = 0.77$. If instead $(\Gamma, \mathbf{u}) = (3, \mathbf{u}^\circ)$, then the rank sum statistic for unrelated health conditions is still 483.5, but its expectation is now 457.5, and its variance is 874.3, yielding a deviate of $0.88 = (483.5 - 457.5)/\sqrt{874.3}$. In other words, the observed data on unrelated health conditions are consistent with the absence of hidden bias, $\Gamma = 1$, but they are also quite consistent with a hidden bias $(\Gamma, \mathbf{u}) = (3, \mathbf{u}^\circ)$ that would alter the study's conclusion about the effects of methylmercury on chromosome damage. Though the test found no evidence of hidden bias, it did not reduce the sensitivity to bias of the primary comparison; that is, it did not reject the point $(\Gamma, \mathbf{u}) = (3, \mathbf{u}^\circ)$ of maximum sensitivity.

As noted in §6.2.4, the test for hidden bias based on the unaffected outcome $\mathbf{y}$ has greatest power against an unobserved covariate $\mathbf{u}$ strongly related to $\mathbf{y}$. On the other hand, as seen in Chapter 4, the test of the null hypothesis of no treatment effect on the response $\mathbf{r}$ is most sensitive to bias from an unobserved covariate $\mathbf{u}$ strongly related to $\mathbf{r}$. If $\mathbf{y}$ has only a weak relationship with $\mathbf{r}$, as is true of the relationship between unrelated health conditions and the percentage of Cu cells, then the test for bias based on $\mathbf{y}$ will have little power against values of $\mathbf{u}$ that matter most for $\mathbf{r}$. For reducing sensitivity to hidden bias, the ideal $\mathbf{y}$ would be unaffected by the treatment but otherwise would be strongly related to the response $\mathbf{r}$. This ideal $\mathbf{y}$ bears some resemblance to a useful type of covariate, namely, a "baseline measure" or "pretest score," which is a measure of outcome of interest recorded prior to treatment, so a "baseline measure" is unaffected by the treatment but is often highly correlated with the outcome.

6.2.6 Should Adjustments Be Made for Unaffected Outcomes?

Should an unaffected outcome be viewed as another covariate? Should adjustments be made for unaffected outcomes in just the same way that adjustments are made for covariates?

The goal is compare subjects who were comparable *prior* to treatment. An outcome is, by definition, measured *after* treatment. Adjustments for unaffected outcomes render people comparable prior to treatment only under special and restrictive circumstances, that is, under assumptions that may be wrong and are often difficult to justify. This is discussed formally and in detail in Rosenbaum (1984b).

A bias visible in an unaffected outcome may only be the faint trace of a much larger imbalance in an unobserved covariate. Indeed, there is a sense in which, if an imbalance in an unaffected outcome is produced by an imbalance in an unobserved covariate, then the covariate imbalance is always at least as large as the outcome imbalance. See Rosenbaum (1989b, §4.3) for formal discussion. Removing the outcome imbalance may do little to remove the imbalance in the unobserved covariate.

6.3 An Outcome With Known Direction of Effect

6.3.1   Using a Nonnegative Effect to Test for Hidden Bias

In many cases, a treatment can have no effect or can have the effect of increasing a particular outcome $\mathbf{Y}$, but it cannot plausibly decrease it. In §6.1.2, radiation from nuclear fallout might have no effect on the frequency of childhood cancers other than leukemia, or it might increase them, but it is hard to imagine that fallout is preventing cancers. In Problem 2, occupational exposures to benzene might have no effect on total mortality or might increase mortality, but it is hard to imagine that benzene exposures prevent death.

In the terminology of §2.5, if the treatment has an effect on $\mathbf{Y}$, then $\mathbf{Y}$ takes the value $\mathbf{y_z}$ if $\mathbf{Z} = \mathbf{z}$ where each $\mathbf{y_z}$ is fixed. There is no interference between units if $y_{zsi} = y_{Tsi}$ whenever $z_{si} = 1$, and $y_{zsi} = y_{Csi}$ whenever $z_{si} = 0$, and there is a nonnegative effect if $y_{Tsi} \geq y_{Csi}$ for all (s, i). Let $T^* = t^*(\mathbf{Z}, \mathbf{Y})$ be an effect increasing statistic, for instance, any of the statistics in §2.4.3.

If the treatment has a nonnegative effect on $\mathbf{Y}$, then the statistic T^* will tend to be larger than if the treatment had no effect on $\mathbf{Y}$. Compare T^* to its distribution in the absence of hidden bias under the null hypothesis of no effect on $\mathbf{Y}$; that is, proceed exactly as in §6.2.1. If T^* is large, falling in the extreme upper tail of this distribution, then two explanations are possible, namely, a positive effect on $\mathbf{Y}$ or hidden bias. However, if T^* is small, falling in the extreme lower tail of this distribution, then this cannot be due to a nonnegative effect on $\mathbf{Y}$, so there is evidence of hidden bias. More precisely, if T^* is small, calculate the tail probability of a value as smaller or smaller than the observed T^* assuming the absence of hidden bias and no effect of the treatment on $\mathbf{Y}$. Report this tail probability as the significance level for testing the hypothesis that hidden biases are absent. These highly intuitive ideas are discussed formally in §6.4.

This logic was used in §6.1.2 in connection with childhood cancers other than leukemia in high- and low-exposure counties in Utah. Under the premise that higher levels of radiation from fallout might cause but would not prevent cancers, the finding that high-exposure counties had lower frequencies of other childhood cancers was taken to indicate the presence of hidden bias.

6.3.2 Did the Test Reduce Sensitivity to Bias?

In §6.3.1, an outcome $\mathbf{Y}$ with a nonnegative effect was used to test the hypothesis that there is no hidden bias; that is, $\Gamma = 1$. To use $\mathbf{Y}$ to test the hypothesis $H_0 : (\Gamma, \mathbf{u}) = (\Gamma^\circ, \mathbf{u}^\circ)$ with $(\Gamma^\circ, \mathbf{u}^\circ)$ specified, proceed exactly as in §6.2.4, comparing T^* to its distribution determined from (6.2) under the hypothesis of no effect on $\mathbf{Y}$. As in §6.3.1, if T^* is large compared to this distribution, there are two explanations: a positive effect on $\mathbf{Y}$ or an incorrect value of $(\Gamma^\circ, \mathbf{u}^\circ)$. However, if T^* is small compared to this distribution, it cannot be due to a nonnegative effect on $\mathbf{Y}$, so there is evidence against the hypothesized value $(\Gamma^\circ, \mathbf{u}^\circ)$. These considerations are stated formally in §6.4.

As in §6.2.4, the typical use of this procedure is to test whether T^* has rendered implausible the point of greatest sensitivity $(\Gamma^\circ, \mathbf{u}^\circ)$ for the outcome $\mathbf{R}$ of primary interest.

6.4 *The Behavior of T^* with Nonnegative Effects

In §6.3, the significance level for the simple hypothesis of no effect on $\mathbf{Y}$ was used to test the composite hypothesis of a nonnegative effect. This section looks at the relevant technical details. Let $T^* = t^*(\mathbf{Z}, \mathbf{Y})$. Under the model (4.6), the chance that $T^* \geq c$ is

$$\alpha_c = \sum_{\mathbf{z} \in \Omega} [t^*(\mathbf{z}, \mathbf{y_z}) \geq c] \frac{\exp(\gamma \mathbf{z}^T \mathbf{u})}{\sum_{\mathbf{b} \in \Omega} \exp(\gamma \mathbf{b}^T \mathbf{u})}. \tag{6.4}$$

Let $\mathbf{a}$ be any one fixed treatment assignment, $\mathbf{a} \in \Omega$. Suppose we observed $\mathbf{Z} = \mathbf{a}$ and, consequently, $\mathbf{Y} = \mathbf{y_a}$ and $T^* = t^*(\mathbf{a}, \mathbf{y_a})$. The hypothesis that the treatment had no effect on $\mathbf{Y}$ would say that $\mathbf{Y} = \mathbf{y_a}$, no matter how treatments $\mathbf{Z}$ were assigned. The significance level, say $\alpha_{c,\mathbf{a}}$, for testing this hypothesis would be

$$\alpha_{c,\mathbf{a}} = \sum_{\mathbf{z} \in \Omega} [t^*(\mathbf{z}, \mathbf{y_a}) \geq c] \frac{\exp(\gamma \mathbf{z}^T \mathbf{u})}{\sum_{\mathbf{b} \in \Omega} \exp(\gamma \mathbf{b}^T \mathbf{u})}. \tag{6.5}$$

The following proposition says that when the treatment has a nonnegative effect, the statistic T^* is stochastically larger than all of its permutation distributions under the null hypothesis of no effect. In particular, when the treatment has a nonnegative effect, there is at most a 5% chance that T^*

will fall in the lower 5% tail of its permutation distribution under the hypothesis of no effect. The proof is the same as the proof of Proposition 4 in §2.9.

Proposition 24 *If $t^*(\cdot, \cdot)$ is effect increasing and the treatment has a nonnegative effect on $\mathbf{Y}$, then $\alpha_c \geq \alpha_{c,\mathbf{a}}$ for all c and all $\mathbf{a} \in \Omega$.*

If $\gamma = 0$, then

$$
\alpha_c = \frac{|\{\mathbf{z} \in \Omega : t^*(\mathbf{z}, \mathbf{y_z}) \geq c\}|}{K}, \qquad \alpha_{c,\mathbf{a}} = \frac{|\{\mathbf{z} \in \Omega : t^*(\mathbf{z}, \mathbf{y_a}) \geq c\}|}{K}, \quad (6.6)
$$

and $\alpha_{c,\mathbf{a}}$ is the usual significance level for a randomization test.

6.5   Bias of Known Direction

6.5.1   Does Disability Insurance Discourage Work?

The Social Security Disability Program provides financial support to disabled workers and their families. Concern is often expressed that disability insurance could provide a disincentive to work. For example, Parsons (1980, p. 130) wrote:

> The recent increase in nonparticipation in the labor force of prime-aged males can apparently be largely explained by the increased generosity of social welfare transfers, particularly Social Security disability payments.

Would most recipients of Social Security Disability Insurance (DI) work if they did not receive benefits? In a methodologically interesting study, Bound (1989) examined this question using a severely biased comparison in which the direction of the bias was clear; see also Parsons (1991) and Bound (1991). Specifically, Bound compared recipients of DI, say $Z_{si} = 0$, to individuals who had applied for but were denied benefits, say $Z_{si} = 1$. The treatment is the *denial* of benefits to an applicant. In 1978, for noninstitutionalized men aged 45 to 64, these two groups were similar in terms of several demographic characteristics (Bound 1989, Table 2), including median age of 55.6 for $Z_{si} = 1$ versus 58.3 for $Z_{si} = 0$, years of education of 9.2 versus 9.1, percentage nonwhite of 13.2% versus 12.4%, and percentage married of 74.3% versus 79.9%. Nonetheless, the comparison is biased because benefits tend to be granted to individuals who are more seriously ill and who are less able to work, and Bound (1989, p. 484) documents that the "rejected applicants are healthier and more capable of work than those who were accepted." So when rejected applicants are compared to applicants who received benefits, one expects to see greater participation in the workforce by rejected applicants for two reasons: they are more fit for work, and they lack the disincentive effect of the benefit. In other words, Bound is arguing that the direction of bias is known for the outcome **R** of primary interest, namely, workforce participation. Is knowing the direction of the bias helpful?

Bound looked at a variety of measures of workforce participation, but for current purposes, it suffices to consider the percentage employed among workers who applied for disability insurance. For noninstitutionalized men aged 45 to 64 in 1978, Bound (1989, Table 2) found that 28.7% of rejected applicants ($Z_{si} = 1$) and 2.3% of recipients of DI were employed. Bound's argument is essentially the following. The difference, 28.7 − 2.3, might reflect a disincentive effect of DI, or it might reflect the superior health of rejected applicants, or it might reflect both to some degree. In other words, the difference 28.7 − 2.3 might overestimate the effect that denying benefits would have had on workers who actually received benefits, but it seems unlikely to underestimate the effect. If one accepted this argument, one would conclude that most recipients of DI would not work if benefits were denied; after all, when benefits were denied to a healthier group of workers, most of them did not return to work.

Notice that knowing the direction of bias is helpful in some contexts but not in others. Bound (1989) found estimates of the disincentive effect that were much smaller than other studies had found. It was therefore relevant that the small size of these estimated effects could not be attributed to bias. Had Bound found larger estimates than others had found, knowing that the bias tended to make the estimates larger would have weakened rather than strengthened the case for his estimates when compared to those obtained by others.

Bound's (1989) paper prompted a comment by Parsons (1991) with a reply by Bound (1991). Briefly, the discussion concerned whether or not there were additional hidden biases, beyond the bias of known direction caused by denying benefits to healthier individuals.

6.5.2   *Bias of Known Direction: Formal Statement

This section provides one simple, formal statement of the argument used in §6.5.1, so that a bias of known direction implies an inference that is biased in a known direction. The inference is based on a statistic, $t(\cdot, \cdot)$, which is both arrangement-increasing and effect increasing, two properties shared by most statistics discussed in this book. Also, the distribution of treatment assignments is governed by (4.6) with $\gamma \geq 0$. For brevity and simplicity of discussion, assume the treatment has an additive effect, $\mathbf{R} = \mathbf{r}_C + \tau\mathbf{Z}$, so interest focuses on τ; however, this is by no means a critical assumption. The discussion here builds upon §6.4.

Consider testing the hypothesis $H_0 : \tau = \tau_0$ by computing $T = t(\mathbf{Z}, \mathbf{R} - \tau_0\mathbf{Z})$ from the adjusted responses, $\mathbf{R} - \tau_0\mathbf{Z}$, and comparing T to its null distribution under (4.6). In general, the adjusted response is $\mathbf{R} - \tau_0\mathbf{Z} =$

$\mathbf{r}_C + (\tau - \tau_0)\mathbf{Z}$, so for each $\mathbf{z} \in \Omega$ write $\mathbf{y_z} = \mathbf{r}_C + (\tau - \tau_0)\mathbf{z}$ for the adjusted response that would have been observed under treatment assignment $\mathbf{z}$. If the hypothesized effect is too small, that is, if $\tau > \tau_0$, then for every $\mathbf{z} \in \Omega$, the statistic computed from the adjusted responses is too large, that is, $t(\mathbf{z}, \mathbf{y_z}) \geq t(\mathbf{z}, \mathbf{r}_C)$, because $t(\cdot, \cdot)$ is effect increasing, and $(y_{zsi} - r_{Csi})(2z_{si} - 1) = z_{si}(\tau - \tau_0)(2z_{si} - 1) \geq 0$ for both $z_{si} = 0$ and $z_{si} = 1$. It follows that if $\tau > \tau_0$, then

$$\sum_{\mathbf{z} \in \Omega}[t(\mathbf{z}, \mathbf{y_z}) \geq k]\frac{\exp(\gamma \mathbf{z}^T \mathbf{u})}{\sum_{\mathbf{b} \in \Omega}\exp(\gamma \mathbf{b}^T \mathbf{u})}$$

$$\geq \sum_{\mathbf{z} \in \Omega}[t(\mathbf{z}, \mathbf{r}_C) \geq k]\frac{\exp(\gamma \mathbf{z}^T \mathbf{u})}{\sum_{\mathbf{b} \in \Omega}\exp(\gamma \mathbf{b}^T \mathbf{u})} = \zeta(\mathbf{r}_C, \mathbf{u}), \text{ say.}$$

Here, $\zeta(\mathbf{r}_C, \mathbf{u})$ is the chance that $t(\mathbf{z}, \mathbf{r}_C) \geq k$ due to bias alone. Several technical conclusions follow from this. After stating the technical conclusions, they are then interpreted in terms of the bias of the inference. First, $t(\mathbf{Z}, \mathbf{R} - \tau_0\mathbf{Z}) \geq t(\mathbf{Z}, \mathbf{r}_C)$, so $t(\mathbf{Z}, \mathbf{R} - \tau_0\mathbf{Z})$ is stochastically larger than $t(\mathbf{Z}, \mathbf{r}_C)$ whenever $\tau > \tau_0$. Moreover, by Proposition 17, $\zeta(\mathbf{r}_C, \mathbf{u})$ is arrangement-increasing. Finally, if u_{si} had a conditional probability distribution given r_{Csi} so that u_{si} and r_{Csi} were positively related (formally, if $\Pr(\mathbf{u}|\mathbf{r}_C) = \prod_{s=1}^{S} \prod_{i=1}^{n_s} \Pr(u_{si}|r_{Csi})$, where $\Pr(u_{si}|r_{Csi})$ is TP_2), then the chance that $t(\mathbf{Z}, \mathbf{r}_C) \geq k$ is greater than or equal to the chance under a uniform randomized experiment; see Rosenbaum (1989a) for details, and Krieger and Rosenbaum (1994) for a generalization.

If the hypothesis $H_0 : \tau = \tau_0$ is tested using a rank test such as Wilcoxon's rank sum or signed rank test whose randomization distribution rejects at level α when $T \geq k$, then the following several interpretations follow when τ_0 is too small, that is, when $\tau > \tau_0$. First, no matter what pattern of biases $\mathbf{u}$ are present, the value τ_0 that is too small is more likely to be rejected than the true value τ. Second, if u_{si} and r_{Csi} are positively related, then the chance of rejecting both τ_0 and τ is at least α. Third, the stronger the positive relationship between $\mathbf{r}_C$ and $\mathbf{u}$, the more likely rejection becomes.

In Bound's (1989) study, one continuous measure of workforce participation was the *change* in earnings, that is, earnings in the year after application for DI benefits minus average earnings in the two years before application. Notice that this is the *change* in earnings, so most changes are negative, and the more negative the change, the greater the decline in earnings. In this case, r_{Csi} is the change in earnings if DI benefits are granted, $r_{Tsi} = r_{Csi} + \tau$ is the change in earnings if benefits are denied, and τ is the disincentive effect, which is presumably nonnegative. Also, u_{si} measures fitness for work, larger values indicating greater fitness. Presumably, $\gamma \geq 0$, so that greater fitness for work, u_{si}, is associated with a greater chance that benefits will be denied, $Z_{si} = 1$. Presumably, greater fitness for work, u_{si}, is nonnegatively associated with a larger, that is, less negative, decline in earnings, r_{Csi}. Suppose that we test the hypothesis that the disincentive effect is $H_0 : \tau = \tau_0$, versus $H_A : \tau > \tau_0$ when in fact the true disincentive effect is larger, $\tau > \tau_0$, and the test refers a conventional rank test to its randomization distribution, rejecting at level α when the randomization distribution is correct. If bias due to $\mathbf{u}$ would have led us to reject the true τ, then it would also lead us to reject τ_0's that are too small, and the probability that this will happen is at least α and it increases as the strength of association between $\mathbf{u}$ and $\mathbf{r}_C$ increases. If a small disincentive effect τ_0 looks plausible—if the conventional test fails to reject it—then this failure to reject cannot be attributed to hidden bias due to $\mathbf{u}$. Of course, similar considerations apply to confidence intervals and Hodges–Lehmann point estimates derived from the test. Bound's estimates of the disincentive effect are smaller than those found by others, but the small size of these estimates cannot be due to hidden biases of the type just described, because they push the estimate in the opposite direction.

6.6   Bibliographic Notes

Campbell (1969) and Campbell and Stanley (1963) discuss the strengths and weaknesses of various study designs, including the design in §6.1.2. Cook, Campbell, and Peracchio (1990, pp. 546–7 and pp. 561–564), Cook (1991), Shadish and Cook (1999, Table 1) and Shadish, Cook and Campbell (2002) refer to unaffected outcomes as "nonequivalent dependent variables." Disease specificity in §6.1.3 has been widely discussed; see, for example, Yerushalmy and Palmer (1959), Lilienfeld (1959), Sartwell (1960), Hill (1965), Susser (1973), and Rothman (1986) to sample a range of opinion. The use of known effects to test for hidden bias is discussed in Rosenbaum (1984a, 1989a,b). The link in §6.2.5 between a test for bias using one outcome and a sensitivity analysis for another is discussed in Rosenbaum (1992).

6.7   Problems

1. Petitti, Perlman, and Sidney (1986) commented on the contradictory results of several studies concerning the relationship between postmenopausal estrogen use and heart disease. After adjusting for several covariates, they found that the risk of death from cardiovascular disease for users of estrogen was half the risk of nonusers; however, they found the same difference for deaths from accidents, homicides, and suicide. They write:

There is no biologically plausible reason for a protective effect of postmenopausal estrogen use on mortality from accidents, homicide, and suicide. We believe that our results are best explained by the assumption that postmenopausal estrogen users in this cohort are healthier than those who had no postmenopausal estrogen use, in ways that have not been quantified and cannot be adjusted for. The selection of healthier women for estrogen use in this population is not necessarily a characteristic shared by other populations.

Do you agree?

Consider also Kreiger, Kelsey, Holford, and O'Connor (1982).

2. Infante, Rinsky, Wagoner, and Young (1977) found a statistically significant excess of leukemia deaths among benzene workers when compared to the general population, but a statistically significant deficit of deaths overall. If one were willing to assume that working with benzene might cause disease or death but would not prevent them, then what would you conclude about the general health of benzene workers compared to the population as a whole? Could this explain the excess of leukemia deaths? Why or why not?

3. What is the smallest value of Γ such that the expected value of T^* in §6.2.2 would equal the observed value of 483.5 for some $\mathbf{u} \in U$? (*Hint*: Use the methods of Chapter 4. *Answer*: $\Gamma = 1.73$ and $\mathbf{u}$ contains 15 ones and 24 zeros ordered in the same way as $\mathbf{y}$ in Table 6.3, that is, $u_i, = 0$ when $y_i = 0$ and $u_i = 1$ when $y_i > 0$.)

4. How is the calculation in Problem 3 above related to the range of Hodges–Lehmann estimates of effect that one could calculate for $\mathbf{y}$ using the procedure in §4.6.8? (*Hint*: What is the smallest Γ such that the interval of Hodges–Lehmann estimates includes zero?)

5. Why is there no largest value of Γ such that the expected value of T^* in §6.2.2 equals the observed value of 483.5 for some $\mathbf{u} \in U$? (*Hint*: Review §4.7.5.)

6.8 References

Beck, H. L. and Krey, P. W. (1983) Radiation exposures in Utah from Nevada nuclear tests. *Science*, **220**, 18–24.

Bound, J. (1989) The health and earnings of rejected disability insurance applicants. *American Economic Review*, **79**, 482–503.

Bound, J. (1991) The health and earnings of rejected disability insurance applicants: Reply. *American Economic Review*, **81**, 1427–1434.

Campbell, D. T. (1966) Pattern matching as an essential in distal knowing. In: *The Psychology of Egon Brunswik*, Kenneth Hammond, ed., New York: Holt, Rinehart and Winston, pp. 81–106.

Campbell, D. T. (1969) Reforms as experiments. *American Psychologist*, **24**, 409–429.

Campbell, D. T. and Stanley, J. (1963) *Experimental and Quasi Experimental Designs for Research.* Chicago: Rand McNally.

Cook, T. D. (1991) Clarifying the warrant for generalized causal inferences in quasi-experimentation. In: *Evaluation and Education at Quarter Century*, M. W. McLaughlin and D. Phillips, eds., NSSE 1991 Yearbook, pp. 115–144.

Cook, T. D., Campbell, D. T. and Peracchio, L. (1990) Quasi Experimentation. In: *Handbook of Industrial and Organizational Psychology*, M. Dunnette and L. Hough, eds., Palo Alto, CA: Consulting Psychologists Press, Chapter 9, pp. 491–576.

D'Abadie, C. and Proschan, F. (1984) Stochastic versions of rearrangement inequalities. In *Inequalities in Statistics and Probability*, Y. L. Tong, ed., Hayward, CA: Institute of Mathematical Statistics, pp. 4–12.

Donohue, J. J. and Levitt, S. D. (2001) Legalized abortion and crime. *Quarterly Journal of Economics,* **116**, 379–420.

Gastwirth, J. (1988) *Statistical Reasoning in Law and Public Policy.* New York: Academic.

Hill, A. B. (1965) The environment and disease: Association or causation? *Proceedings of the Royal Society of Medicine*, **58**, 295–300.

Hollander, M., Proschan, F., and Sethuraman, J. (1977) Functions decreasing in transposition and their applications in ranking problems. *Annals of Statistics*, **5**, 722–733.

Infante, P., Rinsky, R., Wagoner, J., and Young, R. (1977) Leukemia in benzene workers. *Lancet*, July 9, 76–78.

Kreiger, N., Kelsey, J., Holford, T., and O'Connor, T. (1982) An epidemiologic study of hip fracture in postmenopausal women. *American Journal of Epidemiology*, **116**, 141–148.

Krieger, A. M. and Rosenbaum, P. R. (1994) A stochastic comparison for arrangement increasing functions. *Combinatorics, Probability and Computing*, **3**, 345–348.

Land, C. E. (1979) The hazards of fallout or of epidemiologic research. *New England Journal of Medicine*, **300**, 431–432.

Legorreta, A. P., Silber, J. H., Costantino, G. N., Kobylinski, R. W., and Zatz, S. L. (1993) Increased cholecystectomy rate after the introduction of laparoscopic cholecystectomy. *Journal of the American Medical Association*, **270**, 1429–1432.

Lilienfeld, A. (1959) On the methodology of investigations of etiologic factors in chronic diseases–Some comments. *Journal of Chronic Diseases*, **10**, 41–46.

Lyon, J. L., Klauber, M. R., Gardner, J. W., and Udall, K. S. (1979) Childhood leukemias associated with fallout from nuclear testing. *New England Journal of Medicine*, **300**, 397–402.

Mittleman, M. A., Maclure, M., Sherwood, J. B., Mulry, R. P., Tofler, G. H., Jacobs, S. C., Friedman, R., Benson, H., and Muller, J. E. (1995) Triggering of acute myocardial infaction onset by episodes of anger. *Circulation*, **92**, 1720–1725.

Parsons, D. O. (1980) The decline of male labor force participation. *Journal of Political Economy*, **88**, 117–134.

Parsons, D. O. (1991) The health and earnings of rejected disability insurance applicants: Comment. *American Economic Review*, **81**, 1419–1426.

Petitti, D., Perlman, J., and Sidney, S. (1986) Postmenopausal estrogen use and heart disease. *New England Journal of Medicine*, **315**, 131–132.

Rosenbaum, P. R. (1984a) From association to causation in observational studies. *Journal of the American Statistical Association*, **79**, 41–48.

Rosenbaum, P. R. (1984b) The consequences of adjustment for a concomitant variable that has been affected by the treatment. *Journal of the Royal Statistical Society*, Series **A**, **147**, 656–666.

Rosenbaum, P. R. (1989a) On permutation tests for hidden biases in observational studies: An application of Holley's inequality to the Savage lattice. *Annals of Statistics*, **17**, 643–653.

Rosenbaum, P. R. (1989b) The role of known effects in observational studies. *Biometrics*, **45**, 557–569.

Rosenbaum, P. R. (1992) Detecting bias with confidence in observational studies. *Biometrika*, **79**, 367–374.

Ross, H. L., Campbell, D. T., and Glass, G. V. (1970) Determining the social effects of a legal reform: The British "Breathalyser" crackdown of 1967. *American Behavioral Scientist*, **13**, 493–509.

Rothman, K. (1986) *Modern Epidemiology.* Boston: Little, Brown.

Sackett, D. (1979) Bias in analytic research. *Journal of Chronic Diseases*, **32**, 51–68.

Sartwell, P. (1960) On the methodology of investigations of etiologic factors in chronic diseases: Further comments. *Journal of Chronic Diseases*, **11**, 61–63.

Shadish, W. R. and Cook, T. D. (1999) Design rules: More steps toward a complete theory of quasi-experimentation. *Statistical Science*, **14**, 294–300.

Shadish, W. R., Cook, T. D., and Campbell, D. T. (2002) *Experimental and Quasi-Experimental Designs for Generalized Causal Inference.* Boston: Houghton-Mifflin.

Skerfving, S., Hansson, K., Mangs, C., Lindsten, J., and Ryman, N. (1974) Methylmercury-induced chromosome damage in man. *Environmental Research*, **7**, 83–98.

Susser, M. (1973) *Causal Thinking in the Health Sciences: Concepts and Strategies in Epidemiology.* New York: Oxford University Press.

Trichopoulos, D., Katsouyanni, K., Zavitsanos, X., Tzonou, A., and Dalla-Vorgia, P. (1983) Psychological stress and fatal heart attack: The Athens 1981 earthquake natural experiment. *Lancet*, 26 February, 441–444.

Yerushalmy, J. and Palmer, C. (1959) On the methodology of investigations of etiologic factors in chronic diseases. *Journal of Chronic Diseases*, **10**, 27–40.

7
Multiple Reference Groups in Case-Referent Studies

7.1 Multiple Reference Groups

7.1.1 What Are Multiple Reference Groups?

A case-referent study compares the frequency or intensity of exposure to the treatment among cases and among referents or noncases who are free of the disease; see §3.3. If referents or noncases are selected from several different sources, then the study has several distinct referent groups.

Case-referent studies with several referent groups are common. A few examples follow. Notice, in particular, the types of referent groups used. Gutensohn, Li, Johnson, and Cole (1975) compared the frequency of tonsillectomy among cases of Hodgkin's disease to two groups of referents, namely, spouses of cases and siblings of cases. The Collaborative Group for the Study of Stroke in Young Women (1973) compared the use of oral contraceptives by cases of stroke among young women aged 15 to 44 with the use of oral contraceptives among two referent groups matched for age and race, namely, women discharged from the same hospital for some other disease and women who were neighbors of the cases. In a study of subacute sclerosing panencephalitis in young children, Halsey, Modlin, Jabbour, Dubey, Eddins, and Ludwig (1980) used playmates of cases as one referent group and children admitted to the same hospital as another. Kreiger, Kelsey, Holford, and O'Connor (1982) compared estrogen use among female cases of hip fracture with estrogen use in two referent groups, namely, trauma and nontrauma referents, that is, other women from the same hospitals who were admitted for traumas such as fractures or sprains other than hip fractures and women who were admitted for disorders that did not involve a physical trauma.

The use of multiple referent groups in case-referent studies is different from the use of multiple control groups. The case group and several referent groups are each exposed to the treatment and each exposed to the control in varying degrees. In contrast, a study with multiple control groups has several groups of subjects who did not receive the treatment. A case-referent study may have a single referent group and several control groups, and one will be discussed in §8.1.4. In principle, a case-referent study could have both multiple referent groups and multiple control groups. Notice that the term "control group" is sometimes used loosely to describe any group used for comparison, but in this book a control group refers specifically to subjects who did not receive the treatment.

7.1.2   An Example: A Study of Reye's Syndrome with Four Referent Groups

Aspirin bottles contain the following warning: "Children and teenagers should not use this medicine for chicken pox or flu symptoms before a doctor is consulted about Reye syndrome, a rare but serious illness reported to be associated with aspirin." Reye's syndrome typically occurs in children under the age of 15 following an upper respiratory tract infection, and it often leads to death from cerebral damage. Hurwitz (1989) surveys the evidence linking Reye's syndrome and aspirin.

Hurwitz, Barrett, Bregman et al. (1985) compared the use of aspirin by cases of Reye's syndrome with its use in four referent groups, namely, referents from the case's emergency room, school, community, and inpatients at the case's hospital. This study was not the first to link aspirin with Reye's syndrome, but from a strictly methodological view, it is the most interesting. Earlier studies had convinced the US Surgeon General and the American Academy of Pediatrics to advise against giving children aspirin for chicken pox or flu, and that appears to have been very good advice, supported by the new US Public Health Service study by Hurwitz et al. (1985), which attempted to address criticisms of earlier studies. Observational studies are often subject to criticism, at times warranted criticism, and some caution in asserting conclusions may be appropriate. However, caution and uncertainty do not, by themselves, favor the null hypothesis of no treatment effect over other hypotheses; rather, caution in the face of uncertainty may sometimes merit strong action to warn of a possible hazard.

In their study, all subjects, both cases and referents, had to have had a "respiratory, gastrointestinal, or chicken pox antecedent illness within 3 weeks before hospitalization." In other words, the population being described consists of children who have had one of these illness within 3

TABLE 7.1. Counts of Aspirin Use by Cases and Referents.

	Aspirin	No Aspirin	% Using Aspirin
Cases	28	2	93%
Emergency room	7	18	28%
Inpatient	5	17	23%
School	24	17	59%
Community	30	27	53%

TABLE 7.2. Odds Ratios for Aspirin Use.

	School	Community	Emergency Room	Inpatient
Cases	9.9	12.6	36.0	47.6
School		1.3	3.6	4.8
Community			2.9	3.8
Emergency room				1.3

weeks. The study matched cases to referents on the basis of age, race, date of illness, and the specific type of antecedent illness. In addition, individual referent groups were matched in other ways; specifically, emergency room and inpatient referents came from the same hospital as the case, school referents came from the same school or day care center, and community referents were selected by random digit telephone dialing from the same community as the case. Table 1 in their article shows that cases and referents were similar in terms of age, race, gender, and antecedent illness. (The comparisons that follow are based on the published data and ignore the matching, but for comparisons done both ways, the matched and unmatched analyses agree with each other; compare Tables 7.2 and 7.3 and the results reported in the first paragraph of page 852 in Hurwitz et al. 1985.)

Table 7.1 gives the frequencies of aspirin use by case and referent groups in Hurwitz et al. (1985). Of the 30 cases, 28 had used aspirin. In each referent group, the relative frequency of aspirin use was much lower.

Tables 7.2 and 7.3 are calculated from Table 7.1. Table 7.2 gives the odds ratio comparing aspirin use in each pair of groups. For instance, the odds of using aspirin were 9.92 times higher in the case group than in the school control group. Table 7.3 gives the usual chi-square statistic with continuity correction for a 2×2 table for each pair of groups. Each chi-square has one degree of freedom, and is significant at the 0.05 level or 0.01 level if greater than 3.84 or 6.63, respectively.

Two patterns are apparent in Tables 7.2 and 7.3. First, cases used aspirin substantially and significantly more than did all four referent groups. If the study were free of biases, Cornfield's (1951) result in §3.3.2 would

TABLE 7.3. Chi-Squares for Aspirin Use.

	School	Community	Emergency Room	Inpatient
Cases	9.0	12.9	22.4	24.3
School		0.1	4.6	6.0
Community			3.3	4.6
Emergency room				0.0

suggest that aspirin use increases the risk of Reye's syndrome by 9.9 to 47.6 times. The second pattern concerns differences between the referent groups. School and community referents have similar frequencies of aspirin use, the odds ratio being 1.3. Emergency room and inpatient referents have similar frequencies of aspirin use, again with an odds ratio of 1.3. However, school and community referents are three to five times more likely to use aspirin than emergency room or inpatient referents, and three of these four comparisons are significant at the 0.05 level.

The differences in aspirin use among the referent groups are inconsistent with a study that is free of selection and hidden biases. If there were no bias, the frequency of aspirin use in each referent group should differ only by chance from the frequency of aspirin use among all children without Reye's syndrome in the relevant population of children having recently had one of the antecedent illnesses. However, in contrast with the study of nuclear fallout and childhood leukemia in §6.1.2, the differences among the referent groups, while statistically significant, are much smaller than each of the four differences between the cases and each referent group. There is evidence of hidden bias, but the biases appear to be much smaller than the ostensible treatment effect.

7.1.3 Selection Bias and Hidden Bias in Case-Referent Studies

Recall from §3.3 that a synthetic case-referent study begins with the population of M subjects in §3.2.1 and then draws a random sample without replacement of subjects who are cases and a random sample of subjects who are not cases, possibly after stratification on the observed covariates $\mathbf{x}$. In some studies, the sample of cases includes all of the cases in the population; this is a trivial sort of random sampling, but it is random sampling nonetheless. In such a synthetic case-referent study, the reference group is the sample of subjects who are not cases. Synthetic case-referent studies are sometimes conducted and they form a simple model for the situation in which the process of selecting cases and referents neither introduces additional biases nor reduces bias. In a synthetic case-referent study free of hidden bias, §3.3 showed that the odds ratio estimates the population odds ratio and the usual Mantel–Haenszel test appropriately tests the null hypothesis of no effect of the treatment or exposure. Section §4.4 considered the test of no effect and its sensitivity to hidden bias in a synthetic case-referent study in which there is no selection bias.

Most case-referent studies are not synthetic case-referent studies. In particular, a study with several different referent groups is, by definition, not a synthetic case-referent study, since the several reference groups are, by definition, not random samples from the relevant population of subjects who are not cases. The purpose of this section is to discuss case-referent studies in which the reference group is not a random sample or stratified random sample of subjects who are not cases. This section summarizes findings based on the considerations in the appendix, §7.3, which should be consulted for formal proof.

Hidden bias may be introduced when treatments or exposures are assigned to the M subjects in the population. Selection bias may be introduced when cases and referents are selected into the study from the cases and referents in the population. The relationship between selection and hidden bias can be subtle. A careful choice of referent group may remove a hidden bias that would have been present had the entire population been available for study. A poor choice of referent group may introduce a selection bias into a study that was free of hidden bias, that is, a bias that would not have been present in the entire population.

In the discussion that follows, the model for hidden bias, (4.1), (4.2), or (4.6), is assumed to hold in the population of M subjects before cases and referents are selected. The intent is to test the null hypothesis of no treatment effect, so the null hypothesis is tentatively assumed to hold for the purpose of testing it. The null hypothesis of no treatment effect says that whether or not a subject is a case is unaffected by the treatment or exposure. More precisely, $\mathbf{r}$ is the N-dimension vector with $r_{si} = 1$ if the ith subject in stratum s is a case and $r_{si} = 0$ if this subject is not a case, and the null hypothesis H_0 says that $\mathbf{r}$ would be unaltered if the pattern of exposures $\mathbf{Z}$ were changed.

The null hypothesis H_0 refers to $\mathbf{r}$, that is, to whether or not a subject is a case. In §7.1.2, $\mathbf{r}$ distinguishes cases of Reye's syndrome from children without Reye's syndrome. In many case-referent studies, other variables besides $\mathbf{r}$ are used in selecting referents, and some of these may be outcomes in the sense that they describe subjects after exposure to the treatment. This turns out to be an important consideration in distinguishing hidden bias and selection bias. As has been true throughout Chapters 4 and 6, the term hidden bias refers to an unobserved covariate $\mathbf{u}$, that is, a variable describing subjects prior to treatment so that $\mathbf{u}$ is unaffected by the treatment.

Under H_0, if subjects are selected based on quantities that are unaffected by the treatment, such as the response $\mathbf{r}$, the observed covariate $\mathbf{x}$, the unobserved covariate $\mathbf{u}$, or some other unaffected outcome $\mathbf{y}$, then the form of the model (4.6) and the value of γ are unchanged. In this situation, if there is no hidden bias in the sense that $\gamma = 0$, then conventional methods such as the Mantel–Haenszel test may be used to test H_0, and if hidden bias is possible in the sense that γ might not equal zero, then the methods in Chapter 4 may be used to appraise sensitivity to bias. In this specific sense, if cases and referents are selected based on quantities that are unaffected by the treatment, there is no selection bias.

Consider instead an outcome that is affected by the treatment, say $Y = y_Z$, where as in §6.3 the value of this outcome changes if the treatment assignment Z changes. For example, see Problem 1. If subjects are selected using the affected outcome $Y = y_Z$, then the selection may introduce a bias where none existed previously, and it may distort the distribution of Z, so the model (4.6) can no longer be used to study sensitivity to hidden bias. For instance, if the treatment had a positive effect on $Y = y_Z$, then selecting referents with high values of Y would produce a referent group in which too many referents had been exposed to the treatment. Under H_0, membership in the referent group should not change with the treatment or exposure. In the hypothetical study in §3.3.3 of smoking and lung cancer, a referent group of cardiac patients would lead to selection bias because smoking causes cardiac disease. Selecting cardiac patients as referents means selecting subjects who have higher frequencies of smoking, more exposure to the treatment, because exposure to the treatment caused some of their cardiac disease.

Although selecting subjects based on quantities unaffected by the treatment does not introduce selection bias in the test of H_0, it may alter the quantity or pattern of hidden bias. That is, even if selection is based on fixed quantities unaffected by the treatment, the selected subjects may differ systematically from the unselected subjects in their values of the unobserved covariate u_{si}. Indeed, an investigator who uses siblings or neighbors as referents is, presumably, trying to reduce hidden bias without introducing selection bias, that is, to produce matched sets that are more homogeneous than the population in terms of certain covariates u that are not explicitly measured, without distorting the frequency of exposure to the treatment. Though often a sensible strategy, an attempt to reduce hidden bias by the choice of referent can at times backfire, increasing rather than reducing the amount of hidden bias, and perhaps this happened with the inpatient referents in §7.1.2. Let u be a binary variable indicating whether or not the adult caring for a child is aware of the advice against aspirin use by the Surgeon General and the American Academy of Pediatrics. It is conceivable that a case tends to differ more from an inpatient referent in terms of u than a case would typically differ from a referent selected at random from children who are not cases.

To summarize: Selecting referents based on characteristics unaffected by the treatment does not introduce selection bias in the test of H_0, but it may decrease or increase the quantity of hidden bias. Selecting referents based on characteristics affected by the treatment can introduce selection biases that distort the distribution of treatments or exposures $\mathbf{Z}$, thereby invalidating both:

(i) tests that would have been appropriate in the absence of hidden bias; and

(ii) sensitivity analyses that would have been appropriate in the presence of hidden bias.

Again, formal proofs are given in the appendix, §7.3.

7.2   Matched Studies with Two Referent Groups

7.2.1   An Example: Breast Cancer and Age at First Birth

Table 7.4 describes data collected by Lilienfeld, Chang, Thomas and Levin (1976) and reported by Liang and Stewart (1987). The cases in this study were women in Baltimore who were diagnosed with primary malignant breast cancer. Each case was matched to one or two female referents on the basis of age and race. A hospital referent had been admitted to the same hospital as the case. A neighbor referent came from the same neighborhood as the case. There were 409 cases, of whom 195 were matched with both a hospital and a neighbor referent, 164 were matched only to a hospital referent, and 50 were matched only to a neighbor referent. The exposure of interest here is the age of first birth, ≤ 24 or ≥ 25, where women who had never given birth were classified as ≥ 25.

Each count in Table 7.4 is a matched set, the total count being 409 matched sets, rather than $(3 \times 195) + (2 \times 164) + (2 \times 50) = 1013$ women. For instance, there are 47 matched sets in which the case and both referents first gave birth at an age ≤ 24. This format is the correct format for presenting matched comparisons because it indicates who is matched to whom, information that is necessary for an appropriate analysis.

For a moment, ignore the distinction between hospital and neighbor referents and compare the age at first birth of cases and referents using the Mantel–Haenszel statistic. There are $S = 409$ strata defined by the matched pairs and matched triples. There are a total of 209 cases who first gave birth at an age of 25 or greater, where 172.3 were expected in the absence of hidden bias, leading to a significance level < 0.0001. If there is no hidden bias, there is strong evidence of a higher risk of breast cancer among women 25 or older at first birth.

A first step is to test for hidden bias by comparing the frequency of older first births in the two referent groups. McNemar's test is used to compare the age at first birth in the two referent groups for the 195 matched triples. In the absence of hidden bias, matched hospital and neighbor referents

TABLE 7.4. Age at First Birth for Cases of Breast Cancer and Matched Referents.

Hospital Referent	Neighbor Referent	Case ≤ 24	Case ≥ 25
≤ 24	≤ 24	47	32
≤ 24	≥ 25	17	31
≥ 25	≤ 24	18	21
≥ 25	≥ 25	8	21
≤ 24	None	58	53
≥ 25	None	24	29
None	≤ 24	17	12
None	≥ 25	11	10

should have similar frequencies of first births at ≤ 24 years of age. There are $87 = 17 + 31 + 18 + 21$ discordant pairs, that is, hospital/neighbor pairs in which one woman first gave birth at an age ≤ 24 and the other at an age ≥ 25. In $48 = 17 + 31$ of these discordant pairs, the hospital referent first gave birth at an age ≤ 24 and in the remaining 39 pairs the neighbor referent first gave birth at an age ≤ 24. McNemar's test compares this 48 versus 39 split of the 87 pairs with a binomial distribution with 87 independent trials each having probability 0.5 of success, finding little or no evidence that the frequency of first births at ages ≤ 24 differs in the two referent groups.

In parallel with the discussion in §6.2.3, it may be shown that the above test comparing the two referent groups has the following properties. First, in testing for hidden bias, the test has the correct level; that is, if $\gamma = 0$, then 5% of the time the test will falsely reject the hypothesis that $\gamma = 0$ with a significance level of 0.05 or less. Second, for $\gamma > 0$, the power of the test increases steadily as the values of u are rearranged to make the two referent groups more dissimilar; that is, the power function of the appropriate one-sided test is arrangement-increasing in the pair of vectors containing the u's and the binary indicators distinguishing hospital and neighbor referents. The test can hope to distinguish $\gamma = 0$ from $\gamma > 0$ if the referent groups are quite different in their typical values of u.

7.2.2 Partial Comparability of Cases and Multiple Referent Groups

All of the case-referent studies discussed in this chapter have the following features in common. A new (or incident) case of disease is made available to the investigator. This case is then matched to one or more referents who are similar in certain ways, for instance, a new patient at the same hospital of the same age and race. Then the case is matched to one or more additional referents who are similar to the case in different ways, for instance, a neighbor of the same age and race. This strategy will be called

TABLE 7.5. Notation for a Matched Set with Two Referent Groups.

	Case	Referent Group 1	Referent Group 2
Size	1	I_s	J_s
Subscripts	$(s,1)$	$(s,2),\ldots,(s,I_s+1)$	$(s,I_s+2),\ldots,(s,n_s)$

partial comparability. Each referent group is comparable to the case in some ways but not in others. In Gutensohn, Li, Johnson, and Cole (1975), the cases and their spouses have similar home environments and diets as adults, while cases and siblings have similar childhood environments and diets and have some genetic similarities.

Consider testing the null hypothesis of no treatment or exposure effect, H_0, by comparing cases with two referent groups, the situation with more than two groups being similar. Matched set s contains $n_s \geq 2$ subjects, the first subject is the case, so $r_{s1} = 1$, the next $I_s \geq 0$ subjects are referents from the first group, and the last $J_s \geq 0$ subjects are referents from the second group, so $n_s = 1 + I_s + J_s$ and $r_{si} = 0$ for $i = 2,\ldots,n_s$. This is summarized in Table 7.5 for matched set s. In the example in §7.2.1, $n_s = 3$, $I_s = 1$, $J_s = 1$, for 195 matched sets s; $n_s = 2$, $I_s = 1$, $J_s = 0$, for 164 more matched sets, and $n_s = 2$, $I_s = 0$, $J_s = 1$, for 50 matched sets.

Under H_0 and in the absence of selection bias, a subject does not change from one referent group to another based on exposure to the treatment Z_{si}, nor does exposure to the treatment cause cases of disease, so r_{si} and the order of subjects is fixed under H_0 not varying with $\mathbf{Z}$.

To express partial comparability, write the unobserved covariate u_{si} as a weighted sum of three unobserved covariates

$$u_{si} = \psi_0 v_{0si} + \psi_1 v_{1si} + \psi_2 v_{2si}, \tag{7.1}$$

where

$$\psi_k \geq 0, \qquad 1 = \sum \psi_k, \qquad 0 \leq v_{ksi} \leq 1 \quad \text{for all } k, s, i.$$

The first thing to notice about (7.1) is that it is not a new assumption, but rather a reexpression of the old assumption in §4.2 that $\mathbf{u} \in U$. Any $\mathbf{u} \in U$ may be written as $\mathbf{u} = \psi_0 \mathbf{v}_0 + \psi_1 \mathbf{v}_1 + \psi_2 \mathbf{v}_2$ for some $\mathbf{v}_k \in U$ for $k = 1, 2, 3$, and some $\psi_k \geq 0$, $1 = \sum \psi_k$, and any $\mathbf{u}$ written in this way is an element of U, so (7.1) is the same as the assumption that $\mathbf{u} \in U$. So far, (7.1) is nothing new.

To express partial comparability, require

$$v_{1si} = v_{1s1} \quad \text{for } i = 2,\ldots,I_s+1,$$

and

$$v_{2si} = v_{2s1} \quad \text{for } i = I_s+2,\ldots,n_s, \tag{7.2}$$

which says that the referents in group 1 have the same value of v_{1si} as the case and the referents in group 2 have the same value of v_{2si} as the case, though v_{0si} may differ for cases and referents. The first thing to note is that, taken together, (7.1) and (7.2) are not a new assumption, just a reexpression of the assumption that $\mathbf{u} \in U$. Once again, any $\mathbf{u} \in U$ may be written to satisfy (7.1) and (7.2) by taking $\psi_0 = 1$, $\psi_1 = 0$, $\psi_2 = 0$, and $v_0 = \mathbf{u}$, and any $\mathbf{u}$ that satisfies (7.1) and (7.2) is an element of U. So far, nothing new has been assumed.

The sensitivity analysis considers a range of values of the parameter $(\gamma, \psi_0, \psi_1, \psi_2)$ under the model (4.6). Taking $(\gamma, \psi_0, \psi_1, \psi_2) = (\gamma, 1, 0, 0)$ gives exactly the sensitivity analysis in Chapter 4, because in this case (7.2) does not restrict $\mathbf{u}$ in any way and $u_{si} = v_{0si}$. In other words, $(\gamma, \psi_0, \psi_1, \psi_2) = (\gamma, 1, 0, 0)$ signifies that partial comparability did nothing to make either type of referent similar to the case in terms of the unobserved covariate u. If instead $(\gamma, \psi_0, \psi_1, \psi_2) = (\gamma, 0, 1, 0)$ then $u_{si} = v_{1si}$ and, using (7.2), the case and the first referent group have identical values of u, so there is no hidden bias using the first reference group. In the example of §7.2.1, this would mean that cases and matched hospital referents are comparable in terms of u. In the same way, $(\gamma, \psi_0, \psi_1, \psi_2) = (\gamma, 0, 0, 1)$ signifies that cases and the second referent group have identical values of u; for instance, that cases and neighbors have identical values of u. If $(\gamma, \psi_0, \psi_1, \psi_2) = (\gamma, 1/3, 1/3, 1/3)$ then the bias is split equally between the uncontrolled v_0 and the partially controlled v_1 and v_2 so partial comparability has reduced but not eliminated hidden bias for both groups.

In other words, the sensitivity considers several possibilities, including the possibility that partial comparability did nothing to reduce hidden bias, that it eliminated bias for one referent group but did nothing for the other, or that it reduced but did not eliminate bias for one or both groups.

7.2.3   *Sensitivity Analysis with Two Matched Referent Groups

The procedure for sensitivity analysis with two reference groups is similar to that in §4.4.4 and 4.4.5 for case-referent studies; however, the sensitivity bounds now reflect the partial comparability of the referent groups. Since the first subject in each matched set is the case, so $r_{s1} = 1$ and $r_{si} = 0$ for $i = 2, \ldots, n_s$, it follows that a sign-score statistic has the form $T = \sum_s d_s \sum_i r_{si} Z_{si} = \sum_s d_s Z_{s1}$. Most often, T is the Mantel–Haenszel statistic with $d_s = 1$ for each s and $T = \sum Z_{s1}$ is the number of exposed cases, but the statistic could give different weights d_s to different cases as in §4.4.5.

In the notation of §4.4, $B_s = Z_{s1}$ and $p_s = \text{prob}(Z_{s1} = 1|\mathbf{m})$. To conduct the sensitivity analysis, bounds are needed on p_s under (7.1) and (7.2), say

$P_s^- \leq p_s \leq P_s^+$. For the Mantel–Haenszel statistic, the bounds are used in (4.20) for a large sample approximation; alternatively, a formula analogous to (4.10) is used for exact calculations. If there are varying weights d_s, then (4.21) is used.

The bounds require some preliminary notation. Let $\boldsymbol{\eta}_s^+$ be the vector of dimension $n_s - 1 = I_s + J_s$ whose first I_s coordinates equal $\exp(\gamma\psi_1)$ and whose last J_s coordinates equal $\exp(\gamma\psi_2)$. In parallel, let $\boldsymbol{\eta}_s^-$ be the vector of dimension $n_s - 1 = I_s + J_s$ whose first I_s coordinates equal $\exp(\gamma\psi_0 + \gamma\psi_2)$ and whose last J_s coordinates equal $\exp(\gamma\psi_0 + \gamma\psi_1)$.

The mth elementary symmetric function of an n-dimensional argument $\mathbf{a}$, with n positive coordinates, is the sum of all products of m coordinates of $\mathbf{a}$, that is,

$$\mathrm{SYM}_m(\mathbf{a}) = \sum_{\mathbf{b} \in \Psi_m} \prod_j a_j^{b_j},$$

where Ψ_m is the set containing the $\binom{n}{m}$ vectors $\mathbf{b}$ with 1 or 0 coordinates such that $\sum_{j=1}^n b_j = m$. This function is defined for $m = 0, 1, \ldots, n$. For instance, $\mathrm{SYM}_0(\mathbf{a}) = 1$, $\mathrm{SYM}_1(\mathbf{a}) = a_1 + \cdots + a_n$, $\mathrm{SYM}_2(\mathbf{a}) = a_1 a_2 + a_1 a_3 + \cdots + a_{n-1} a_n$, and $\mathrm{SYM}_n(\mathbf{a}) = a_1 a_2 \ldots a_n$. For $m = 1, \ldots, n$, write $g_m(\mathbf{a}) = \mathrm{SYM}_m\{\mathbf{a}\} / \mathrm{SYM}_{m-1}\{\mathbf{a}\}$, and notice that $g_m(\mathbf{a})$ is defined for vectors $\mathbf{a}$ with strictly positive coordinates.

The bounds, $P_s^- \leq p_s \leq P_s^+$, on $p_s = \mathrm{prob}(Z_{s1} = 1|\mathbf{m})$ are $P_s^- = P_s^+ = 1$ if $m_s = n_s$, $P_s^- = P_s^+ = 0$ if $m_s = 0$, and otherwise if $0 < m_s < n_s$,

$$P_s^+ = \frac{\exp(\gamma)}{\exp(\gamma) + g_{m_s}(\boldsymbol{\eta}_s^+)}, \qquad P_s^- = \frac{1}{1 + g_{m_s}(\boldsymbol{\eta}_s^-)}. \qquad (7.3)$$

These expressions are derived in the appendix, §7.4.

To illustrate, consider the example in §7.2.1. Consider one of the 195 matched sets with both a hospital and a patient referent, so $n_s = 3$, $I_s = 1$, $J_s = 1$. In this matched set, the number of women who gave birth at age 25 or above could be $m_s = 0, 1, 2$, or 3. We want bounds, P_s^- and P_s^+, on the probability $p_s = \mathrm{prob}(Z_{s1} = 1|\mathbf{m})$ that the case in this matched set was among the women who gave birth at age 25 or above. Given $m_s = 0$, the case certainly did not give birth at age 25 or above, so in this case $P_s^- = P_s^+ = 0$. Similarly, if all of the women in this matched set gave birth at 25 or above, so $m_s = 3$, then $P_s^- = P_s^+ = 1$. The two cases just considered, $m_s = 0$ or $m_s = n_s$, concern concordant matched sets that contribute only a constant to the test statistic and so do not affect the inference. Now $\boldsymbol{\eta}_s^+ = [\exp(\gamma\psi_1), \exp(\gamma\psi_2)]$ and $\boldsymbol{\eta}_s^- = [\exp\{\gamma(\psi_0 + \psi_2)\}, \exp\{\gamma(\psi_0 + \psi_1)\}]$. If $m_s = 1$, then only one of the three women gave birth at 25 or above, so $\mathrm{SYM}_1(\boldsymbol{\eta}_s^+) = \exp(\gamma\psi_1) + \exp(\gamma\psi_2)$, $\mathrm{SYM}_0(\boldsymbol{\eta}_s^+) = 1$, and

$$P_s^+ = \frac{\exp(\gamma)}{\exp(\gamma) + \exp(\gamma\psi_1) + \exp(\gamma\psi_2)},$$

and similarly

$$P_s^- = \frac{1}{1 + \exp\{\gamma(\psi_0 + \psi_2)\} + \exp\{\gamma(\psi_0 + \psi_1)\}}.$$

If $\gamma = 0$, so there is no hidden bias, then $P_s^- = P_s^+ = 1/3$ in the two expressions just given. If partial comparability did nothing to reduce bias, so $\gamma \neq 0$, $\psi_0 = 1$, and $\psi_1 = \psi_2 = 0$, then $P_s^+ = \exp(\gamma)/\{\exp(\gamma) + 2\}$ and $P_s^- = 1/\{1 + 2\exp(\gamma)\}$ in the expressions just given, and these are the bounds from §4.4.4.

In the same way, if two of the three women gave birth at 25 or above, $m_s = 2$, $\text{SYM}_2(\boldsymbol{\eta}_s^+) = \exp\{\gamma(\psi_1 + \psi_2)\}$, so

$$P_s^+ = \frac{\exp(\gamma)}{\exp(\gamma) + \dfrac{\exp\{\gamma(\psi_1 + \psi_2)\}}{\exp(\gamma\psi_1) + \exp(\gamma\psi_2)}},$$

and similarly

$$P_s^- = \frac{1}{1 + \dfrac{\exp\{\gamma(2\psi_0 + \psi_1 + \psi_2)\}}{\exp\{\gamma(\psi_0 + \psi_2)\} + \exp\{\gamma(\psi_0 + \psi_1)\}}}.$$

If $\gamma = 0$, then $P_s^- = P_s^+ = 2/3$ in the two expressions just given. If $\gamma \neq 0$, $\psi_0 = 1$, and $\psi_1 = \psi_2 = 0$, then $P_s^+ = 2 \cdot \exp(\gamma)/\{2 \cdot \exp(\gamma) + 1\}$ and $P_s^- = 2 \cdot \exp(\gamma)/\{2 \cdot \exp(\gamma) + \exp(2\gamma)\}$, as in §4.4.4.

The pairs in §7.2.1 with a single referent have $n_s = 2$. For these pairs, $P_s^- = P_s^+ = 0$ if $m_s = 0$ and $P_s^- = P_s^+ = 1$ if $m_s = 2$. For $n_s = 2$, $m_s = 1$ with just a hospital referent,

$$P_s^- = \frac{\exp(\gamma\psi_1)}{\exp(\gamma\psi_1) + \exp(\gamma)}, \qquad P_s^+ = \frac{\exp(\gamma)}{\exp(\gamma) + \exp(\gamma\psi_1)}; \qquad (7.4)$$

see Problem 3. In these expressions, notice that $P_s^- = P_s^+ = 1/2$ if $\psi_1 = 1$. In other words, if there is substantial hidden bias in the sense that γ is large, but if all of that bias is in the variable v_{1si} for which cases and hospital referents are comparable, then there is no bias in pairs containing a case and a hospital referent. Similarly, with just a neighbor referent,

$$P_s^- = \frac{\exp(\gamma\psi_2)}{\exp(\gamma\psi_2) + \exp(\gamma)}, \qquad P_s^+ = \frac{\exp(\gamma)}{\exp(\gamma) + \exp(\gamma\psi_2)}.$$

7.2.4 Example: Sensitivity Analysis for Breast Cancer and Age at First Birth

Table 7.6 gives the results of the sensitivity analysis for the study in §7.2.1 which compared cases of breast cancer to hospital and neighbor referents.

TABLE 7.6. Sensitivity Analysis for Significance Levels (P) for Breast Cancer and Age at First Birth.

	Interpretation	Γ	ψ_0	ψ_1	ψ_2	P
A	No hidden bias	1	1	0	0	0.0001
B	No reduction in bias	1.5	1	0	0	0.0197
C	No reduction in bias	2	1	0	0	0.48
D	Hospital referents free of bias	2	0	1	0	0.0007
E	Neighbor referents free of bias	2	0	0	1	0.025
F	Bias partly removed	2	1/3	1/3	1/3	0.0485
G	Hospital referents free of bias	3	0	1	0	0.0113
H	Neighbor referents free of bias	3	0	0	1	0.36

The sensitivity analysis considers a range of assumptions about the hidden biases affecting each of the referent groups. Among other possibilities, the table considers the situations in which both groups are affected by the same biases, the hospital referents are free of bias but the neighbors are not, or neighbors are free of bias but the hospital referents are not.

The table gives the sensitivity parameters $(\Gamma, \psi_0, \psi_1, \psi_2)$, where $\Gamma = \exp(\gamma)$, together with the upper bound on the significance level for the Mantel–Haenszel test comparing the age at first birth for cases and referents. The lower bound on the significance level is less than 0.0001 in each situation and so is not included in the table. The parameter Γ measures the total quantity of hidden bias, so in this regard, situations C through F are similar, and situations G and H are similar. The parameters (ψ_0, ψ_1, ψ_2), where $1 = \psi_0 + \psi_1 + \psi_2$, indicate the degree to which hidden bias is controlled by using referents from the same hospital or neighborhood. With $(\psi_0, \psi_1, \psi_2) = (1, 0, 0)$, the use of hospital and neighbor referents did nothing to make the cases and referents comparable in terms of the unobserved covariate u_{si}—this gives the sensitivity analysis discussed in Chapter 4. With $(\psi_0, \psi_1, \psi_2) = (0, 1, 0)$, the hospital referents are free of hidden bias, because cases and hospital referents have the same value of u_{si}. With $(\psi_0, \psi_1, \psi_2) = (0, 0, 1)$, the neighbor referents are free of hidden bias. With $(\psi_0, \psi_1, \psi_2) = (1/3, 1/3, 1/3)$, the bias is partially reduced in both referent groups.

Of the 409 cases of breast cancer, 209 first gave birth at age 25 or older. In the absence of hidden bias, situation A in Table 7.6, there is a single significance level < 0.0001 from the usual Mantel–Haenszel test, indicating increased risk of breast cancer. This comparison is insensitive to all patterns of hidden bias of magnitude $\Gamma = 1.5$, situation B, but it becomes sensitive for $\Gamma = 2$, situation C. An unobserved covariate associated with twice the odds of an older first birth could explain the apparently higher risk of breast cancer among such women. However, a bias of magnitude $\Gamma = 2$ or $\Gamma = 3$ could not explain the apparently higher risk if the bias affected only the neighbors and not the hospital referents, situations D and G. A bias of magnitude $\Gamma = 2$ could not explain the increased risk if it affected only hospital referents and not neighbors, situation E, but a larger bias $\Gamma = 3$ with the same pattern could explain the increased risk, situation H.

In short, an unobserved covariate that doubled the odds of a late first birth could explain the observed increase in risk of breast cancer if it affected both referent groups equally. A bias of this magnitude affecting only one referent group could not explain the increased risk. In this sense, partial comparability may reduce sensitivity to hidden bias.

7.2.5 Computations Illustrated

Table 7.7 shows some of the computations leading to situation D in Table 7.6; that is, $(\Gamma, \psi_0, \psi_1, \psi_2) = (2, 0, 1, 0)$. This is the situation in which there is hidden bias of magnitude $\Gamma = 2$, but it does not affect hospital referents. Rows 1 to 4 in Table 7.7 describe the $195 = 47 + 67 + 60 + 21$ matched triples, rows 5 to 7 describe the $164 = 58 + 77 + 29$ pairs with only a hospital referent, and rows 8–10 describe the $50 = 17 + 23 + 10$ pairs with only a neighbor referent. Rows 1, 4, 5, 7, 8, and 10 describe concordant matched sets in which either all n_s women gave birth at an age ≤ 24 or all gave birth at an age ≥ 25. While these sets may be included in the computations, they contribute only a constant to the test statistic and therefore do not affect the results. Row 2 describes the 67 matched triples in which two women had first births at ages ≤ 24 and one had a first birth at an age ≥ 25. In 32 of these matched sets, the case of breast cancer was the one woman whose first birth occurred at an age ≥ 25. In these 67 matched sets, depending on the value of the unobserved covariate $\mathbf{u}$, the probability that the case will have age ≥ 25 at first birth could be as low as $P_s^- = 0.25$ or as high as $P_s^+ = 0.40$. These probabilities are based on the formula (7.3) in §7.2.3. Summing from $s = 1$ to $s = 409$ in formula (4.20) gives the deviate used to obtain the significance level in Table 7.6.

Row 6 of Table 7.7 concerns the 77 discordant matched pairs in which there is a hospital referent but no neighbor referent. Note that $P_s^- = P_s^+ = 1/2$ in these pairs, because the hospital referents are assumed free of bias in this calculation.

7.3 *Appendix: Selection and Hidden Bias

This appendix discusses the impact of certain types of nonrandom selection of subjects for comparison. The section serves two purposes. First, it provides the basis for the assertions in §7.1.3 concerning the relationship

TABLE 7.7. Computations Illustrated.

n_s	I_s	J_s	m_s	#Matched Sets	#Sets With Case ≥ 25	P_s^-	P_s^+
3	1	1	0	47	0	0.00	0.00
3	1	1	1	67	32	0.25	0.40
3	1	1	2	60	52	0.60	0.75
3	1	1	3	21	21	1.00	1.00
2	1	0	0	58	0	0.00	0.00
2	1	0	1	77	53	0.50	0.50
2	1	0	2	29	29	1.00	1.00
2	0	1	0	17	0	0.00	0.00
2	0	1	1	23	12	0.33	0.67
2	0	1	2	10	10	1.00	1.00

between selection bias and hidden bias in case-referent studies. Second, many observational studies entail several comparisons with subsets of the subjects selected in various ways, and the appendix considers the relationship among these different comparisons.

Assume that hidden bias is expressed by the model (4.1), (4.2), or (4.6). Also, to test the null hypothesis of no treatment effect, H_0, assume that the treatment does not affect the fixed responses $\mathbf{r}$.

Divide the units in subclass s into two mutually exclusive and exhaustive groups, G_{I_s} and G_{E_s}, based on any fixed feature of the units, so $G_{I_s} \cup G_{E_s} = \{1, \ldots, n_s\}$ and $G_{I_s} \cap G_{E_s} = \emptyset$. The division may be based on the unobserved covariate $\mathbf{u}$ or on an unaffected outcome $\mathbf{y}$, or under H_0, on the unaffected responses $\mathbf{r}$ themselves, or on a combination of all three, possibly with the aid of a table of random numbers. Suppose that the units in G_{I_s} are included in a comparison but the units in G_{E_s} are excluded. Write $\dot{\mathbf{Z}}$ for the coordinates of $\mathbf{Z}$ in $G_{I_s}, s = 1, \ldots, S$, and write $\ddot{\mathbf{Z}}$ for the coordinates in $G_{E_s}, s = 1, \ldots, S$, with a similar notation for other quantities. For instance, write

$$\dot{m}_s = \sum_{i \in G_{I_s}} Z_{si} \quad \text{and} \quad \dot{\mathbf{m}} = \begin{bmatrix} \dot{m}_1 \\ \vdots \\ \dot{m}_s \end{bmatrix},$$

so $\dot{m}_s$ of the $\dot{n}_s$ subjects included in G_{I_s} from stratum s received the treatment. Also, $\dot{\Omega}$ contains all vectors with binary coordinates of dimension $\dot{N} = \dot{n}_1 + \cdots + \dot{n}_s$ with $\dot{m}_s$ ones among the $\dot{n}_s$ coordinates for stratum s.

Starting from (4.1) or (4.2), in parallel with §4.2.3,

$$\mathrm{pr}(\dot{\mathbf{Z}} = \dot{\mathbf{z}}|\dot{\mathbf{m}}) = \frac{\exp(\gamma \dot{\mathbf{z}}^T \dot{\mathbf{u}})}{\sum_{\mathbf{b} \in \dot{\Omega}} \exp(\gamma \dot{\mathbf{b}}^T \dot{\mathbf{u}})}, \qquad \dot{\mathbf{u}} \in [0, 1]^{\dot{N}}. \tag{7.5}$$

The same distribution (7.5) can be obtained beginning with (4.6) instead of beginning with (4.1) or (4.2) by conditioning on $\mathbf{m}$ and $\ddot{\mathbf{Z}}$. Note that $\mathbf{m}$ and $\ddot{\mathbf{Z}}$ determine $\dot{\mathbf{m}}$. Using the definition of conditional probability applied to (4.6) and simplifying using $\mathbf{z}^T\mathbf{u} = \dot{\mathbf{z}}^T\dot{\mathbf{u}} + \ddot{\mathbf{z}}^T\ddot{\mathbf{u}}$ gives

$$\text{pr}(\dot{\mathbf{Z}} = \dot{\mathbf{z}} \mid \dot{\mathbf{m}}, \ddot{\mathbf{Z}}) = \frac{\exp(\gamma\dot{\mathbf{z}}^T\dot{\mathbf{u}})}{\sum_{\dot{\mathbf{b}}\in\dot{\Omega}} \exp(\gamma\dot{\mathbf{b}}^T\dot{\mathbf{u}})}, \qquad \dot{\mathbf{u}} \in [0, 1]^{\dot{N}}. \qquad (7.6)$$

In other words, the distribution in (7.5) and (7.6) arises either by starting with (4.2) and ignoring the excluded units, or by starting with (4.6) and "setting aside" the excluded units by conditioning on their treatment assignments, $\ddot{\mathbf{Z}}$, so they no longer enter the permutation distribution.

The models in (7.5) and (4.6) have the same form and the same value of the parameter γ, but selection of subjects changes $\mathbf{u}$ to $\dot{\mathbf{u}}$. If there is no hidden bias to start with, that is, if $\gamma = 0$, then there is no hidden bias after selection, and the treatments $\dot{\mathbf{Z}}$ are uniformly distributed on $\dot{\Omega}$, so the methods in Chapter 3 may be used. If there is hidden bias to start with, that is, if $\gamma \neq 0$, then after selection the methods in Chapter 4 may be used to appraise sensitivity to bias. However, as selection changes $\mathbf{u}$ to $\dot{\mathbf{u}}$, the quantity and pattern of hidden bias may be different after selection. For instance, if only siblings of cases are used as referents, and if the unobserved covariate describes childhood diet, then $\dot{\mathbf{u}}$ may be more nearly constant within matched sets than $\mathbf{u}$ is.

The situation is entirely different if subjects are selected based on a quantity affected by the treatment, say $\mathbf{Y} = \mathbf{y_Z}$. This may distort the distribution of treatments or exposures $\mathbf{Z}$, so model (7.5) does not hold and the methods of Chapter 4 cannot be used. Even if there is no hidden bias in the sense that $\gamma = 0$, after selection based on $\mathbf{y_Z}$, the distribution of $\dot{\mathbf{Z}}$ will not generally be uniform on $\dot{\Omega}$, so there will be selection bias in testing H_0 though there is no hidden bias.

7.4   *Appendix: Derivation of Bounds for Sensitivity Analysis

This appendix derives the bounds $P_s^- \leq p_s \leq P_s^+$ on $p_s = \text{prob}(Z_{s1} = 1\mid\mathbf{m})$ in §7.2.3 assuming partial comparability expressed by (7.1) and (7.2) and $\gamma \geq 0$. The process is slightly different from that in Chapter 4 because the v_{ksi} are subject to the constraint (7.2). Write $\zeta_{si} = \exp(\gamma u_{si}) = \exp\{\gamma(\psi_0 v_{0si} + \psi_1 v_{1si} + \psi_2 v_{2si})\}$. Recall the notation in Table 7.5. Write

$$\zeta_s = [\underbrace{\zeta_{s2}, \dots, \zeta_{s,I_{s+1}}}_{I_s} \underbrace{\zeta_{s,I_{s+2}}, \dots, \zeta_{s,n_s}}_{J_s}]$$

for the $n_s - 1$ dimensional vector describing the $n_s - 1$ referents in matched set s. To calculate $\text{prob}(Z_{s1} = 1|\mathbf{m})$, take term s in the product in (4.6), and sum over all $\mathbf{z}_s \in \Omega_s$ such that $z_{s1} = 1$, which is a sum over $\binom{n_s-1}{m_s-1}$ terms. Then

$$\text{prob}(Z_{s1} = 1|\mathbf{m}) = \frac{\zeta_{s1} \cdot \text{SYM}_{m-1}(\boldsymbol{\zeta}_s)}{\zeta_{s1} \cdot \text{SYM}_{m-1}(\boldsymbol{\zeta}_s) + \text{SYM}_m(\boldsymbol{\zeta}_s)} = \frac{\zeta_{s1}}{\zeta_{s1} + g_m(\boldsymbol{\zeta}_s)},$$
(7.7)

where $g_m(\mathbf{a}) = \text{SYM}_m\{\mathbf{a}\}/\text{SYM}_{m-1}\{\mathbf{a}\}$ and $g_m(\mathbf{a})$ is defined for vectors a with strictly positive coordinates.

In obtaining bounds on $\text{prob}(Z_{sl} = 1|\mathbf{m})$ in (7.7), the following fact about ratios of symmetric functions is useful. See Mitrinovic (1970, §2.15.3, Theorem 1, p. 102) for a proof of the first part of Lemma 25.

Lemma 25 *The ratio $g_m(\mathbf{a})$ is monotone-increasing in each coordinate of* a. *Hence,* $\text{prob}(Z_{s1} = 1|\mathbf{m})$ *is increasing in ζ_{s1} and decreasing in ζ_{si} for $i \geq 2$.*

Because of the constraints (7.2), the ζ_{si} are linked and cannot be changed arbitrarily. In particular, changing either v_{1s1} or v_{2s1} affects several ζ_{si}. As a result, the monotonicity in Lemma 25 is not sufficient to determine the bounds $P_s^- \leq p_s \leq P_s^+$. As will be seen in the proof of Proposition 27 below, Lemma 26 is needed to determine the extreme values of v_{1s1} or v_{2s1}.

Define $\mathbf{f}(\alpha, \beta)$ as the $n_s - 1$ dimensional vector

$$\mathbf{f}(\alpha, \beta) = [\underbrace{\alpha, \dots, \alpha}_{I_s} \ \underbrace{\beta, \dots, \beta}_{J_s}].$$

Lemma 26 *The function*

$$\frac{v\alpha\beta}{v\alpha\beta + g_m\{\mathbf{f}(\alpha, \beta)\}}$$
(7.8)

is monotone-increasing in α and in β for $\alpha > 0$, $\beta > 0$, for every fixed $v > 0$.

Proof. It will be shown that (7.8) is monotone-increasing in α, the proof for β being similar. Proving that (7.8) is monotone-increasing in α is the same as proving that $g_m\{\mathbf{f}(\alpha, \beta)\}/\alpha$ is monotone-decreasing in α. Using the definition of SYM_m, it follows that if $\theta > 0$, then $g_m(\theta\mathbf{a}) = \theta g_m(\mathbf{a})$. Let $\alpha \leq \alpha^*$ so the task is to show $g_m\{\mathbf{f}(\alpha, \beta)\}/\alpha \geq g_m\{\mathbf{f}(\alpha^*, \beta)\}/\alpha^*$. Let $\theta = \alpha^*/\alpha \geq 1$. Then, as required,

$$\frac{g_m\{\mathbf{f}(\alpha, \beta)\}}{\alpha} = \frac{\theta g_m\{\mathbf{f}(\alpha, \beta)\}}{\theta\alpha} = \frac{g_m\{\mathbf{f}(\theta\alpha, \theta\beta)\}}{\theta\alpha}$$

$$\geq \frac{g_m\{\mathbf{f}(\theta\alpha, \beta)\}}{\theta\alpha} = \frac{g_m\{\mathbf{f}(\alpha^*, \beta)\}}{\alpha^*},$$

where the inequality follows from the monotonicity of $g_m(\mathbf{a})$. ∎

Proposition 27 *Under partial comparability expressed by (7.1) and (7.2),*

$$\frac{1}{1 + g_m[\mathbf{f}\{\exp(\gamma\psi_0 + \gamma\psi_2), \exp(\gamma\psi_0 + \gamma\psi_1)\}]} \leq \mathrm{prob}(Z_{s1} = 1|\mathbf{m})$$

$$\leq \frac{\exp(\gamma)}{\exp(\gamma) + g_m[\mathbf{f}\{\exp(\gamma\psi_1), \exp(\gamma\psi_2)\}]}. \tag{7.9}$$

Proof. Under the constraint (7.2),

$$
\begin{aligned}
\exp\{\gamma(\psi_1 v_{1s1} + \psi_2 v_{2s1})\} &\leq \zeta_{s1} \leq \exp\{\gamma(\psi_0 + \psi_1 v_{1s1} + \psi_2 v_{2s1})\} \\
\exp(\gamma\psi_1 v_{1s1}) &\leq \zeta_{si} \leq \exp\{\gamma(\psi_0 + \psi_1 v_{1s1} + \psi_2)\} \\
&\quad \text{for} \quad i = 2, \ldots, I_s + 1, \tag{7.10} \\
\exp(\gamma\psi_2 v_{2s1}) &\leq \zeta_{si} \leq \exp\{\gamma(\psi_0 + \psi_1 + \psi_2 v_{2s1})\} \\
&\quad \text{for} \quad i = I_s + 2, \ldots, n_s.
\end{aligned}
$$

Using Lemma 25, $\mathrm{prob}(Z_{s1} = 1|\mathbf{m})$ is increased by raising ζ_{s1} to its upper bound and reducing ζ_{si} for $i \geq 2$ to its lower bound in (7.10). If ζ_{s1} is at its upper bound and ζ_{si} for $i \geq 2$ is at its lower bound in (7.10), then

$$\mathrm{prob}(Z_{s1} = 1|\mathbf{m}) = \frac{v\alpha\beta}{v\alpha\beta + g_m\{\mathbf{f}(\alpha, \beta)\}} \tag{7.11}$$

with $v = \exp(\gamma\psi_0)$, $\alpha = \exp(\gamma\psi_1 v_{1s1})$, and $\beta = \exp(\gamma\psi_2 v_{2s1})$; hence, using Lemma 26, (7.11) is maximized when $v_{1s1} = v_{2s1} = 1$, giving the upper bound in (7.9) upon using $1 = \psi_0 + \psi_1 + \psi_2$. If ζ_{s1} is at its lower bound and ζ_{si} for $i \geq 2$ is at its upper bound in (7.10), then (7.11) holds with $v = \exp\{-\gamma(2\psi_0 + \psi_1 + \psi_2)\}$, $\alpha = \exp\{\gamma(\psi_0 + \psi_1 v_{1s1} + \psi_2)\}$, and $\beta = \exp\{\gamma(\psi_0 + \psi_1 + \psi_2 v_{2s1})\}$, so the minimum value of (7.11) occurs with $v_{1s1} = v_{2s1} = 0$. ∎

7.5   Bibliographic Notes

There are two books about case-referent studies by Breslow and Day (1980) and Schlesselman (1982) and, in addition, most epidemiology texts discuss case-referent studies and selection bias in detail; see Kelsey, Thompson, and Evans (1986), Kleinbaum, Kupper, and Morgenstern (1982), Lilienfeld and Lilienfeld (1980), MacMahon and Pugh (1970), Miettinen (1985a), Rothman (1986), Rothman and Greenland (1998), and also the articles by Cornfield (1951), Holland and Rubin (1988), Mantel (1973), Mantel and Haenszel (1959), and Prentice and Breslow (1978). Gastwirth (1988) discusses the studies of Reye's syndrome. The use of more than one reference group is briefly discussed in most texts and is discussed in greater detail by Cole (1979), Fairweather (1987), Kelsey, Thompson, and Evans (1986, pp. 160–163), Liang and Stewart (1987), and Rosenbaum (1987). The technical material in §7.2 through §7.4 is largely based on Rosenbaum (1991).

7.6 Problems

1. A subject's knowledge of the investigator's hypothesis. Weiss (1994) discusses whether it is appropriate to exclude from a case referent study those subjects who were aware of the study's hypothesis. He writes:

   > It is likely that for many persons who develop an illness, knowledge of a hypothesis concerning its etiology is obtained *after* the diagnosis .... In some instances, knowledge of the hypothesis could occur more commonly among exposed than nonexposed cases: inquiries by medical personnel into the history of exposure to a possible etiologic factor may be of relatively greater salience to exposed persons ....
   >
   > [... excluding knowledgeable subjects] will create bias if the acquisition of knowledge of the etiologic hypothesis by cases is related to their exposure status.

   Express Weiss's correct argument in the terminology of §7.1.3, that is, in terms of an affected outcome $\mathbf{Y} = \mathbf{y_Z}$ used in selecting subjects.

2. Bad tequila in Acapulco. In 1985, the *Journal of Chronic Diseases* published an exchange led by Olli Miettinen (1985b) with dissent and discussion by James Schlesselman, Alvan Feinstein, and Olav Axelson. Miettinen writes:

   > As an illustration, consider the hypothetical example of testing the hypothesis that a cause of traveller's diarrhea is the consumption of tequila (a Mexican drink), with cases derived from a hospital in Acapulco, Mexico, over a defined period of time. What might be the proper [referent] group?

   Answer Miettinen's question. What is the treatment? What is the outcome? What is $\pi_{[j]}$? Consider the binary variable indicating whether a subject was in Acapulco during the defined period of time. Is this binary variable a covariate? How would $\pi_{[j]}$ vary with this binary variable? What would be the consequence of matching on this variable? Would this automatically preclude certain referent groups? Read the published discussion and compare your answer to the answers given there.

3. A derivation. Obtain (7.4) from (7.3) using the observation that $\psi_1 = 1 - \psi_0 - \psi_2$ from (7.1).

7.7   References

Breslow, N. and Day, N. (1980) *The Analysis of Case-Control Studies.* Volume 1 of *Statistical Methods in Cancer Research.* Lyon, France: International Agency for Research on Cancer of the World Health Organization.

Cole, P. (1979) The evolving case-control study. *Journal of Chronic Diseases,* **32**, 15–27.

Collaborative Group for the Study of Stroke in Young Women (1973) Oral contraception and increased risk of cerebral ischemia or thrombosis. *New England Journal of Medicine,* **288**, 871–878.

Cornfield, J. (1951) A method of estimating comparative rates from clinical data: Applications to cancer of the lung, breast and cervix. *Journal of the National Cancer Institute,* **11**, 1269–1275.

Fairweather, W. (1987) Comparing proportion exposed in case-control studies using several control groups. *American Journal of Epidemiology,* **126**, 170–178.

Gastwirth, J. (1988) *Statistical Reasoning in Law and Public Policy.* New York: Academic.

Gutensohn, N., Li, F., Johnson, R., and Cole, P. (1975) Hodgkin's disease, tonsillectomy and family size. *New England Journal of Medicine,* **292**, 22–25.

Halsey, N., Modlin, J., Jabbour, J., Dubey, L., Eddins, D., and Ludwig, D. (1980) Risk factors in subacute sclerosing panencephalitis: A case-control study. *American Journal of Epidemiology,* **111**, 415–424.

Herbst, A., Ulfelder, H., and Poskanzer, D. (1971) Adenocarcinoma of the vagina: Association of maternal stilbestrol therapy with tumor appearance in young women. *New England Journal of Medicine,* **284**, 878–881.

Holland, P. and Rubin, D. (1988) Causal inference in retrospective studies. *Evaluation Review,* **12**, 203–231.

Hurwitz, E. (1989) Reye's syndrome. *Epidemiologic Reviews,* **11**, 249–253.

Hurwitz, E. S., Barrett, M. J., Bregman, D., Gunn, W. J., Pinsky, P., Schonberger, L. B., Drage, J. S., Kaslow, R. A., Burlington, D. B., and Quirman, G. V. (1985) Public health service study on Reye's syndrome and medications. *New England Journal of Medicine,* **313**, 14, 849–857.

Kelsey, J., Thompson, W., and Evans, A. (1986) *Methods in Observational Epidemiology.* New York: Oxford University Press.

Kleinbaum, D., Kupper, L., and Morgenstern, H. (1982) *Epidemiologic Research.* Belmont, CA: Wadsworth.

Kreiger, N., Kelsey, J., Holford, T., and O'Connor, T. (1982) An epidemiologic study of hip fracture in postmenopausal women. *American Journal of Epidemiology,* **116**, 141–148.

Liang, K. and Stewart, W. (1987) Polychotomous logistic regression methods for matched case-control studies with multiple case or control groups. *American Journal of Epidemiology,* **125**, 720–730.

Lilienfeld, A., Chang, L., Thomas, D., and Levin, M. (1976) Rauwolfia derivatives and breast cancer. *Johns Hopkins Medical Journal,* **139**, 41–50.

Lilienfeld, A. and Lilienfeld, D. (1980) *Foundations of Epidemiology* (second edition). New York: Oxford University Press.

MacMahon, B. and Pugh, T. (1970) *Epidemiology: Principles and Methods.* Boston: Little, Brown.

Mantel, N. (1973) Synthetic retrospective studies and related topics. *Biometrics,* **29**, 479–486.

Mantel, N. and Haenszel, W. (1959) Statistical aspects of retrospective studies of disease. *Journal of the National Cancer Institute,* **22**, 719–748.

Miettinen, O. (1985a) *Theoretical Epidemiology.* New York: Wiley.

Miettinen, O. (1985b) The "case-control" study: Valid selection of subjects (with discussion) *Journal of Chronic Diseases,* **38**, 543–548.

Mitrinovic, D. S. (1970) *Analytic Inequalities.* Berlin: Springer-Verlag.

Prentice, R. and Breslow, N. (1978) Retrospective studies and failure time models. *Biometrika,* **65**, 153–158.

Rosenbaum, P. R. (1987) The role of a second control group in an observational study (with Discussion). *Statistical Science,* **2**, 292–316.

Rosenbaum, P. R. (1991) Sensitivity analysis for matched case-control studies. *Biometrics,* **47**, 87–100.

Rothman, K. (1986) *Modern Epidemiology.* Boston: Little, Brown.

Rothman, K. and Greenland, S. (1998) *Modern Epidemiology.* Philadelphia: Lippincott-Raven.

Schlesselman, J. (1982) *Case-Control Studies.* New York: Oxford University Press.

Weiss, N. (1994) Should we consider a subject's knowledge of the etiologic hypothesis in the analysis of case-control studies? *American Journal of Epidemiology,* **139**, 247–249.

8
Multiple Control Groups

8.1 The Role of a Second Control Group

8.1.1 Observational Studies with More than One Control Group

An observational study has multiple control groups if it has several distinct groups of subjects who did not receive the treatment. In a randomized experiment, every control is denied the treatment for the same reason, namely, the toss of a coin. In an observational study, there may be several distinct ways that the treatment is denied to a subject. If these several control groups have outcomes that differ substantially and significantly, then this cannot reflect an effect of the treatment, since no control subject received the treatment. It must reflect, instead, some form of bias.

Multiple control groups arise in several ways. In some contexts, a subject may become a control either because the treatment was not offered or because, though offered, it was declined. Some years ago, the Educational Testing Service sought to evaluate the impact of the College Board's Advanced Placement (AP) Program which offers high-school students the opportunity to earn college credit for advanced courses taken in high school. In such a study, a student may become a control in either of two ways: the student may decline to participate in the AP Program, or the student may attend a school that does not offer the AP Program. What hidden biases might be present with these two sources of controls? Schools offering the AP Program may have greater resources or have more college-bound students than other schools, and these characteristics are often found in school districts with higher family incomes. Possibly, students in schools offering the AP Program would have better outcomes even if the AP Program itself had no effect. On the other hand, if the AP Program is available in a school, students who decline to participate may be less motivated for academic work than those who do participate. Here, too, if the AP Program had no effect, greater motivation among AP participants might possibly lead to better outcomes. The use of both control groups provides little protection in this study because the most plausible hidden biases lead to patterns of responses that resemble treatment effects. In the end, a single control group was used, namely, matched students in the same school who declined to participate.

A related but different situation arises with compensatory programs, that is, programs such as Head Start that are intended to assist disadvantaged students. Campbell and Boruch (1975) discuss biases that may affect studies of compensatory programs. As with the AP Program, if a compensatory program is offered, students who decline to participate might be less motivated. In contrast with the AP Program, admission to a compensatory program often depends on need, the program being offered to students with the greatest need. After matching or adjustment for observed covariates, such as test scores and grades prior to admission into the program, if the compensatory program had no effect, one might expect that program participants, being better motivated, would outperform students who decline to participate, but would not perform as well as students who were not offered the program because of less need.

Generally, the goal is to select control groups so that, in the observed data, an actual treatment effect would have an appearance that differed from the most plausible hidden bias. Arguably, this is true in the case of the compensatory program but not in the case of the AP Program.

Multiple control groups also arise when controls are selected from several existing groups of subjects who did not receive the treatment. Seltser and Sartwell (1965) studied the possibility that X-ray exposure was causing deaths among radiologists. They compared members of the Radiological Society of North America, a professional association of radiologists, to four other specialty societies. In two of these four societies, the American Academy of Ophthalmology and Otolaryngology and the American Association of Pathologists and Bacteriologists, most society members would normally use X-rays vastly less often than radiologists do. For the period 1945 to 1954, Seltser and Sartwell (1965, Table 7) calculated age-adjusted mortality rates per 1000 person-years of 16.4 for the radiologists, 12.5 for the pathologists and bacteriologists, and 11.9 for the ophthalmologists and otolaryngologists. In other words, in these two control groups the mortality rates are similar, as they should be in the absence of hidden bias. Both groups had substantially lower mortality than among the radiologists, consistent with a possible effect of radiation exposure.

TABLE 8.1. Road Workers Compared to Two Control Groups, Office and Quarry Workers.

Renal Function	Road	Office	Quarry
Abnormal	24	4	2
Normal	68	39	36
Total	92	43	38

Multiple control groups are sometimes formed from several groups that received different treatments that were believed to be the same in ways that matter for the outcome under study. When formed in this way, differences among the control groups may reflect either hidden bias or unanticipated differences in the effects of the control treatments. The problems in §8.5 discuss the possibility that contraception using the intrauterine device causes ectopic pregnancy. There are five comparison groups defined by other forms of contraception.

If subjects who receive negligible doses of a treatment are distinguished from those who actively avoid or abstain from the treatment, then two control groups are produced. In §8.1.4, an example is discussed that distinguishes abstention from alcohol and extremely low consumption of alcohol.

As noted in Chapter 7, a case-referent study may have one or more referent groups and one or more control groups. A referent is not a case. A control did not receive the treatment. A case-referent study of alcohol consumption might have two referent groups, say neighbors and siblings of cases, and two control groups, say abstainers and those who are not abstainers but who report virtually no alcohol consumption. Such a design would offer several opportunities to look for selection and hidden biases.

8.1.2 An Example: Occupational Exposure to Hydrocarbons and Kidney Disease

To investigate whether occupational exposures to hydrocarbons cause renal disorders, Douglas and Carney (1998) compared 92 road workers exposed to asphalt or bitumen fumes to two control groups, namely, "38 hard rock quarry workers not occupationally exposed to hydrocarbons, and 43 office workers also not exposed to hydrocarbons." They identified renal function abnormalities using blood and urine tests, with results given in Table 8.1. In fact, 3 of the 24 road workers with abnormal kidney function had multiple abnormalities, whereas none of the controls had multiple abnormalities.

Using Fisher's exact test for a 2×2 table without hidden bias, the slightly higher rate of renal abnormalities among office workers when compared to quarry workers is not significant in either a one-sided ($P = .40$) or two-sided ($P = 2 \times .40 = .80$) test, so the control groups do not differ significantly. However, the rate of renal abnormalities is significantly higher among road workers than among office workers ($P = 0.018$) and higher than among

TABLE 8.2. Cytogenetic Damage in Pb-Zn Miners and Two Control Groups.

	Miners	Housewives	Distant
SCE ≥ 9	22	0	0
SCE < 9	95	57	61
Total	117	57	61

quarry workers $(P = 0.004)$ in one-sided tests, and of course it is also higher than the combined control group $(P = .0009)$.

The comparison of roadworkers with the combined control group is insensitive to a bias of $\Gamma = 1.8$ with upper bound on the significance level of $P = .044$, but becomes sensitive at $\Gamma = 2$ with upper bound $P = .072$. These exact calculations use the extended hypergeometric distribution, as in §4.4.1, Corollary 15.

8.1.3 An Example: Cytogenetic Damage in Miners

Miners are often exposed to a variety of potentially hazardous substances, including radon gas. Marjan Bilban (1998), a researcher in Slovenia, compared cytogenetic damage of male miners at a lead and zinc mine to two control groups. Because virtually all males in the vicinity of the mine had worked with lead or zinc in some capacity, one control group consisted of housewives from the vicinity of the mine who had no occupational exposure to the mine or its products. The other control group, 86% male, came from the general population farther from the mine; it is labeled "distant controls." So the first control group lived in the same vicinity as the miners, but differed in gender, while the second control group was more similar in terms of gender, but lived elsewhere.

Several measures of cytogenetic damage were obtained from blood lymphocytes, including the average number of sister chromatid exchanges (SCE). The frequencies of SCE of 9 or more are given in Table 8.2. Notice that SCE ≥ 9 was not uncommon among miners, but was never seen in either control group. Fisher's exact test for a 2×2 table will be used to compare the miners to each control group and to the combined control group with 118 subjects. The sensitivity analysis is given Table 8.3, which reports upper bounds on the one-sided significance levels from Fisher's exact test, obtained using the extended hypergeometric distribution, as discussed in §4.4.1, Corollary 15. Comparing the miners to the combined group, the upper bounds on the significance levels for $\Gamma = 6$ and 7 are, respectively, 0.045 and 0.068.

Neither control group is ideal. Although one would not anticipate a strong relationship between gender and SCE, some differences are certainly possible. Environmental hazards vary from place to place, and the distant controls may have been exposed to a different pattern of environmental hazards besides lead and zinc from the mine. Nonetheless, similar results

TABLE 8.3. Sensitivity Analysis for Significance Levels Comparing Pb-Zn Miners to Controls.

Γ	Housewives	Distant	Combined
1	.00059	.00039	.0000025
2	.014	.011	.00041
3	.051	.042	.0036

TABLE 8.4. Alcohol and Mortality.

	Group	Dead	Alive
Control groups {	Abstain	7	6
	None	27	71
Treated groups {	Lowest	14	38
	Middle	13	20
	Highest	21	20

were obtained for local housewives and distant controls, and the comparison is insensitive to small and moderate biases. To explain the higher level of SCE among miners as a bias, one would need to postulate substantial biases affecting both control groups in a similar manner.

8.1.4   An Example: Are Small Doses of Alcohol Beneficial?

Petersson, Trell, and Kristenson (1982) reviewed six studies suggesting that moderate consumption of alcohol may have beneficial effects in preventing some cardiac disease, and then they conducted a similar study with the additional feature that there were two control or "no alcohol" groups. Alcohol use was graded based on the response to ten questions. Nine of the questions described situations and asked about drinking in relation to those situations. One question asked about alcohol abstention. One control group consisted of men who described themselves as abstaining from all consumption of alcohol. The other control group consisted of men who did not say they abstained from alcohol, but who answered "no" to all nine questions about drinking behavior.

Between 1974 and 1979, the study obtained questionnaire responses from 7725 male residents of Malmö, Sweden. By 1980, there had been 127 deaths among these 7725 men. Each death was matched with two referents who were alive in 1980, the referents being matched to the death based on age and date of entry into the study. In principle, the analysis should make use of the matching, but this is not possible from the published data, so for this example, the matching is ignored. This study is, almost, a synthetic case-referent study in the sense of §3.3.2; see also §7.1.3. Petersson, Trell, and Kristenson (1982)'s data are given in Table 8.4. (In this table, their 0

TABLE 8.5. Odds Ratios for Alcohol and Mortality.

	Abstain	None	Low	Middle
None	0.33			
Low	0.32	0.97		
Middle	0.56	1.71	1.76	
High	0.90	2.76	2.85	1.62

TABLE 8.6. Chi-Squares for Alcohol and Mortality.

	Abstain	None	Low	Middle
None	2.6			
Low	2.3	0.0		
Middle	0.31	1.1	0.9	
High	0.02	6.1	4.8	0.6

is called "None," their 1 is called "Lowest," their 2 is called "Middle" and their ≥ 3 is called "Highest.")

Table 8.5 contains odds ratios comparing the groups in Table 8.4. In this table, the odds for the row are divided by the odds for the column, so for None/Abstain the odds ratio is $(27/71)/(7/6) = 0.33$. Table 8.6 gives the chi-square statistics for a 2×2 table, with the continuity correction; they have one degree of freedom and are significant at 0.05 or 0.025 if greater than 3.84 or 5.02, respectively. For instance, 2.60 is the chi-square for the 2×2 table Abstain/None $\times$ Dead/Alive.

When High alcohol consumption is compared to None or Low, the odds ratios are above two and significant at 0.05, consistent with greater mortality among those who consume larger quantities of alcohol. However, the High group does not differ significantly from the Abstain group, and the odds ratio is actually slightly less than one. In fact, the odds ratios suggest the greatest mortality is found among abstainers, though the differences are not quite statistically significant. In particular, comparing the two control groups, the odds of death are slightly more than three times greater in the Abstain group than in the None group, though again this difference is not significant at 0.05. Notice that there are only 13 men in the Abstain group.

Of the abstainers, Petersson, Trell, and Kristenson write: "Most of these men, however, had chronic disease as the reason for their abstention, or even a past history of alcoholism. Increased mortality in nondrinkers may create a false impression of a preventive effect of any versus no daily drinking in relation to general and cardiovascular health." In other words, though earlier studies had claimed a beneficial effect of moderate alcohol consumption, Petersson, Trell, and Kristenson are raising doubts about whether that is an actual effect of alcohol, because moderate consumption appears better than abstention but no better and perhaps worse than negligible doses.

8.1.5 An Example: Availability and Refusal of Treatment; Addressing Self Selection

A treatment may be available to some people and not to others, and may be refused by some people when available. For instance, many treatments are available in some places or organizations—cities, states, corporations, health plans—and not in other, ostensibly similar, places or organizations. In many contexts, people have the right to refuse treatment, and they often do. Several problems are evident. First, different places or organizations may actually be different in ways we neither anticipate nor measure, and this may create a hidden bias. Second, individuals who refuse treatment may differ from those who accept it, and this may create a hidden bias. Moreover, because one does not refuse a treatment that is not offered, the people from the place where the treatment is not available do not divide themselves into those who accept and those who reject treatment; the self selection operates in one place, not in the other.

If people were randomly assigned to places or organizations, the situation would resemble the randomized design proposed by Zelen (1979). In that design, patients are divided at random into two groups, and the experimental treatment is offered to patients in the first group, some of whom refuse it. Zelen (1979) proposed comparing the two random groups, keeping the randomization intact by looking not at the effect of the treatment, but rather at the effect of the offer of the treatment. An alternative analysis that also keeps the randomization intact is the instrumental variable analysis in §5.4. If one believed the assignment to place or organization was free of hidden bias, and if one believed the exclusion restriction, then similar analyses, but with adjustments for observed covariates, might be performed in an observational study.

In observational studies, there is random assignment neither in the assignment to places or organizations nor in the decision to accept or refuse treatment. If the only reason outcomes differ in the three groups is the actual effect of the treatment, then the treated group is expected to differ from the two comparison groups which are expected not to differ from each other. If the only reason the three groups differ is a bias introduced by the decision to accept or refuse treatment, then the people from the second place should "fall between"—that is, be a statistical mixture of—the treated subjects and those who refused treatment. If the only reason the three groups differ is that people from the second place are not comparable to the others, then one expects them to differ from the other two groups that are expected not to differ from each other. Obviously, the situation could be more complicated, with groups differing for more than one rea-

TABLE 8.7. Cognitive Therapy for Early Psychosis.

	COPE	Refused	Unavailable
Before Treatment SANS	17.3	17.7	20.8
After Treatment SANS	12.9	16.6	23.1
Before Treatment BDI	8.5	5.2	5.4
After Treatment BDI	7.5	2.7	4.2

son. Because these three simple explanations predict different patterns of outcomes, one analysis—perhaps one of several—will compare the groups to see how they differ.

An example occurs in a study by Jackson, McGorry, Edwards, et al. (1998) of cognitive therapy (COPE) given to 40 patients for early psychosis. Their program was available only in Victoria, Australia, so one comparison group (Unavailable) consisted of 14 ostensibly similar patients from other areas. The other comparison group (Refused) consisted of 20 patients who refused the program of cognitive therapy. For two outcomes, the Scale for the Assessment of Negative Symptoms (SANS) and the Beck Depression Inventory (BDI), before and after treatment, the means for their data are displayed in Table 8.7. For both outcomes, lower scores signify less severe symptoms. Before treatment, the differences among the groups were not significant at the 0.05 level. After treatment, the COPE group was significantly better on SANS than the Unavailable group. After treatment, the COPE group was significantly worse on the BDI than the Refused group. The published data do not permit a more detailed analysis, and the sample sizes are small, so the means are unstable and the absence of significance has limited meaning. In particular, one is hesitant to make much of the interesting pattern in which the Refused group had better final outcomes than the Unavailable group. In larger samples, the design used by Jackson et al. (1998) would have provided more insight into possible biases than a study with either comparison group alone.

8.1.6   An Example: What Does It Mean to Not Receive Treatment?

Zabin, Hirsch, and Emerson (1989) studied the effects of having an abortion on the education, psychological status, and subsequent pregnancies of black teenage women in Baltimore. So the treatment is having an abortion, but what is the control? A woman will have an abortion only if she is pregnant and does not want the child; without the abortion but with good health, she will have a child. So one might compare a teen having an abortion to a teen having a child. Of course, one teen has a child and the other does not. Is this the correct comparison? Perhaps abortions do great psychological damage, but not as much damage as being the teenage mother of an unwanted child. Moreover, it is sometimes asserted, correctly or otherwise, that if abortions were illegal, young women would be more careful about avoiding unwanted pregnancies. If one believed both that making abortions illegal prevented abortions, and also that removing abortion as a legal option prevented most unwanted pregnancies, then one might think it is more appropriate to compare a teen who had an abortion to a teen who was not pregnant. Perhaps there is a little merit in each of these two arguments. From a scientific perspective, however, the preference is clear: A scientist would like to know about all three teens, the one who had an abortion, the one who had a child, and the one who was not pregnant.

Notice that powerful self selection effects are at work, and hidden biases are likely. The decision by a young pregnant woman to have an abortion or have a child may reflect many characteristics that are incompletely measured, such as the degree to which she is determined to complete high school. Sexually active teenagers who do not use adequate birth control may tend to differ in many ways from those who do not become pregnant.

Zabin, Hirsch, and Emerson (1989) derived two control groups from 360 young black women who came for pregnancy tests at two clinics in Baltimore. The "treated" group consisted of women who were pregnant and had an abortion. Women who were pregnant and had a child formed the first control group. Women whose pregnancy test revealed they were not pregnant formed the second control group. Because the second control group consisted of women who suspected they might be pregnant, it is likely to be less biased than a control group of women who simply were not pregnant. The second control group did not have to decide whether to have an abortion or a child, so it is not subject to the self selection effect that formed the treated group and the first control group. The purposeful decision to have an abortion or a child creates a self selection effect. If that selection effect were the only force producing differences among the three groups, then the negative pregnancy test group, being a mixture unaffected by the selection, should be found "between" the treated group and the first control group. If the only force producing differences among the groups were the psychological effects of having an abortion, then the two control groups should have similar outcomes. If the only force producing differences among the groups is the presence of the newborn child, then the abortion and negative pregnancy test groups should have similar outcomes. Three very simple theories should produce three recognizably different outcomes. Obviously, several forces may be at work simultaneously to produce other patterns of outcomes.

Zabin, Hirsch, and Emerson studied many outcomes with varied results. As one illustration, consider "negative educational change" two years after the pregnancy test. In the abortion group, 17.8% had a negative educational change, whereas in the child-bearing group it was 37.3% and in the negative pregnancy test group it was 37.4%. Despite the ambiguities in this study, it would be hard to argue that this pattern of results supports a claim that abortions cause negative educational change at two years.

The two control groups together provide more information about hidden biases than either group provides alone, in part because the control groups systematically vary certain important unobserved differences.

8.1.7   An Example: Using a Second Control Group as Partial Replication of an Unanticipated Finding

A second control group is sometimes the basis for internal replication of part of a study. One might do this, for example, if treated subjects are hard to find, but potential controls are plentiful and inexpensive. A second control group, not used in the initial stages of data analysis, provides a few of the many benefits of a replication, including an increase in sample size, and some data that are unaffected by what was learned from initial analyses using the first control group.

For instance, Kim, McConnell, and Greenwood (1977) used a second control group in this way. They were interested in the effects of a particular financial maneuver—the formation of a captive finance subsidiary—on the value of a company's bonds. This maneuver "... creates a new class of security holders with claims that are superior to those of the old bondholders," (p. 797), so it might be expected to depress the value of the bonds. Kim, McConnell, and Greenwood confined attention to firms with long-term debt whose stock traded on the New York Stock Exchange (NYSE). Twenty-four such firms formed a captive finance subsidiary between 1940 and 1971, so treated firms are in extremely limited supply. Each of these firms might have one or several actively traded bonds. Each bond of each treated firm was matched to a control bond of another NYSE firm on the basis of Moody's bond rating, term-to-maturity, coupon interest rate, and coupon interest payment dates. Many control bonds were available. They found a substantial decline in value of the bonds of treated firms when compared to control bonds, a decline that began about seven months before the formation of the subsidiary and that ended about two months after, because (pp. 804-805): "Apparently, information about the impending formation of the finance subsidiary began to reach the market (and stockholders and bondholders began to react to the information) about seven months prior to the actual incorporation."

One might worry, of course, that while a decline in bond values was anticipated, the specific timing of the decline was not. Many different patterns of changes in bond values might, after the fact, be explained as consistent with the theory that the formation of the subsidiary caused a substantial decline in bond values. For instance, if the decline had taken place entirely after the formation of the subsidiary, then this too would have seemed consistent with the theory. A theory that is consistent with many different patterns of data is only weakly corroborated when one of those patterns is seen; however, if that one pattern reappeared upon replication, then the corroboration would be much stronger. A replication with entirely new data was not feasible since no additional treated firms were available. Unable to replicate the treated group, but in an effort to address the issue as well as the available data would allow, Kim, McConnell, and Greenwood (1977) formed a second control group of matched bonds, finding that "the two control groups yielded approximately equal monthly returns over the period ... ."

In a sense, the second control group addresses half the problem. To be fully convincing, one would want to independently replicate a pattern first suggested by the comparison of the treated group and the first control group. The second control group provides the opportunity to independently replicate half of the pattern, but provides no replication of the pattern in the treated group. In short, when treated units are extremely scarce, a second control group may provide an incomplete replication of patterns suggested by the data.

8.2 Selecting Control Groups: Systematic Variation and Bracketing

The goal in selecting control groups is to distinguish treatment effects from the most plausible systematic biases. If this is to be done successfully, the pattern of outcomes anticipated if the treatment has an effect must differ from the pattern anticipated from a hidden bias. The principles of "systematic variation" and "bracketing" are intended to ensure that this is so.

In a thoughtful discussion of control groups in behavioral research, Campbell (1969) quotes Bitterman's (1965) discussion of control by systematic variation. Bitterman's work is in a challenging field, comparative psychology, which seeks to "study the role of the brain in learning [by comparing] the learning of animals with different brains" [Bitterman (1965, p. 396)]. Bitterman (1965, pp. 399–400) writes:

> Another possibility to be considered is that the difference between the fish and rat which is reflected in these curves is not a difference in learning at all, but a difference in some confounded variable—sensory, motor, or motivational. Who can say, for example, whether the sensory and motor demands made upon the two animals in these experiments were exactly the same? Who can say whether the fish were just as hungry as the rats? ...

> I do not, of course, know how to arrange a set of conditions for the fish which will make sensory and motor demands exactly equal to those which are made upon the rat in some given ex-

perimental situation. Nor do I know how to equate drive level or reward value in the two animals. Fortunately, however, meaningful comparisons are still possible, because for *control by equation* we may substitute what I call control by *systematic variation*. Consider, for example, the hypothesis that the difference between the curves ... is due to a difference, not in learning, but in degree of hunger. The hypothesis implies that there is a level of hunger at which the fish *will* show progressive improvement, and, put in this way, the hypothesis becomes easy to test. If, despite the widest possible variation in hunger, progressive improvement fails to appear in the fish, we may reject the hunger hypothesis.

In the terminology of this book, Bitterman's "control by equation" is analogous to controlling confounding due to an observed covariate x by matching subjects with the same or similar values of x as in Chapters 3 and 10; then differences in outcomes cannot be due to differences in x. "Control by systematic variation" concerns a variable u which is not recorded. If, however, two control groups can be formed so that u is much higher in one group than in the other, and if the groups do not differ materially in their outcomes, then this is consistent with the claim that differences in u are not responsible for differences in outcomes observed between treated and control groups. The principle of systematic variation leads to the following advice. When designing an observational study, identify the most plausible hidden biases and find control groups that sharply vary their levels.

For instance, if one were concerned that physical exercise might be related to kidney abnormalities, then in the study by Douglas and Carney (1998), the use of both quarry and office workers as controls is likely to systematically vary the amount of physical exercise that occurs on the job. Presumably, quarry workers are closer to the treated group, the road workers, in that both jobs entail physical exercise. Since the road workers had more kidney abnormalities than both control groups, with quarry workers having the fewest abnormalities, differences in exercise on the job are not a plausible explanation of the pattern of kidney abnormalities.

A step beyond systematic variation of u is "bracketing," as discussed by Campbell (1969). Bracketing seeks two control groups such that, in the first group, u tends to be higher than in the treated group and, in the second, u tends to be lower than in the treated group. The goal is two control groups that are farther apart from each other in terms of u than they are from the treated group. Possibly this is true of the example in §7.1.1 of a compensatory educational program. In the absence of a program effect, matched controls who decline to participate might be expected to underperform the program group, while matched controls who are not eligible because of less need might be expected to outperform the program group. On the other hand, with a large positive treatment effect and no hidden bias, the program group would tend to outperform both matched control groups. Bracketing yields a study design in which treatment effects and plausible biases are likely to have different appearances in observable data.

8.3 Comparing Outcomes in Two Control Groups

8.3.1 A Model for Assignment to One of Several Groups

In earlier chapters, the notation and model described two groups, a treated and a control group. This section discusses a model for assignment to one of two or more groups. As will be seen, it is closely connected both to the model discussed in earlier chapters and to randomized experiments with two or more groups. Specifically, the model generalizes (4.2) to permit more than two groups. Also, when there is no hidden bias, the model includes the distribution of treatment assignments in completely randomized experiments and randomized block experiments.

As always, there are n_s subjects in stratum s defined by the observed covariate $\mathbf{x}_s$, with $N = n_1 + \cdots + n_s$, and the ith of these n_s subjects has an unobserved covariate u_{si}, with $0 \leq u_{si} \leq 1$. If each subject falls in one of K groups, $K \geq 2$, rather than a single treated group or a single control group, write $G_{sik} = 1$ if the ith subject in stratum s is in group k, and write $G_{sik} = 0$ otherwise, with $1 = \sum_{k=1}^{K} G_{sik}$ for each (s, i). With fixed covariates $\mathbf{x}_s$ and u_{si}, consider the following multinomial logit model (Cox, 1970, §7.5) for $(G_{si1}, \ldots, G_{siK})$:

$$\text{prob}(G_{sik} = 1) = \frac{\exp\{\xi_k(\mathbf{x}_s) + \delta_k u_{si}\}}{\sum_{j=1}^{K} \exp\{\xi_j(\mathbf{x}_s) + \delta_j u_{si}\}}, \tag{8.1}$$

where the N vectors $(G_{si1}, \ldots, G_{siK})$ are mutually independent.

Suppose that we wish to compare two specific groups, j and k, $j \neq k$, and so select all subjects with $(G_{sij} + G_{sik}) = 1$, that is, all subjects who belong to one of these two groups. Then

$$\text{prob}(G_{sij} = 1 | G_{sij} + G_{sik} = 1)$$

$$= \frac{\exp\{\xi_j(\mathbf{x}_s) + \delta_j u_{si}\}}{\exp\{\xi_j(\mathbf{x}_s) + \delta_j u_{si}\} + \exp\{\xi_k(\mathbf{x}_s) + \delta_k u_{si}\}}$$

$$= \frac{\exp\{\kappa(\mathbf{x}_s) + \gamma u_{si}\}}{1 + \exp\{\kappa(\mathbf{x}_s) + \gamma u_{si}\}}, \tag{8.2}$$

where

$$\kappa(\mathbf{x}_s) = \xi_j(\mathbf{x}_s) - \xi_k(\mathbf{x}_s), \qquad \gamma = \delta_j - \delta_k, \qquad \text{and} \quad 0 \leq u_{si} \leq 1,$$

which is identical in form to (4.2). In other words, if the model (8.1) describes assignment to all K groups, then when comparing any two fixed groups, say j and k, the model and methods of earlier chapters may be used directly, recognizing only that the $\kappa(\mathbf{x}_s)$ and the γ appropriate for comparing two groups j and k will be different from the $\kappa(\mathbf{x}_s)$ and the γ for comparing two other groups j' and k'. In particular, if there are only $K = 2$ groups, then models (8.1) and (4.2) are essentially the same.

Write $m_{sk} = \sum_{i=1}^{n_s} G_{sik}$ for the number of subjects assigned to group k in stratum s. Also write $\boldsymbol{\delta} = (\delta_1, \dots, \delta_K)^T$, so there is no hidden bias if $\boldsymbol{\delta} = \mathbf{0}$.

8.3.2   *The Model and Randomized Block Experiments with Several Groups

This section, which may be skipped, describes the relationship between the model (8.1) and a randomized block experiment. The conclusion is that if there is no hidden bias in the sense that $\boldsymbol{\delta} = \mathbf{0}$, then the distribution of assignments G_{sik} under (8.1) is the same as in a randomized block experiment if one conditions on a sufficient statistic for the $\xi_k(\mathbf{x}_s)$. This is analogous to the conclusion in §3.2 for two groups. The notation introduced and results in this section are not used in later sections. The section is intended solely as further motivation for the model (8.1).

Write $\mathbf{G}$ for the $N \times K$ matrix containing the G_{sik}, and write $\mathbf{M}$ for the $S \times K$ matrix containing the $m_{sk} = \sum_i^{n_s} G_{sik}$. Conditioning on $\mathbf{M}$ fixes the number of subjects in group k in stratum s. Then

$$\text{prob}(\mathbf{G} = \mathbf{g}|\mathbf{M}) = \frac{\exp(\mathbf{u}^T \mathbf{g} \boldsymbol{\delta})}{\sum^* \exp(\mathbf{u}^T \mathbf{g}^* \boldsymbol{\delta})}, \tag{8.3}$$

where $\sum^*$ is a sum is over all $\prod \binom{n_s}{m_{s1} \dots m_{sK}}$ possible $\mathbf{g}^*$ such that:

(i) $g_{sik}^* = 0$ or $g_{sik}^* = 1$;

(ii) $1 = \sum_k g_{sik}^*$; and

(iii) $m_{sk} = \sum_{i=1}^{n_s} g_{sik}^*$ for each s, k.

If there is no hidden bias in the sense that $\boldsymbol{\delta} = \mathbf{0}$, then (8.3) is constant or uniform, assigning the same probability to each possible $\mathbf{g}$. If there is just one stratum, $S = 1$, then this uniform distribution of treatment assignments is the same as the distribution used to derive the Kruskal–Wallis (1952) test. If there are several strata, $S \geq 2$, but each $m_{sk} = 1$, so one subject gets each treatment in each stratum, then this uniform distribution is the same as the distribution used to derive Friedman's (1937) test. These two standard tests are discussed in many texts, for instance, Hollander and Wolfe (1973) or Lehmann (1975).

In short, the model (8.1) is familiar in two senses. First, as shown in §8.3.1, in comparing groups two at a time, the model leads back to the methods used to compare treated and control groups in earlier chapters. Second, when all groups are considered together, if there is no hidden bias, the model leads back to familiar randomization tests comparing several groups.

8.3.3 Power of the Test Comparing Outcomes in Two Control Groups

Consider a study in which group $k = 1$ is the treated group and groups $k = 2, \ldots, K$ are the $K - 1$ control groups, with $K > 2$. What does it mean to say that group 1 is a treatment group and groups $k = 2, \ldots, K$ are control groups? The answer *defines* multiple control groups, distinguishing a study with multiple control groups from a study that simply has several different treatments. Consider the response R_{si} of the ith subject in stratum s under the K possible group assignments for this subject. To say that groups $k = 2, \ldots, K$ are control groups is to say that the response of this subject, R_{si}, would be the same no matter which control group received this subject, though R_{si} might be higher or lower if the subject received the treatment by being assigned to group 1. Notice carefully that this is a statement about the responses of individual subjects under different treatments, not a statement about the observed responses in the several groups. This is, again, the definition of multiple control groups. If the response of an individual would change depending upon the control group to which that subject was assigned, then the differences between the control groups affect the response, so they are not really control groups but rather active treatments with varied effects.

Pick two control groups, $j \geq 2$ and $k \geq 2$, $j \neq k$. Suppose that $G_{sik} + G_{sij} = 1$ so the ith subject in stratum s is in one of these two control groups. Then the response R_{si} of this subject is the same if $G_{sik} = 1$ or if $G_{sij} = 1$; that is, conditionally given that $G_{sik} + G_{sij} = 1$, the response R_{si} is fixed. Also, $\mathrm{prob}(G_{sik} = 1 | G_{sij} + G_{sik} = 1)$ is given by (8.2), which has the form used in Chapter 4 to compare two stratified or matched groups. Focus attention on the subjects in these two groups, setting aside subjects in other groups. Write $\tilde{\mathbf{u}}$ and $\tilde{\mathbf{r}}$ for the vectors of dimension $\sum_s (m_{sj} + m_{sk})$ containing the u_{si} and R_{si} for subjects in groups j and k, and write $\tilde{\mathbf{Z}}$ for the vector of the same dimension, where $\tilde{Z}_{si} = 1$ if $G_{sij} = 1$ and $\tilde{Z}_{si} = 0$ if $G_{sik} = 1$. Note that $\tilde{\mathbf{r}}$ is fixed as $\tilde{\mathbf{Z}}$ varies since groups j and k are control groups. Also, let $\tilde{\Omega}$ be the set of possible values of $\tilde{\mathbf{Z}}$, so each $\tilde{\mathbf{Z}} \in \tilde{\Omega}$ has m_{sj} ones and m_{sk} zeros among its $m_{sj} + m_{sk}$ coordinates for stratum s.

Contrast the responses of the subjects in groups j and k using an arrangement-increasing statistic $T = t(\tilde{\mathbf{Z}}, \tilde{\mathbf{r}})$, for instance, any of the statistics in §2.4.3. For example, if $m_{sj} + m_{sk} = 1$ for each s then T might be the signed rank or McNemar statistic comparing the pairs of subjects from groups j and k. Let

$$\alpha = \frac{1}{|\widetilde{\Omega}|} \sum_{\widetilde{\mathbf{z}} \in \widetilde{\Omega}} [t(\widetilde{\mathbf{z}}, \widetilde{\mathbf{r}}) \geq a]$$

and

$$\beta(\widetilde{\mathbf{r}}, \widetilde{\mathbf{u}}) = \sum_{\widetilde{\mathbf{z}} \in \widetilde{\Omega}} [t(\widetilde{\mathbf{z}}, \widetilde{\mathbf{r}}) \geq a] \frac{\exp(\gamma \widetilde{\mathbf{z}}^T \widetilde{\mathbf{u}})}{\sum_{\mathbf{b} \in \widetilde{\Omega}} \exp(\gamma \mathbf{b}^T \widetilde{\mathbf{u}})},$$

where $\gamma = \delta_j - \delta_k$ as in (8.2). Using arguments exactly parallel to those in §6.2.3, the following points are established. All statements about probabilities refer to conditional probabilities given the $G_{sik} + G_{sij}$.

- *The comparison of responses in two control groups tests the null hypothesis that there is no hidden bias.* A test that rejects the hypothesis of no hidden bias—that is, the hypothesis that $\boldsymbol{\delta} = \mathbf{0}$—when $T \geq a$ has level α. For instance, the rank sum test or the signed rank test would have their usual null distribution discussed in Chapter 2. This is true because $\boldsymbol{\delta} = \mathbf{0}$ implies $\gamma = 0$.

- *The test can have power only if there is systematic variation of the unobserved covariate.* The power of the test is $\beta(\widetilde{\mathbf{r}}, \widetilde{\mathbf{u}})$, which can differ from α only if $\gamma = \delta_j - \delta_k \neq 0$. If $\delta_j > \delta_k$ then group j will tend to have higher values of u than group k, and similarly if $\delta_j < \delta_k$ then group j will tend to have lower values of u, but if $\delta_j = \delta_k$ they will tend to have the same distribution of u. In the terminology of §8.2, a comparison of control groups j and k can hope to detect a hidden bias only if these two groups exhibit systematic variation of u.

- *The power of this test for hidden bias is greatest if the unobserved covariate u and the responses of controls are strongly related.* This is highly desirable, since an unobserved covariate strongly related to the response is one that does the most to distort inferences about treatment effects. More precisely, the power function $\beta(\widetilde{\mathbf{r}}, \widetilde{\mathbf{u}})$ is arrangement-increasing. This is analogous to Proposition 23 in §6.2.3.

- *Bracketing.* There is bracketing in the sense of §8.2 if $\delta_j > \delta_1 > \delta_k$ for in this case the distribution of u in the treated group tends to fall below the distribution in group j but above that in group k.

- *Unbiased tests.* If the unobserved covariate and the response under the controls are positively related, then the test is an unbiased test of $H_0 : \delta_j = \delta_k$ against $H_A : \delta_j > \delta_k$; for specifics, see Rosenbaum (1989a, §6).

In short, the behavior of the power function $\beta(\widetilde{\mathbf{r}}, \widetilde{\mathbf{u}})$ provides some formal justification for the principles of systematic variation and bracketing. See also the Problems in §8.5 for the relationship between the power $\beta(\widetilde{\mathbf{r}}, \widetilde{\mathbf{u}})$ and the sensitivity of tests for treatment effects.

8.4   Bibliographic Notes

The general use of multiple control groups to detect hidden biases in observational studies is discussed in detail by Campbell (1969), Rosenbaum (1984, 1987) and Shadish, Cook and Campbell (2002). The use of two control groups is common; it may be the most commonly used device to detect hidden biases. A few additional examples follow. Roghmann and Sodeur (1972) studied the effects of military service on authoritarian attitudes, comparing the West German Army to two control groups, namely, members of a German Catholic fraternity and West German students. In a study of the effects of constant noise on hearing, Taylor, Pearson and Mair (1965) compared jute weavers to two control groups: workers in the jute industry not exposed to noise and school teachers. In a study of the effects of the sequential use of obstetric forceps and vacuum extractors in vaginal deliveries, Ezenagu, Kakaria, and Bofill (1999) compared such deliveries to two groups, one using just forceps, the other using just a vacuum extractor. In a study of the psychological symptoms of women requesting removal of breast implants, Wells et al. (1997) compared such patients to two control groups, namely breast cancer patients and healthy controls. A study of 17 individuals with a rare genetic resistance to parvovirus B19 by Brown et al. (1994) had two treated and two control groups, using data from blood banks and current blood samples. Weston and Mansinghka (1971) compared the financial performance of 63 large conglomerate firms to two control groups, one comprised of industrial firms, the other comprised of a mixture of firms of different types. Another use of multiple control groups is to examine the impact of measurement procedures separated from effects of the treatment; see Solomon (1949) and Payne (1951) for general discussion, and see Berk, Lenihan, and Rossi (1980) for some discussion of an example. The power and unbiasedness of tests for hidden bias in §8.3.3 are discussed in Rosenbaum (1989a). The Advanced Placement Program example in §8.1 is discussed in Rosenbaum (1987). See also the Bibliographic Note, §7.5, concerning multiple referent groups in case-referent studies. An alternative strategy uses pretreatment or baseline measures of the response as one "control," and both pretreatment and posttreatment measures of response for untreated controls as another; see Rosenbaum (2001) for detailed discussion.

8.5 Problems

1. IUD and ectopic pregnancy: What does it mean to not receive treatment? Rossing, Daling, Voigt, Stergachis, and Weiss (1993) conducted a case-referent study of the possible increase in the risk of tubal or ectopic pregnancy caused by use of an intrauterine devise as a contraceptive. Can you spot an important issue that will arise in such a study? (*Hint:* An increased risk compared to *what?*)

2. IUD and ectopic pregnancy: Compared to what? Rossing et al. (1993, p. 252) write:

   There is considerable evidence to indicate that, among women who conceive a pregnancy, users of an intrauterine device (IUD) are more likely than nonusers to have implantation occur outside the uterus ... [In studies of] nonpregnant, sexually active women ... both increased and decreased risks have been reported. In the two studies in which the comparison group was restricted to noncontracepting, nonpregnant women, however, current IUD users were observed to be at a reduced risk of ectopic pregnancy.

   Why do the studies disagree with each other?

3. IUD and ectopic pregnancy: Several comparisons. Rossing et al. (1993, p. 252) collected data on female members of the Group Health Cooperative of Puget Sound who developed an ectopic pregnancy between 1981 and 1986; these were the cases. Referents were females from the same cooperative, selected using age and county of residence, with certain exclusions. Table 8.8 gives their data comparing cases and referents for IUD users and five other contraceptive groups. What conclusion do you draw?

4. IUD and ectopic pregnancy: Are comparison groups always control groups? Table 8.8 compares IUD users to five comparison groups. Are these control groups in the sense defined in §8.3.3? Why or why not?

5. Dose and response: The pattern of point estimates. The example in §8.1.4 had two control groups and it also had three treated groups at three doses or levels of alcohol consumption. Lilienfeld and Lilienfeld (1980, p. 309) write:

   If a factor is of causal importance in a disease, then the risk of developing the disease should be related to the degree of exposure to the factor; that is, a *dose-response relationship* should exist.

TABLE 8.8. Tubal Pregnancy and IUD.

Contraceptive method	Cases	Referents
IUD	18	60
Oral	13	202
Barrier	19	263
Sterilization	31	153
Rhythm, withdrawal, other	11	33
None	157	120

Ignoring the Abstain group, does there appear to be increasing risk of death with increasing alcohol consumption in Table 8.5?

6. Dose and response: Could the pattern, or lack of pattern, be due to chance? Excluding the Abstain group, some of the odds ratios in Table 8.5 differ significantly and others do not; see Table 8.6. Given this, what would you say about the presence or absence of a dose response relationship. Compare your thoughts with the sensible recommendations of Maclure and Greenland (1992, p. 103).

7. Dose and response: Can hidden bias produce a pattern of increasing response with increasing dose? Concerning dose and response, Weiss (1981, p. 488) writes:

   ... one or more confounding factors can be related closely enough to both exposure and disease to give rise to [a dose response relationship] in the absence of cause and effect.

   Suggest a simple linear regression model with a continuous dose Z, a continuous response R, and an unobserved covariate u such that a dose response relationship is produced in the absence of a treatment effect. That is, build a simple model illustrating Weiss's point.

8. Dose and response: Sensitivity analysis with continuous or discrete doses of treatment. Find an exponential family model relating discrete or continuous doses Z to an unobserved covariate u that reduces to the logit model (4.2) when there are just two doses. Use this model to develop a method of sensitivity analysis similar to that in Chapter 4 when treatments come in dose levels. (Solution: Rosenbaum (1989b), Gastwirth, Krieger, and Rosenbaum (1998).)

9. Power to detect bias and sensitivity: Power against the worst u. Suppose the treatment has no effect, and that there are two control groups and a binary response $\mathbf{r}$. Suppose further that the one-sided Mantel–Haenszel statistic is used twice. First, in a test for hidden bias, the two control groups are compared to each other. Second, in a test for

a treatment effect, the treated group is compared to the two control groups, ignoring the distinction between the two control groups. Define the notation so each test rejects when its test statistic is large in the positive direction. What value $\mathbf{u} \in U$ of the unobserved covariate leads to greatest sensitivity in the test for a treatment effect in §4.4.6? What value $\widetilde{\mathbf{u}}$ of the unobserved covariate leads to greatest conditional power $\beta(\widetilde{\mathbf{r}}, \widetilde{\mathbf{u}})$ in the test for hidden bias in §8.3.3? What is the relationship between $\mathbf{u}$ and $\widetilde{\mathbf{u}}$? Why is this highly desirable?

10. Did the test for hidden bias reduce sensitivity to hidden bias? The bracketed case. Suppose there is a treated group, $k = 1$, and two control groups, $k = 2$ and $k = 3$, such that the two control groups are known to bracket the treated group, in the sense of §8.3.3, so that $\delta_2 > \delta_1 > \delta_3$. Continuing Question 9: How would you investigate whether the test for bias has reduced sensitivity to bias? That is: How would you conduct an analysis similar to that in §6.2.4? What value $\mathbf{u}^\circ$ would be tested? How is $\mathbf{u}^\circ$ related to the value $\widetilde{\mathbf{u}}$ that maximizes the power $\beta(\widetilde{\mathbf{r}}, \widetilde{\mathbf{u}})$? Why is this, too, highly desirable?

11. Did the test for hidden bias reduce sensitivity to hidden bias? The unbracketed case. Why is bracketing important in Problem 10? What can happen without bracketing? (Hint: Consider the case of $\delta_1 > \delta_2 = \delta_3$.)

8.6 References

Berk, R. A., Lenihan, K. J., and Rossi, P. H. (1980) Crime and poverty: Some experimental evidence from ex-offenders. *American Journal of Sociology*, **45**, 766–786.

Bilban, M. (1998) Influence of the work environment in a Pb-Zn mine on the incidence of cytogenetic damage in miners. *American Journal of Industrial Medicine*, **34**, 455–463.

Bitterman, M. (1965) Phyletic differences in learning. *American Psychologist*, **20**, 396–410.

Brown, K. E., Hibbs, J. R., Gallinella, G., Anderson, S. M., Lehman, E. D., McCarthy, P., and Young, N. S. (1994) Resistance to parvovirus B19 infection due to lack of virus receptor (erythrocyte P antigen). *New England Journal of Medicine*, **330**, 1192–1196.

Campbell, D. (1969) Prospective: Artifact and Control. In *Artifact in Behavioral Research*, R. Rosenthal and R. Rosnow, eds., New York: Academic, pp. 351–382.

Campbell, D. and Boruch, R. (1975) Making the case for randomized assignment to treatments by considering the alternatives: Six ways in which quasi-experimental evaluations in compensatory education tend to underestimate effects. In: *Evaluation and Experiment*, C. Bennett and A. Lumsdaine, eds., New York: Academic, pp. 195–296.

Campbell, D. and Stanley, J. (1963) *Experimental and Quasi-Experimental Designs for Research*. Chicago: Rand McNally.

Cox, D. R. (1970) *The Analysis of Binary Data*. London: Methuen.

Douglas, D. and Carney, G. (1998) Exposure to asphalt or bitumen fume and renal disease. *Occupational and Environmental Medicine*, **55**, 645–646.

Ezenagu, L. C., Kakaria, R., Bofill, J. A. (1999) Sequential use of instruments at operative vaginal delivery: Is it safe? *American Journal of Obsetrics and Gynecology*, **180**, 1446–1449.

Friedman, M. (1937) The use of ranks to avoid the assumption of normality implicit in the analysis of variance. *Journal of the American Statistical Association*, **32**, 675–701.

Gastwirth, J. L., Krieger, A. M., and Rosenbaum, P. R. (1998) Dual and simultaneous sensitivity analysis for matched pairs. *Biometrika*, **85**, 907–920.

Hollander, M. and Wolfe, D. (1973) *Nonparametric Statistical Methods*. New York: Wiley.

Jackson, H., McGorry, P., Edwards, J., Hulbert, C., Henry, L., Francey, S., Maude, D., Cocks, J., Power, P., Harrigan, S., and Dudgeon, P. (1998) Cognitively-oriented psychotherapy for early psychosis (COPE): Preliminary results. *British Journal of Psychiatry*, **172**, 93–100.

Kim, E. H., McConnell, J. J., and Greenwood, P. R. (1977) Capital structure rearrangements and me-first rules in an efficient capital market. *Journal of Finance*, **32**, 789–810.

Kruskal, W. and Wallis, W. (1952) Use of ranks in one-criterion variance analysis. *Journal of the American Statistical Association*, **47**, 583–621.

Lehmann, E. L. (1975) *Nonparametrics: Statistical Methods Based on Ranks*. San Francisco: Holden-Day.

Lilienfeld, A., Chang, L., Thomas, D., and Levin, M. (1976) Rauwolfia derivatives and breast cancer. *Johns Hopkins Medical Journal*, **139**, 41–50.

Lilienfeld, A. and Lilienfeld, D. (1980) *Foundations of Epidemiology* (second edition). New York: Oxford University Press.

Maclure, M. and Greenland, S. (1992) Tests for trend and dose-response: Misinterpretations and alternatives. *American Journal of Epidemiology*, **135**, 96–104.

Payne, S. L. (1951) The ideal model for controlled experiments. *Public Opinion Quarterly*, **15**, 557–562.

Petersson, B., Trell, E., Kristenson, H. (1982) Alcohol abstention and premature mortality in middle aged men. *British Medical Journal*, **285**, 1457–1459.

Roghmann, K. and Sodeur, W. (1972) The impact of military service on authoritarian attitudes: Evidence from West Germany. *American Journal of Sociology*, **78**, 418–433.

Rosenbaum, P. R. (1984) From association to causation in observational studies. *Journal of the American Statistical Association*, **79**, 41–48.

Rosenbaum, P. R. (1987) The role of a second control group in an observational study (with Discussion). *Statistical Science*, **2**, 292–316.

Rosenbaum, P. R. (1989a) On permutation tests for hidden biases in observational studies: An application of Holley's inequality to the Savage lattice. *Annals of Statistics*, **17**, 643–653.

Rosenbaum, P. R. (1989b) Sensitivity analysis for matched observational studies with many ordered treatments. *Scandinavian Journal of Statistics*, **16**, 227–236.

Rosenbaum, P. R. (2001) Stability in the absence of treatment. *Journal of the American Statistical Association*, **96**, 210–219.

Rossing, M., Daling, J., Voigt, L., Stergachis, A., and Weiss, N. (1993) Current use of an intrauterine device and risk of tubal pregnancy. *Epidemiology*, **4**, 252–258.

Seltser, R. and Sartwell, P. (1965) The influence of occupational exposure to radiation on the mortality of American radiologists and other medical specialists. *American Journal of Epidemiology*, **81**, 2–22.

Shadish, W. R. and Cook, T. D. (1999) Design rules: More steps toward a complete theory of quasi-experimentation. *Statistical Science*, **14**, 294–300.

Shadish, W. R., Cook, T. D., and Campbell, D. T. (2002) *Experimental and Quasi-Experimental Designs for Generalized Causal Inference*. Boston: Houghton-Mifflin.

Solomon, R. (1949) An extension of control group design. *Psychological Bulletin*, 137–150.

Taylor, W., Pearson, J., and Mair, A. (1965) Study of noise and hearing in jute weaving. *Journal of the Acoustical Society of America*, **38**, 113–120.

Weiss, N. (1981) Inferring causal relationships: Elaboration of the criterion of "dose-response." *American Journal of Epidemiology*, **113**, 487–490.

Wells, K. E., Roberts, C., Daniels, S. M., Hann, D., Clement, V., Reintgen, D., and Cox, C. E. (1997) Comparison of psychological symptoms of woman requesting removal of breast implants with those of breast cancer patients and healthy controls. *Plastic and Reconstructive Surgery*, **99**, 680–685.

Weston, J. F. and Mansinghka, S. K. (1971) Tests of the efficiency performance of conglomerate firms. *Journal of Finance*, **26**, 919–936.

Zabin, L. S., Hirsch, M. B., and Emerson, M. R. (1989) When urban adolescents choose abortion: Effects on education, psychological status, and subsequent pregnancy. *Family Planning Perspectives*, **21**, 248–255.

Zelen, M. (1979) A new design for randomized clinical trials. *New England Journal of Medicine*, **300**, 1242–1245.

9

Coherence and Focused Hypotheses

9.1 Coherent Associations

9.1.1 What Is Coherence?

The 1964 US Surgeon General's report, *Smoking and Health* (Bayne-Jones et al. 1964, p. 20), lists five criteria for judgment about causality, the fifth being "the coherence of the association." A single sentence defines coherence (Bayne-Jones et al. 1964, p. 185): "A final criterion for the appraisal of causal significance of an association is its coherence with known facts in the natural history and biology of the disease." There follows a long discussion of the many ways in which the association between smoking and lung cancer is coherent. Per capita consumption of cigarettes had, at that time, been increasing, and the incidence of lung cancer was also increasing. Men, at that time, smoked much more than women and had a much higher incidence of lung cancer. And so on. To this, Sir Austin Bradford Hill (1965, p. 10) adds: "... I regard as greatly contributing to coherence the histopathological evidence from the bronchial epithelium of smokers and the isolation from cigarette smoke of factors carcinogenic for the skin of laboratory animals." The pattern of associations in §1.2 between smoking and cardiovascular disease would also be described as coherent. Coherence is discussed by Susser (1973, pp. 154–162) and more critically by Rothman (1986, p. 19). MacMahon and Pugh (1970, p. 21) use the phrase "consonance with existing knowledge" in place of coherence. Coherence is related to Fisher's "elaborate theory," as discussed in §1.2.

Typically, coherence is defined briefly, if at all, and then illustrated with examples, often compelling examples. This chapter offers a definition of coherence. The definition is reasonably close to traditional usage of the term, and it has consequences for design and inference. The purpose is to distinguish coherence from efforts to detect hidden biases, as described in Chapters 6 through 8. Though both are useful, coherence and detection differ in their logic and their evidence.

In detecting hidden biases, the goal is to collect data so that an actual treatment effect would be visibly different from the most plausible hidden biases. Control groups, referent groups, or outcomes with known effects are selected with this goal in mind; see, in particular, §§6.2.3, 7.2.2, 8.1, and 8.3.3. The many tables in Campbell and Stanley (1963) concern the ability of various research designs to distinguish treatment effects from biases of various kinds. Detection concerns distinguishing biases and effects.

Coherence is different. There is no reference to particular biases, no assurance that certain specific biases would be visibly different from treatment effects. A coherent pattern of associations is one that is, at each of many points, in harmony with existing knowledge of how the treatment should behave if it has an effect. In a coherent association, an elaborate theory describing the treatment and its effects closely fits equally elaborate data. What does such a close fit say about hidden biases?

9.1.2   Focused Hypotheses About Treatment Effects

Sir Karl Popper writes:

> It is easy to obtain confirmations, or verifications, for nearly every theory—if we look for confirmations.... Confirmations should count only if they are the result of *risky predictions*; that is to say, if, unenlightened by the theory in question, we should have expected an event that was incompatible with the theory—an event that would have refuted the theory [Popper 1965, p. 36] ... as scientists we do not seek highly probable theories but ... powerful and improbable theories [Popper 1965, p. 58].

> A theory will be said to be better corroborated the more severe the [evaluations] it has passed.... [An evaluation is] more severe the greater the probability of failing it (the absolute or prior probability as well as the probability in the light of what I call our 'background knowledge', that is to say, knowledge which, by common agreement, is not questioned while [evaluating] the theory under investigation) [Popper 1983, p. 244].

In §1.1, Fisher's elaborate theories are risky predictions because the form of the anticipated treatment effect is tightly and narrowly specified. If such an elaborate theory is confirmed at each of many opportunities, then one senses there is more dramatic evidence against the null hypothesis of no treatment effect and in favor of the elaborate theory of an effect. Indeed, statistical theory supports this view in the sense that hypothesis tests against ordered alternatives often provide more dramatic evidence against a null hypothesis; see, for instance, Jonckheere (1954), Page (1963), Barlow, Bartholomew, Bremner, and Brunk (1972), and Robertson, Wright, and Dykstra (1988). In Chapter 4, dramatic evidence was typically less sensitive to hidden bias. Perhaps a coherent association, an association that matches an elaborate theory, is less sensitive to hidden bias.

9.2   Signed Rank Statistics for Coherent Predictions

9.2.1   Predictions About Multivariate Outcomes in Matched Pairs

There are $2S$ subjects in S matched pairs and the ith subject in pair s exhibits a K-dimensional response, $\mathbf{R}_{si}$, $s = 1, \ldots, S$, $i = 1, 2$. The coherent hypothesis makes predictions about the direction of the effect of the treatment on each of the K responses. Instead of testing each of the K responses separately, a single test will be performed that is particularly effective when the coherent predictions are correct. If complex, ornate data closely match an equally complex, ornate theory, then the results should be less sensitive to hidden bias—at least, that is the hope and the motivation.

This hope is realized in an example later in §9.3. In that example, the treatment is exposure of operating room personnel to anesthetic gases, and the response $\mathbf{R}_{si}$ consists of $K = 2$ measures of mutagenicity, specifically, the number of sister chromatid exchanges and the number of micronuclei. The coherent theory predicts that anesthetic gases cause mutations, and that different measures of mutagenicity will all point in this same direction. The example serves to illustrate both the strengths and limitations of coherence.

In this section, it is assumed that the response has been recorded or rearranged in such a way that the coherent hypothesis predicts higher responses for treated subjects. More complex predictions can also be tested using partial orders of the type discussed in §2.8.

The model for sensitivity analysis is the same as in §4.3.1. One person in each pair received the treatment, signified by $Z_{si} = 1$, and the other received the control, signified by $Z_{si} = 0$, so $Z_{s1} + Z_{s2} = 1$ for each s, and the distribution of treatment assignments is given by (4.7). Although

(4.7) refers to conditional probabilities given $\mathbf{m} = (m_1, \dots, m_S)$, because $m_s = Z_{s1} + Z_{s2} = 1$ for every s, the conditioning on $\mathbf{m}$ is suppressed in the notation in this section.

9.2.2 The Sum of Several Signed Rank Statistics

The test statistic is the sum of the K Wilcoxon signed rank statistics for the K coordinates, R_{sik}, $k = 1, \dots, K$ of $\mathbf{R}_{si}$. Recall that the responses have been reorganized so that higher responses on each coordinate are anticipated under treatment, so the sum of K signed rank statistics will be large if this anticipation is confirmed.

For each k, rank the absolute differences $|R_{s1k} - R_{s2k}|$ from 1 to S with average ranks for ties, writing q_{sk} for the rank. Also, write $c_{s1k} = 1$ if $R_{s1k} > R_{s2k}$, and $c_{s1k} = 0$ otherwise, and $c_{s2k} = 1$ if $R_{s2k} > R_{s1k}$, and $c_{s2k} = 0$ otherwise, so that a tied pair, with $R_{s1k} = R_{s2k}$, has $c_{s1k} = c_{s2k} = 0$. Then the signed rank statistic for R_{sik} is $T_k = \sum_{s=1}^{S} q_{sk} \sum_{i=1}^{2} c_{sik} Z_{si}$, and the sum of these K signed rank statistics is $T = \sum T_k$.

9.2.3 Coherent Tests in the Absence of Hidden Bias

This section discusses testing the null hypothesis of no effect in the absence of hidden bias, that is, with $\Gamma = 1$, so that $\Pr(Z_{s1} = 1) = 1/2$. In §9.2.4, a sensitivity analysis is developed to address hidden bias, that is, $\Gamma \geq 1$ with $\Pr(Z_{s1} = 1)$ unknown. In §9.2.5, hypotheses other than no effect are discussed.

It is convenient to express T as a constant plus the sum of S independent random variables, each random variable taking just two values. Specifically, because $Z_{s2} = 1 - Z_{s1}$,

$$T = \sum_{s=1}^{S} \left\{ Z_{s1} \sum_{k=1}^{K} c_{s1k}\, q_{sk} + (1 - Z_{s1}) \sum_{k=1}^{K} c_{s2k}\, q_{sk} \right\}$$

$$= \sum_{s=1}^{S} Z_{s1} \sum_{k=1}^{K} (c_{s1k} - c_{s2k})\, q_{sk} + \sum_{s=1}^{S} \sum_{k=1}^{K} c_{s2k}\, q_{sk}$$

$$= \sum_{s=1}^{S} Z_{s1}\, w_s + L, \tag{9.1}$$

where $w_s = \sum_{k=1}^{K} (c_{s1k} - c_{s2k})\, q_{sk}$ and $L = \sum_{s=1}^{S} \sum_{k=1}^{K} c_{s2k}\, q_{sk}$.

In the absence of hidden bias, under the model (4.7) with $\Gamma = 1$, the S distinct Z_{s1}'s are mutually independent with $\Pr(Z_{s1} = 1) = \Pr(Z_{s1} = 0) = 1/2$. Under the null hypothesis of no treatment effect, the responses $\mathbf{R}_{si}$ are fixed, not varying with the Z_{si}'s, so the c_{sik} and q_{sk} are also fixed. It follows from (9.1) that, when $\Gamma = 1$, the null expectation of T is $E(T) = \frac{1}{2}\left(\sum_{s=1}^{S} w_s\right) + L$ and the null variance of T is $var(T) = \frac{1}{4}\sum_{s=1}^{S} w_s^2$. An approximate one-sided significance level is obtained by comparing

$$\frac{T - E(T)}{\sqrt{var(T)}}$$

to the standard Normal distribution.

9.2.4 Sensitivity Analysis for a Coherent Test

The sensitivity analysis for $\Gamma \geq 1$ uses bounds on $\Pr(Z_{s1} = 1)$. Write $\overline{\overline{p}}_s = \Gamma/(1+\Gamma)$ if $w_s \geq 0$ and $\overline{\overline{p}}_s = 1/(1+\Gamma)$ if $w_s < 0$. Similarly, write $\underline{\overline{p}}_s = 1/(1+\Gamma)$ if $w_s \geq 0$ and $\underline{\overline{p}}_s = \Gamma/(1+\Gamma)$ if $w_s < 0$. Notice that $\overline{\overline{p}}_s$ maximizes the chance of positive contributions to T and minimizes the chance of negative contributions, while $\underline{\overline{p}}_s$ minimizes the chance of positive contributions and maximizes the chance of negative ones. Let $\overline{\overline{T}}$ be a random variable that is the sum of S independent terms taking value w_s with probability $\overline{\overline{p}}_s$ and value 0 with probability $1 - \overline{\overline{p}}_s$. Define $\underline{\overline{T}}$ similarly but with $\underline{\overline{p}}_s$ in place of $\overline{\overline{p}}_s$. As a special case of Proposition 13, the following proposition provides sensitivity bounds for the coherent signed rank test under the model (4.7).

Proposition 28 *If the treatment has no effect on $\mathbf{R}_{si}$, then for each fixed $\Gamma \geq 1$,*

$$prob\left(\overline{\overline{T}} \geq a\right) \geq prob\left(T \geq a\right) \geq prob\left(\underline{\overline{T}} \geq a\right) \quad \text{for all } a \text{ and all } \mathbf{u} \in U.$$

Under the null hypothesis of no treatment effect, $\overline{\overline{T}}$ has expectation and variance:

$$E\left(\overline{\overline{T}}\right) = \sum_{s=1}^{S} \overline{\overline{p}}_s w_s + L, \qquad var\left(\overline{\overline{T}}\right) = \sum_{s=1}^{S} \overline{\overline{p}}_s\left(1 - \overline{\overline{p}}_s\right) w_s^2,$$

with analogous formulas for $\underline{\overline{T}}$ but with $\underline{\overline{p}}_s$ in place of $\overline{\overline{p}}_s$. As $S \to \infty$, the upper bound on the tail probability, $prob\left(\overline{\overline{T}} \geq a\right)$, is approximated by:

$$1 - \Phi\left[\frac{\left\{a - E\left(\overline{\overline{T}}\right)\right\}}{\sqrt{var\left(\overline{\overline{T}}\right)}}\right],$$

where $\Phi\left(\cdot\right)$ is the standard Normal cumulative distribution. The sensitivity analysis is illustrated in §9.3.

9.2.5   Coherent Equivalence Tests

Write $\mathbf{r}_{Csi}$ for the K–dimensional response that the ith subject in pair s would exhibit under control, $s = 1, \ldots, S$, $i = 1, 2$. Then the treatment has an additive effect on each coordinate if there is a K–dimensional τ such that the observed response is $\mathbf{R}_{si} = \mathbf{r}_{Csi} + \tau Z_{si}$, $s = 1, \ldots, S$, $i = 1, 2$. Point estimates and confidence intervals for each coordinate of τ may be obtained by the methods of Chapters 3 and 4. How can the coherent signed rank statistic T be used to say something about the entire vector τ? A confidence "interval" for the K–dimensional τ based on the scalar T is not possible, but a simple form of equivalence test is possible. If the null hypothesis that the treatment is highly effective is clearly rejected in favor of smaller effects, is this finding sensitive to hidden biases? Although two-sided equivalence tests are most relevant to bioequivalence testing in pharmaceutical experiments, one-sided equivalence tests are most relevant to sensitivity analysis in observational studies.

The hypothesis $H_0 : \tau = \tau_0$ may be tested by calculating the adjusted response, $\mathbf{R}_{si} - \tau_0 Z_{si}$, which equals the fixed quantity $\mathbf{r}_{Csi}$ when the hypothesis is true. Then the coherent signed rank test is applied to the adjusted responses, $\mathbf{R}_{si} - \tau_0 Z_{si}$, testing the hypothesis of no effect on these adjusted responses.

If the hypothesis $H_0 : \tau = \tau_0$, with each $\tau_k > 0$, asserts that the treatment is quite effective, this hypothesis may be tested against the alternative of smaller effects by rejecting H_0 when T is small. Rejection of H_0 in favor of smaller effects provides strong evidence that the treatment is *not* quite effective. The sensitivity analysis is performed as in §9.2.4. In some contexts, the conclusion that the treatment is *not* quite effective may be quite insensitive to hidden bias: very large biases might be needed to mask a highly effective treatment. An example involving a surgical intervention in urology is given by Li, Propert, and Rosenbaum (2001).

9.3   Example: Do Anesthetic Gases Cause Mutations?

9.3.1   Reduced Sensitivity to Hidden Bias from a Coherent Prediction

When anesthetic gases are used in operating rooms, there is some leakage, so operating room personnel may be exposed repeatedly. In an effort to determine whether such exposures were mutagenic, Hoerauf, Lierz, Wiesner,

TABLE 9.1. Mutagenicity of Anesthetic Gases in Exposed Veterinary Surgeons (E) and Controls (C).

Pair	Ages (E/C)	Gender	E-SCE	E-MN	C-SCE	C-MN
1	52/51	M	8.59	10.75	6.37	6.25
2	26/25	F	9.12	13.00	8.35	6.00
3	31/30	M	12.99	6.75	5.35	7.75
4	27/27	F	7.87	7.50	12.03	6.50
5	26/27	M	8.15	9.75	5.88	3.00
6	31/30	F	13.40	5.50	6.39	5.75
7	35/38	M	9.58	4.00	9.74	3.50
8	33/33	F	10.29	9.75	6.62	9.25
9	40/41	F	10.08	12.50	4.30	9.75
10	27/27	F	11.72	7.50	9.28	10.25

Schroegendorfer, Lierz, Spacek, Brunnberg, and Nusse (1999) matched 10 veterinary surgeons to 10 unexposed veterinary physicians, matching for age and gender. They obtained two measures of mutagenicity, namely, the mean number of sister chromatid exchanges per metaphase (SCE) in cultured lymphocytes, and the number of micronuclei (MN) recorded as micronuclei per 500 binucleated cells. Their data are in Table 9.1.

If the Wilcoxon signed rank test is used assuming no hidden bias, then one obtains standardized deviates of 1.99 and 1.58 for SCE and MN, respectively, with approximate one-sided significance levels 0.023 and 0.057. These two separate tests are sensitive to moderate biases: for $\Gamma = 1.5$, the minimum deviates become 1.46 and 1.04 with maximum approximate one-sided significance levels 0.073 and 0.149. Of course, the pattern for both SCE and MN is in the predicted direction. The coherent signed rank test has a deviate of 2.44 in the absence of hidden bias, with approximate one-sided significance level 0.0073. The coherent test is less sensitive to hidden bias: for $\Gamma = 1.5$ and $\Gamma = 2$, respectively, the minimum standardized deviates are 1.94 and 1.63, with approximate maximum one-sided significance levels 0.026 and 0.052. Use of a coherent alternative has reduced sensitivity to hidden bias.

9.3.2 Details of the Calculations

The calculations leading to the coherent test are shown in Table 9.2. As in Wilcoxon's signed rank test, matched pair differences are calculated for each variable, SCE and MN, separately, the absolute differences are ranked with average ranks for ties, and the signs of the differences are noted. The signs and ranks are combined into a coherent score; for instance, for pair #1, the score is $3 \times (+1) + 8 \times (+1) = 11$. Notice that pair #3 has a slightly negative difference for MN, but the largest positive difference for SCE, so the overall score of 5.5 is somewhat positive.

TABLE 9.2. Calculation of the Coherent Signed-Rank Statistic from Matched Pair Differences (d), Ranks (r), and Signs (s).

Pair	dSCE	dMN	rSCE	rMN	sSCE	sMN	Score
1	2.22	4.50	3	8.0	+1	+1	11.0
2	0.77	7.00	2	10.0	+1	+1	12.0
3	7.64	−1.00	10	4.5	+1	−1	5.5
4	−4.16	1.00	7	4.5	−1	+1	−2.5
5	2.27	6.75	4	9.0	+1	+1	13.0
6	7.01	−0.25	9	1.0	+1	−1	8.0
7	−0.16	0.50	1	2.5	−1	+1	1.5
8	3.67	0.50	6	2.5	+1	+1	8.5
9	5.78	2.75	8	6.5	+1	+1	14.5
10	2.44	−2.75	5	6.5	+1	−1	−1.5

Wilcoxon's signed rank statistic for SCE is $3+2+10+4+9+6+8+5 = 47$ and for MN is $8 + 10 + 4.5 + 9 + 2.5 + 2.5 + 6.5 = 43$, so the coherent signed rank statistic is $47 + 43 = 90$. This equals the sum of the scores, $11 + 12 + 5.5 + -2.5 + 13 + 8 + 1.5 + 8.5 + 14.5 + -1.5 = 70$ plus the constant, $L = 7 + 1 + 4.5 + 1.0 + 6.5 = 20$ which adds back the five ranks with negative signs. In the absence of hidden bias, $\Gamma = 1$, the individual signed rank statistics each have expectation $10(10 + 1)/4 = 27.5$ while the coherent statistic has expectation equal to twice this, namely 55, which equals half the sum of the scores plus L or $70/2 + 20 = 55$. In the absence of hidden bias, $\Gamma = 1$, the variance of the coherent statistic is one fourth of the sum of the squares of the scores, or $\frac{1}{4}\left(11^2 + \ldots + -1.5^2\right) = 205.375$ and the deviate is $(90 - 55)/\sqrt{205.375} = 2.44$.

The sensitivity analysis for $\Gamma > 1$ is similar, but it is important to keep in mind that $\bar{\bar{p}}_s = \Gamma/(1+\Gamma)$ if the score is positive and $\bar{\bar{p}}_s = 1/(1+\Gamma)$ if the score is negative, thereby increasing the chance of a positive score and decreasing the chance of a negative one. For instance, with $\Gamma = 2$, the first pair has $\bar{\bar{p}}_1 = 2/3$ but the fourth pair has $\bar{\bar{p}}_4 = 1/3$. With $\Gamma = 2$, the maximum expectation for the coherent signed rank statistic is the weighted sum of scores plus $L = 20$, namely $\frac{2}{3} \times 11 + \ldots + \frac{1}{3} \times -1.5 + 20 = 68$ and the associated variance is $\Gamma/(1+\Gamma)^2 = \frac{2}{9}$ times the sum of the squares of the scores, $11^2 + \ldots + -1.5^2$, or 182.56, so the minimum deviate is $(90 - 68)/\sqrt{182.56} = 1.63$.

The constant L does not affect the deviate or the significance level, since it is added to both the statistic and its expectation. However, without L, the coherent statistic cannot be described as the sum of the separate signed rank statistics. When computing the deviate for the coherent statistic, L may be dropped from both the statistic and its expectation.

9.4   *Properties of the Coherent Signed Rank Test

Properties of the coherent signed rank test are developed in Rosenbaum (1997) and are briefly and informally outlined here. The coherent test has good power when the treatment affects all of the outcomes to a similar degree, with outcomes that are weakly and symmetrically correlated. In this case, the gain in power and insensitivity to bias can be substantial. Of course, multiple outcomes that are perfectly correlated provide no increase in power. Obviously, the situation is less favorable when only some of the outcomes are affected by the treatment, but gains in power and insensitivity are nonetheless possible when at least half the outcomes are affected and there are at least four outcomes. See Rosenbaum (1997, §4.3) for detailed discussion.

It is possible to compare the coherent test statistic, T, to the individual statistics, T_k, by embedding them both in a family of test statistics. Fix a set of weights, $\lambda_k \geq 0$, $1 = \sum \lambda_k$, and consider the statistic, $G = \sum \lambda_k T_k$. Then $KG = T$ when $\lambda_1 = \ldots = \lambda_K = 1/K$, and the two statistics yield the same deviate, $\{G - E(G)\}/\sqrt{\text{var}(G)}$. Also, $G = T_k$ when $\lambda_1 = 0, \ldots$, $\lambda_{k-1} = 0$, $\lambda_k = 1$, $\lambda_{k+1} = 0, \ldots$, $\lambda_K = 0$. The deviate for the coherent signed rank statistic discarding outcome K is obtained from $G = \sum \lambda_k T_k$ with $\lambda_1 = \ldots = \lambda_{K-1} = 1/(K-1)$, $\lambda_K = 0$, and so on. Consider the set of all possible λ_k's such that $G > 0$. On this set of possible λ_k's, the deviate has a special property, semistrict quasiconcavity, whose practical implications will now be described. See Rosenbaum (1997, Appendix) for formal definitions and proof. Suppose that we are unsure whether to use T_k based on the kth outcome or the coherent statistic without the kth outcome, $T - T_k$, assuming both are positive. If their deviates are unequal, the special property implies that the deviate based on all of T is at least as large as the smaller of the two deviates based on T_k alone or on $T - T_k$. Informally, it may be better to know which outcomes are affected and which are not; however, lacking that knowledge, the coherent test is always better than the poorer of two choices, namely using the kth outcome alone or discarding the kth outcome. As discussed in Rosenbaum (1997, Appendix), the special property leads to many similar results of this kind: the combination of two parts is always better than the weaker of the two parts, and may be better than either part, as in §9.3.

9.5   Some Other Forms of Coherence

In general, coherence specifies a detailed pattern of associations anticipated when the treatment is the cause of the associations. This chapter has emphasized one type of coherence, namely, predictions about the direction of effects for several outcome variables. There are, of course, many other patterns one might anticipate.

In some studies, it is possible to obtain measures of the response before and after treatment. The importance of baseline measures of response is emphasized by Campbell and Stanley (1963), Cook, Campbell, and Peracchio (1990), Allison (1990), Cook and Shadish (1994), and Shadish, Cook and Campbell (2002). For example, there are baseline measures of response in the minimum wage data in §5.10, where employment data were obtained before and after the change in New Jersey's minimum wage. For other examples, see Grevert and Goldstein (1977) and Allison and Long (1990). Baseline measures of response are particularly useful when one anticipates that untreated subjects will remain relatively stable, without systematic increases or decreases in response. A statistical test designed to notice and record this particular coherent pattern can yield reduced sensitivity to hidden bias when the pattern occurs (Rosenbaum 2001). Interestingly, in Grevert and Goldstein (1977), the control group was expected to change and the treated group was expected to remain stable: the treatment, naloxone, was expected to block a decline in tension and anxiety.

Yet another pattern is anticipated in an interesting example discussed by Salzberg (1999). Claiming that a fuel company had submitted fraudulent bills for fuel deliveries to certain City buildings, the City of New York sued. The City's evidence included fuel delivery records for these buildings before, during, and after the years of service by this company, together with parallel data for other buildings never served by this company. The coherent prediction is a rise in fuel bills confined to buildings served by this company during the years the company provided fuel.

Still another pattern was anticipated by Allison and Long (1990) in a careful study of the effects that academic departments have on the research productivity of their faculty. They looked at 179 job changes by chemists, biologists, physicists, and mathematicians, and they measured productivity by publications and citations before and after the job change. They predicted and found increased productivity following moves to higher ranked departments, and decreased productivity following moves to lower ranked departments. See their discussion of the several mechanisms that might account for this pattern.

9.6   Strengths and Limitations of Coherence

As an argument, coherence has two strengths. First, a test against a coherent alternative penalizes a data set that exhibits nonsensical associations, making it more difficult to detect an effect in this case. Second, evidence of an effect may be dramatic in either of two ways: the effect may appear dramatic in size, or it may make numerous predictions each of which is confirmed in data. A test against a coherent alternative provides a basis for appraising the latter possibility.

The strengths may be expressed in another way. If the study were free of hidden bias, $\Gamma = 1$, then 5% of the treatment assignments $\mathbf{z} \in \Omega$ would lead to rejection of the null hypothesis of no treatment effect at the 0.05 level. Which 5% of treatment assignments $\mathbf{z} \in \Omega$ should lead to rejection? The argument of coherence says this 5% should be chosen carefully. Specifically, one should reject for those $\mathbf{z} \in \Omega$ that appear most consistent with the pattern anticipated if the treatment actually caused the effect.

Coherence also has two significant limitations. First, a coherent pattern of associations may result from a coherent pattern of hidden biases. For instance, in §9.3, if veterinary surgeons were exposed to some other mutagen besides anesthetic gases, the same coherent pattern of associations might be found. Similarly in §9.5, as discussed by Allison and Long (1990), their anticipated coherent pattern of productivity changes following academic job changes could be produced by more than one mechanism. As discussed in §9.1.1, unlike efforts to detect hidden bias, coherence is concerned with the pattern of associations that is consistent with a treatment effect, but it does not take active steps to ensure that this pattern differs from what is expected from plausible hidden biases. Contrast coherence with "control by systematic variation" in §§8.2 and 8.3.3. If control groups are selected to systematically vary a relevant unobserved covariate then, as seen in §8.3.3, there are good prospects that any hidden bias that may be due to this covariate will be revealed when the control groups are compared. In contrast, the test against a coherent alternative comes with no promise that it can assist in distinguishing a treatment effect from any particular hidden bias.

The second limitation of coherence is that a summary statistic for several outcomes may be useful as a primary endpoint, but detailed results for each outcome are also important. A test against a coherent alternative concerns evidence against the null hypothesis of no effect and in the direction of a coherent pattern of associations. The evidence against the null hypothesis of no effect of any kind may be strong and insensitive to hidden bias, but the evidence about individual outcomes may be quite varied. Because outcomes differ in their consequences for policy, it is rarely sufficient to know that there is at least some effect on one of the outcomes. A sensitivity analysis against a coherent alternative should generally be accompanied by a sensitivity analysis for individual outcomes.

In short, a sensitivity analysis for a test against a coherent alternative helps to appraise evidence provided by the confirmation of numerous specific predictions. It is not, however, a substitute for sensitivity analyses for individual outcomes nor for active steps to detect hidden bias.

9.7   *Appendix: Arrangement-Increasing Functions of Matrices

In this section, the definition of arrangement-increasing functions is extended to certain partial orders and, in particular, the coherent signed rank statistic is seen to be arrangement-increasing with respect to a partial order. In addition, many other similar statistics are arrangement-increasing with respect to a partial order and they play roles similar to that of the coherent signed rank statistic; see the Problems in §9.9. Among these statistics are: (i) sums of several rank sum statistics, (ii) sums of several aligned rank statistics, (iii) sums of several statistics associated with Fisher's exact test for a 2×2 table, and some statistics of the type described in §2.8.4. Section 2.4.4 discussed arrangement-increasing functions $f(\mathbf{a}, \mathbf{b})$ of two stratified N-dimensional vectors $\mathbf{a}, \mathbf{b}$, that is, two vectors indexed by $s = 1, \ldots, S$, $i = 1, \ldots, n_s$, $N = n_1 + \cdots + n_s$. Such a function was defined by two properties. First, it was permutation-invariant in the sense that interchanging two coordinates from the same stratum in both vectors does not alter the value of the function. Second, interchanging two coordinates from the same stratum in one vector so their order disagrees with the order in the other vector decreases, or at least does not increase, the value of the function.

These two properties of arrangement-increasing functions may be stated formally using "interchange vectors" ι_{sij}. There is an interchange vector for each s, $s = 1, \ldots, S$, and each i, j with $1 \leq i < j \leq n_s$ and ι_{sij} is the N-dimensional column vector with a one in the ith coordinate for stratum s, a negative one in the jth coordinate for stratum s, and a zero in all other coordinates. Let $\mathbf{I}$ denote the $N \times N$ identity matrix. When an N-dimensional vector $\mathbf{b}$ is multiplied by the matrix $\mathbf{I} - \iota_{sij}\iota_{sij}^T$ the result $(\mathbf{I} - \iota_{sij}\iota_{sij}^T)\mathbf{b}$ is $\mathbf{b}$ with its ith and jth coordinates of stratum s interchanged. Then $f(\cdot, \cdot)$ is permutation invariant if, for each $\mathbf{a}$, and $\mathbf{b}$ in the domain of $f(\cdot, \cdot)$, for every interchange vector ι_{sij}, $f(\mathbf{a}, \mathbf{b}) = f[(\mathbf{I} - \iota_{sij}\iota_{sij}^T)\mathbf{a}, (\mathbf{I} - \iota_{sij}\iota_{sij}^T)\mathbf{b}]$. A permutation invariant $f(\cdot, \cdot)$ is arrangement-increasing if $\mathbf{a}^T\iota_{sij}\iota_{sij}^T\mathbf{b} \geq 0$ implies $f(\mathbf{a}, \mathbf{b}) \geq f[\mathbf{a}, (\mathbf{I} - \iota_{sij}\iota_{sij}^T)\mathbf{b}]$.

In the same way, let $\mathbf{A}$ and $\mathbf{B}$ be two matrices each with N stratified rows numbered $s = 1, \ldots, S$, $i = 1, \ldots, n_s$. Let $f(\mathbf{A}, \mathbf{B})$ be a real-valued function of these matrix arguments. Then $f(\cdot, \cdot)$ is permutation invariant if simultaneously interchanging two rows of $\mathbf{A}$ and the same two rows of $\mathbf{B}$ in the same subclass does not change $f(\mathbf{A}, \mathbf{B})$; that is, if for every $\mathbf{A}$ and $\mathbf{B}$ in the domain of $f(\cdot, \cdot)$, and for every interchange vector, $f(\mathbf{A}, \mathbf{B}) = f[(\mathbf{I} - \iota_{sij}\iota_{sij}^T)\mathbf{A}, (\mathbf{I} - \iota_{sij}\iota_{sij}^T)\mathbf{B}]$. Write $\mathbf{C} \geq \mathbf{0}$ if $c_{km} \geq 0$ for each k, m. A permutation-invariant function $f(\cdot, \cdot)$ is arrangement-increasing if $\mathbf{A}^T\iota_{sij}\iota_{sij}^T\mathbf{B} \geq \mathbf{0}$ implies $f(\mathbf{A}, \mathbf{B}) \geq f[\mathbf{A}, (\mathbf{I} - \iota_{sij}\iota_{sij}^T)\mathbf{B})$. In other words, if rows (s, i) and (s, j) of $\mathbf{A}$ and $\mathbf{B}$ are ordered in the same way, then interchanging these two rows of $\mathbf{B}$ reduces, or at least does not increase, $f(\mathbf{A}, \mathbf{B})$.

Let $\mathbf{R}$ be an $N \times K$ matrix of K-dimensional responses for N subjects, with the usual partial order $\precsim$ saying that subject (s, i) has a lower response than subject (s, j) if each of the K responses for subject (s, i) is less than or equal to each of the K responses for subject (s, j). This is the partial order used in §9.2 and the first partial order in §2.8.3. With this partial order, the statistic $t(\mathbf{Z}, \mathbf{R})$ in (2.8) and the coherent signed rank statistic in §9.2 are both arrangement-increasing.

Hollander, Proschan, and Sethuraman (1977) discuss properties of arrangement increasing functions of two vectors, and many of these extend to arrangement-increasing functions of two matrices, sometimes with small modifications; see Rosenbaum (1991, §2). For instance, using an extension of their composition theorem, Proposition 23 in §6.2.3 may be proved for poset statistics. A different application arises when the coherent pattern of associations is produced by an underlying and unobserved dose of treatment; see Rosenbaum (1991, §4) for a practical example.

9.8   Bibliographic Notes

Coherence is discussed with varying terminology and varying degrees of enthusiasm by Hill (1965), MacMahon and Pugh (1970, p. 21), Susser (1973, pp. 154–162, 1991), and Rothman (1986, p. 19), among others. The several references by Campbell, Cook, Shadish, Stanley, and Peracchio discuss coherence and detecting bias, in different terminology, and they emphasize that coherent patterns may result from biases rather than treatment effects. Coherence is also discussed more generally by Davidson (1986) and Thagard (2000). There is a large literature on statistical methods for order restricted inference, though the most widely used methods are for totally ordered outcomes rather than partially ordered outcomes; that is, they concern dose-response relationships rather than coherence of multiple outcomes. See Jonckheere (1954) and Page (1963) for nonparametric methods for totally ordered outcomes; these tests are also discussed in the textbook by Hollander and Wolfe (1973). For comprehensive surveys of order restricted inference with extensive bibliographies, see Barlow, Bartholomew, Bremner, and Brunk (1972) and Robertson, Wright, and Dykstra (1988). Maclure and Greenland (1992) are critical of potential misinterpretation of tests for trend and suggest careful terminology for discussing such tests. There is a large literature on equivalence tests; see Schuirmann (1987) and Hsu, Hwang, Liu, and Ruberg (1994), for example. This chapter is based upon Rosenbaum (1991, 1994, 1997) and Li, Propert, and Rosenbaum (2001).

9.9   Problems

1. **Coherent rank sum statistics.** Suppose that there is a single stratum, $S = 1$, so the s subscript is dropped, and a K–dimensional response, $\mathbf{R}_i$, where higher responses on each coordinate of $\mathbf{R}_i$ are anticipated if the treatment has an effect. In this case, one might compute K separate Wilcoxon rank sum statistics, one for each coordinate of $\mathbf{R}_i$, and add them together to form a single test statistic. In a uniform randomized experiment or in an observational study free of hidden bias, $\Gamma = 1$, what are the null expectation and variance of this sum of rank sum statistics? (Rosenbaum 1991)

2. **Coherent rank sum statistics: Sensitivity analysis.** How could you do a sensitivity analysis for the coherent rank sum statistic in the previous problem? (Hint: Write the statistic in the form (4.26) and apply the general method of §4.6.)

3. **Coherent rank sum statistics: Arrangement increasing property.** Show the coherent rank sum statistic is arrangement increasing in $\mathbf{Z}$ and the response matrix, using ideas from the appendix to this chapter. (Rosenbaum 1991)

4. **Coherent Mann-Whitney statistics: Sensitivity analysis.** How could you do a sensitivity analysis for the statistic in §2.8.4? (Hint: Write the statistic in the form (4.26) and apply the general method of §4.6.) (Rosenbaum 1994)

5. **Coherent signed rank statistics with dose-response.** Suppose that, in pair s, the treatment is applied to the treated subject with fixed dose d_s, while the control remains untreated. Consider the statistic $D = \sum_{k=1}^{K} d_s \sum_{s=1}^{S} q_{sk} \sum_{i=1}^{2} c_{sik} Z_{si}$ which gives greater weight to matched pairs in which the dose is higher. When the doses are all the same, say $d_s = 1$, the statistic D is essentially the same as the coherent signed rank test discussed in §9.2. Obtain the null expectation and variance of D for $\Gamma \geq 1$. (Rosenbaum 1997)

9.10   References

Allison, P. D. (1990) Change scores as dependent variables in regression analyses. In: *Sociological Methodology*, C. C. Clogg, ed., Oxford: Basil Blackwell, pp. 93–114.

Allison, P. D. and Long, J. S. (1990) Departmental effects on scientific productivity. *American Sociological Review*, **55**, 469–478.

Barlow, R., Bartholomew, D., Bremner, J., and Brunk, H. (1972) *Statistical Inference Under Order Restrictions*. New York: Wiley.

Bayne-Jones, S., Burdette, W., Cochran, W., Farber, E., Fieser, L., Furth, J., Hickman, J., LeMaistre, C., Schuman, L., and Seevers, M. (1964) *Smoking and Health: Report of the Advisory Committee to the Surgeon General of the Public Health Service*. Washington, DC: US Department of Health, Education, and Welfare.

Campbell, D. and Stanley, J. (1963) *Experimental and Quasi Experimental Designs for Research*. Chicago: Rand McNally.

Cook, T. D. (1991) Clarifying the warrant for generalized causal inferences in quasi-experimentation. In: *Evaluation and Education at Quarter Century*, M. W. McLaughlin and D. Phillips, eds., NSSE 1991 Yearbook, pp. 115–144.

Cook, T. D., Campbell, D. T. and Peracchio, L. (1990) Quasi Experimentation. In: *Handbook of Industrial and Organizational Psychology*, M. Dunnette and L. Hough, eds., Palo Alto, CA: Consulting Psychologists Press, Chapter 9, pp. 491–576.

Cook, T. D. and Shadish, W. R. (1994) Social experiments: Some developments over the past fifteen years. *Annual Review of Psychology*, **45**, 545–580.

Davidson, D. (1986) A coherence theory of truth and knowledge. In: *Truth and Interpretation*, E. Lepore, ed., Oxford: Blackwell, pp. 307–319.

Grevert, P. and Goldstein, A. (1977) Effects of naloxone on experimentally induced ischemic pain and on mood in human subjects. *Proceedings of the National Academy of Sciences* (Psychology), **74**, 1291–1294.

Hill, A. B. (1965) The environment and disease: Association or causation? *Proceedings of the Royal Society of Medicine*, **58**, 295–300.

Hoerauf, K., Lierz, M., Wiesner, G., Schroegendorfer, K., Lierz, P., Spacek, A., Brunnberg, L., Nusse, M. (1999) Genetic damage in operating room personnel exposed to isoflurane and nitrous oxide. *Occupational and Environmental Health*, **56**, 433–437.

Hollander, M., Proschan, F., and Sethuraman, J. (1977) Functions decreasing in transposition and their applications in ranking problems. *Annals of Statistics*, **5**, 722–733.

Hollander, M. and Wolfe, D. (1973) *Nonparametric Statistical Methods*. New York: Wiley.

Hsu, J. C., Hwang, J. T. G., Liu, H. K., and Ruberg, S. J. (1994) Confidence intervals associated with tests for bioequivalence. *Biometrika*, **81**, 103–114.

Jonckheere, A. (1954) A distribution-free *k*-sample test against ordered alternatives. *Biometrika*, **41**, 133–145.

Li, Y., Propert, K. J. and Rosenbaum, P. R. (2001) Balanced risk set matching. *Journal of the American Statistical Association*, **96**, September, to appear.

Maclure, M. and Greenland, S. (1992) Tests for trend and dose-response: Misinterpretations and alternatives. *American Journal of Epidemiology*, **135**, 96–104.

MacMahon, B. and Pugh, T. (1970) *Epidemiology: Principles and Methods*. Boston: Little, Brown.

Mann, H. and Whitney, D. (1947) On a test of whether one of two random variables is stochastically larger than the other. *Annals of Mathematical Statistics*, **18**, 50–60.

Mantel, N. (1967) Ranking procedures for arbitrarily restricted observations. *Biometrics*, **23**, 65–78.

Page, E. (1963) Ordered hypotheses for multiple treatments: A significance test for linear ranks. *Journal of the American Statistical Association*, **58**, 216–230.

Popper, K. (1965) *Conjectures and Refutations*. New York: Harper & Row.

Popper, K. (1983) *Realism and the Aim of Science*. Totowa, NJ: Rowman and Littlefield.

Robertson, T., Wright, F. T., and Dykstra, R. L. (1988) *Order Restricted Statistical Inference*. New York: Wiley.

Rosenbaum, P. R. (1991) Some poset statistics. *Annals of Statistics*, **19**, 1091–1097.

Rosenbaum, P. R. (1994) Coherence in observational studies. *Biometrics*, **50**, 368–374.

Rosenbaum, P. R. (1997) Signed rank statistics for coherent predictions. *Biometrics*, **53**, 556–566.

Rosenbaum, P. R. (2001) Stability in the absence of treatment. *Journal of the American Statistical Association*, **96**, 210–219.

Rothman, K. (1986) *Modern Epidemiology*. Boston: Little, Brown.

Salzberg, A. (1999) Removable selection bias in quasi-experiments. *The American Statistician*, **53**, 103–107.

Schuirmann, D. J. (1987) A comparison of the two one-sided tests procedure and the power approach for assessing the equivalence of average bioavailability. *Journal of Pharmacokinetics and Biopharmaceutics*, **15**, 657–680.

Shadish, W. R., Cook, T. D., and Campbell, D. T. (2002) *Experimental and Quasi-Experimental Designs for Generalized Causal Inference*. Boston: Houghton-Mifflin.

Skerfving, S., Hansson, K., Mangs, C., Lindsten, J., and Ryman, N. (1974) Methylmercury-induced chromosome damage in man. *Environmental Research*, **7**, 83–98.

Susser, M. (1973) *Causal Thinking in the Health Sciences*. New York: Oxford University Press.

Susser, M. (1991) What is a cause and how do we know one? A grammar for pragmatic epidemiology. *American Journal of Epidemiology*, **133**, 635–648.

Thagard, P. (2000) *Coherence in Thought and Action*. Cambridge, MA: MIT Press.

Wilcoxon, F. (1945) Individual comparisons by ranking methods. *Biometrics*, **1**, 80–83.

10
Constructing Matched Sets and Strata

10.1 Introduction: Propensity Scores, Structures, Algorithms

This chapter discusses the construction of matched sets or strata when there are several, perhaps many, observed covariates $\mathbf{x}$. There are three topics: the propensity score, the form of an optimal stratification, and the construction of optimal matched sets. This introduction summarizes the main issues and findings.

As the number p of covariates increases, it becomes difficult to find matched pairs with the same or similar values of $\mathbf{x}$. Even if each covariate is a binary variable, there will be 2^p possible values of $\mathbf{x}$, so with $p = 20$ covariates there are more than a million possible values of $\mathbf{x}$. If there are hundreds or thousands of subjects and $p = 20$ covariates, it is likely that many subjects will have unique values of $\mathbf{x}$. For this reason, it is sometimes said that matching is not useful when there are many covariates. Actually, the problem is not with matching as a technique but rather with the specific objective of obtaining matched pairs or sets that are homogeneous in $\mathbf{x}$. If the objective is defined in other ways, then matching and stratification are not difficult with many covariates.

There are two objectives of matching and stratification besides matched sets that are homogeneous in $\mathbf{x}$. First, if there is no hidden bias so that it suffices to adjust for $\mathbf{x}$, then strata or matched sets are desired that permit use of the conventional methods in Chapter 2. Second, whether or not there is hidden bias, one would like to compare treated and control groups with similar distributions of **x**, even if matched individuals have differing values of **x**. The second objective is called covariate balance.

The propensity score is a device for constructing matched sets or strata when **x** contains many covariates. If the true propensity score were known, both objectives in the previous paragraph would be attained by matching or stratifying on the propensity score, a single covariate. That is, if strata or matched sets are formed that are homogeneous in the propensity score, even if they are heterogeneous in **x**, then the methods of Chapter 2 are appropriate in the absence of hidden bias, and the observed covariates **x** will tend to balance whether or not there is hidden bias. In the case of the first objective, this was demonstrated in §3.2.5; see also §10.2.1 below. Covariate balance is demonstrated in §10.2.2. In practice, the propensity score is not known and must be estimated. One approach was discussed in §§3.4 and 3.5. For matching and stratification, use of estimated propensity scores is illustrated in §10.2.3 and simulation results are reviewed in §10.4.7.

Section 10.3 discusses optimal stratification. As it turns out, under quite general conditions, the form of an optimal stratification is always the same. This implies that in searching for a good stratification, the search may be confined to stratifications of this form because one is sure to be optimal. This optimal form is called a full matching. It is a matched sample in which each matched set contains either a treated subject and one or more controls or else a control subject and one or more treated subjects. Pair matching is not optimal, and neither is matching with a fixed number of controls. In the simulation results described in §10.4.7, full matching is found to be substantially better than matching with a fixed number of controls. It is easy to see why this happens. When the treated and control groups have different distributions of the observed covariates **x**, there are regions of **x** values with many treated subjects and few controls, and other regions with many controls and few treated subjects. Forcing every treated subject to have the same number of controls creates some poor matched sets.

The construction of optimal matched samples is discussed in §10.4. Various types of matching are illustrated and compared, including matching with a fixed number of controls, matching with a variable number of controls, full matching, and balanced matching. Network flow techniques are used to obtain optimal matched samples, and optimal matching is contrasted with a commonly used alternative, greedy matching.

10.2   The Propensity Score

10.2.1   Definition of the Propensity Score

The propensity score is the conditional probability of receiving the treatment given the observed covariates **x**. Recall from §3.2 the model for treatment assignment with $\pi = \text{prob}(Z_{si} = 1)$ and $0 < \pi_{si} < 1$. For all n_s

subjects in stratum s, define the *propensity score* to be

$$\lambda(\mathbf{x}_s) = \frac{\sum_{i=1}^{n_s} \pi_{si}}{n_s}.$$

Since each π_{si} satisfies $0 < \pi_{si} < 1$, it follows that $0 < \lambda(\mathbf{x}_s) < 1$ for each s.

The propensity score $\lambda(\mathbf{x}_s)$ has the following operational interpretation. Pick a subject at random from stratum s, each subject having probability $1/n_s$ of being selected; then this random subject receives the treatment with probability $\lambda(\mathbf{x}_s)$. That is, pick subject (s, i) with probability $1/n_s$, where (s, i) receives the treatment with probability π_{si}, so the marginal probability that a random subject will receive the treatment is $\sum_{i=1}^{n_s} \pi_{si}/n_s = \lambda(\mathbf{x}_s)$.

The propensity score has two useful properties. The first, the more important of the two properties, was discussed in §3.2.5 and it applies when there is no hidden bias, that is, when $\pi_{si} = \lambda(\mathbf{x}_s)$ for every s and i. If there is no hidden bias, then one need not form strata or matched sets that are homogeneous in $\mathbf{x}_s$; it suffices to obtain strata or matched sets that are homogeneous in $\lambda(\mathbf{x}_s)$. If there is no hidden bias, if the strata are homogeneous in $\lambda(\mathbf{x}_s)$, then the conditional distribution of treatment assignments is uniform, and the statistical methods in Chapter 2 for a randomized experiment may be used; see §3.2.5 for details. Since $\mathbf{x}_s$ may be of high dimension, but $\lambda(\mathbf{x}_s)$ is a number, it is often much easier to find subjects with similar values of $\lambda(\mathbf{x}_s)$ than with similar values of $\mathbf{x}_s$. When there is no hidden bias, when there is only overt bias due to $\mathbf{x}_s$, it suffices to adjust for the propensity score $\lambda(\mathbf{x}_s)$.

The second property applies whether or not there is hidden bias; that is, it applies even if $\pi_{si} \neq \lambda(\mathbf{x}_s)$. Strata or matched sets that are homogeneous in $\lambda(\mathbf{x}_s)$ tend to balance $\mathbf{x}_s$ in the sense that treated and control subjects in the same stratum or matched set tend to have the same distribution of $\mathbf{x}_s$. In an experiment, randomization tends to balance all covariates, observed and unobserved, in the sense that treated and control groups tend to have the same distribution of covariate values. In an observational study, strata or matched sets that are homogeneous in the propensity score $\lambda(\mathbf{x}_s)$ tend to balance the observed covariates $\mathbf{x}_s$, though there may be imbalances in unobserved covariates. The balancing property is demonstrated in the next section.

10.2.2 Balancing Properties of the Propensity Score

Pick a value of the propensity score, say Λ, such that there is at least one s with $\lambda(\mathbf{x}_s) = \Lambda$. There may be several strata s with this same value of the propensity score, $\lambda(\mathbf{x}_s) = \Lambda$, so write n_Λ for the total number of subjects in these strata, that is,

$$n_\Lambda = \sum_{s:\lambda(\mathbf{x}_s)=\Lambda} n_s = \sum{}^* n_s.$$

where $\sum{}^*$ denotes a sum over all s such that $\lambda(\mathbf{x}_s) = \Lambda$. Pick a subject at random from these strata; that is, pick a subject (s, i) from $\{(s, i) : \lambda(\mathbf{x}_s) = \Lambda\}$ where each such subject has the same probability of being selected, namely, $1/n_\Lambda$. Write Z and $\mathbf{X}$ for the treatment assignment and observed covariate for this random subject; that is, if the subject selected is (s, i) then $Z = Z_{si}$ and $\mathbf{X} = \mathbf{x}_s$.

What is the probability that $Z = 1$? To emphasize that the probability refers to a subject randomly selected from the n_Λ subjects with $\lambda(\mathbf{x}_s) = \Lambda$, write the probability as prob$\{Z = 1 \mid \lambda(\mathbf{X}) = \Lambda\}$. If $\lambda(\mathbf{x}_s) = \Lambda$, then the random subject comes from stratum s with probability n_s/n_Λ, so

$$\text{prob}\{Z = 1 \mid \lambda(\mathbf{X}) = \Lambda\} = \sum{}^* \frac{n_s\lambda(\mathbf{x}_s)}{n_\Lambda} = \sum{}^* \frac{n_s\Lambda}{n_\Lambda} = \Lambda,$$

where again $\sum{}^*$ denotes a sum over all s such that $\lambda(\mathbf{x}_s) = \Lambda$.

This same conclusion, prob$\{Z = 1 \mid \lambda(\mathbf{X}) = \Lambda\} = \Lambda$, may be expressed in a slightly different way. If $\lambda(\mathbf{x}_s) = \Lambda$, then subject (s, i) is selected with probability $1/n_\Lambda$, and receives the treatment with probability π_{si}; so

$$\text{prob}\{Z = 1 \mid \lambda(\mathbf{X}) = \Lambda\} = \frac{1}{n_\Lambda} \sum{}^* \sum_{i=1}^{n_s} \pi_{si} = \sum{}^* \frac{n_s\lambda(\mathbf{x}_s)}{n_\Lambda} = \Lambda.$$

Proposition 29 below describes the balancing property of the propensity score. It says that a treated and a control subject with the same value of the propensity score have the same distribution of the observed covariate $\mathbf{X}$. This means that in a stratum or matched set that is homogeneous in the propensity score, treated and control subjects may have differing values $\mathbf{X}$, but the differences will be chance differences rather than systematic differences. Again, this balancing concerns observed covariates, but unlike a randomized experiment, the propensity score does not typically balance unobserved covariates. More precisely, Proposition 29 says the following. Pick a value of the propensity score Λ and pick one subject at random from among the n_Λ subjects with this value of the propensity score; then for this subject, treatment assignment Z is independent of the value of the covariate $\mathbf{X}$ given the value of the propensity score $\lambda(\mathbf{X}) = \Lambda$. The proposition is due to Rosenbaum and Rubin (1983).

Proposition 29 *If $\lambda(\mathbf{x}_s) = \Lambda$, then*

$$\text{prob}\{\mathbf{X} = \mathbf{x}_s \mid \lambda(\mathbf{X}) = \Lambda, Z = 1\} = \text{prob}\{\mathbf{X} = \mathbf{x}_s \mid \lambda(\mathbf{X}) = \Lambda, Z = 0\}.$$

Proof. If $\lambda(\mathbf{x}_s) = \Lambda$, then by Bayes' theorem,

$$\text{prob}\{\mathbf{X} = \mathbf{x}_s \mid \lambda(\mathbf{X}) = \Lambda, Z = 1\}$$
$$= \frac{\text{prob}\{Z = 1 \mid \lambda(\mathbf{X}) = \Lambda, \mathbf{X} = \mathbf{x}_s\}\,\text{prob}\{\mathbf{X} = \mathbf{x}_s \mid \lambda(\mathbf{X}) = \Lambda\}}{\text{prob}\{Z = 1 \mid \lambda(\mathbf{X}) = \Lambda\}}.$$

Now, $\text{prob}\{Z = 1 \mid \lambda(\mathbf{X}) = \Lambda, \mathbf{X} = \mathbf{x}_s\} = \text{prob}\{Z = 1 \mid \mathbf{X} = \mathbf{x}_s\} = \lambda(\mathbf{x}_s) = \Lambda$, and $\text{prob}\{Z = 1 \mid \lambda(\mathbf{X}) = \Lambda\} = \Lambda$, so

$$\text{prob}\{\mathbf{X} = \mathbf{x}_s \mid \lambda(\mathbf{X}) = \Lambda, Z = 1\} = \text{prob}\{\mathbf{X} = \mathbf{x}_s \mid \lambda(\mathbf{X}) = \Lambda\},$$

proving the result. ∎

10.2.3 Matching or Stratifying on an Estimated Propensity Score: An Example

In practice, the propensity score $\lambda(\mathbf{x})$ is unknown. In this section, the propensity score $\lambda(\mathbf{x})$ is estimated using a logit model and the estimate is used in place of the true propensity score.

The example concerns a comparison of coronary bypass surgery and medical or drug therapy in the treatment of coronary artery disease; it is from Rosenbaum and Rubin (1984). There were $N = 1515$ subjects, of whom 590 were surgical patients and 925 were medical patients. The vector $\mathbf{x}$ contained 74 observed covariates describing hemodynamic, angiographic, laboratory, and exercise test results, together with information about patient histories. Each of these 74 covariates showed a statistically significant imbalance in treated and control groups; contrast this with the randomized experiment in §2.1.

These 74 covariates were controlled using five strata formed from an estimated propensity score. The propensity score was estimated using a linear logit model that predicted treatment assignment Z from the observed covariates $\mathbf{x}$; see Cox (1970) for detailed discussion of logit models. The model included some interactions and quadratic terms selected by a sequential process described in Rosenbaum and Rubin (1984). The 1515 patients were divided into five strata, each containing 303 patients, based on the estimated propensity score. The stratum with the highest estimated probabilities of surgery contained 69 medical patients and 234 surgical patients, while the stratum with the lowest estimated probabilities of surgery contained 277 medical patients and 26 surgical patients. The theory in §10.2.2 suggests that, had the true propensity scores been used in place of estimates, medical and surgical patients in the same stratum should have similar distributions of the 74 observed covariates.

Table 10.1 shows the imbalance in five of the 74 variables before and after stratification on the propensity score. The column labeled "Before" contains the square of the usual two-sample t-statistic, that is, the F-ratio,

TABLE 10.1. Covariate Imbalance Before and After Stratification: F-Ratios.

Covariate	Before	After: Main Effect	After: Interaction
Abnormal LV Contraction	51.8	0.4	0.9
Progressing Chest Pain	43.6	0.1	1.4
Left Main Stenosis	22.1	0.3	0.2
Cardiomegaly	25.0	0.2	0.0
Current Nitrate Treatment	31.4	0.1	2.2

comparing the means of each variable among medical and surgical patients. The values in this column are large indicating the medical and surgical patients exhibit statistically significant differences on each of these variables, and in fact on each of the 74 covariates. The last two columns give F-ratios from a two-way 2×5 analysis of variance performed for each variable, the two factors being medical versus surgical and the five propensity score strata. The column labeled "After: Main Effect" is the F-ratio for the main effect of medical versus surgical. The column labeled "After: Interaction" is the F-ratio for the two-way interaction. None of the F-ratios in Table 10.1 is significant at the 0.05 level, and most of the F-ratios are less than one, so there is no indication of systematic imbalances in these covariates. For all 74 covariates, only one of the $2 \times 74 = 148$ F-ratios was significant at the 0.05 level. In a randomized experiment, $0.05 \times 148 = 7.4$ F-ratios significant at 0.05 would have been expected by chance alone. Evidently, these 74 observed covariates are more closely balanced within strata than would have been expected in a randomized experiment with five strata. However, randomization would balance both observed and unobserved covariates, while stratification on the propensity score cannot be expected to balance unobserved covariates.

In this case, a fairly coarse stratification on an estimate of the propensity score did tend to balance 74 observed covariates. Using these five strata, the medical and surgical patients were compared with respect to outcomes such as survival and pain relief. Adjusting for the five strata, there was little or no difference in survival. See Rosenbaum and Rubin (1984) for detailed results.

10.2.4 Propensity Scores with Doses of Treatment

Propensity scores have been described for the case of a treated group compared to a control group. When, instead, there are several doses of treatment, the propensity score may be generalized in either of two directions. Suppose in this section only that the treatment Z can take values $1, 2, \ldots, K$. In the first generalization, the propensity score remains a scalar function of covariates, so that matching and stratification on the scalar propensity score proceed as for two treatment groups; see Joffe and

Rosenbaum (1999). In the second approach, each of the K levels of the treatment Z requires its own propensity score, and strata formed from different propensity scores are not comparable; see Imbens (2000) and Rosenbaum (1987). The first generalization is somewhat easier to use, but the second requires fewer assumptions. A brief outline of the two generalizations follows.

Suppose the levels of Z are ordered or ordinal, so $Z = k + 1$ is a higher dose of treatment than $Z = k$ for each k. A common model for a conditional distribution of an ordinal variable given a vector $\mathbf{x}$ of covariates is McCullagh's (1980) ordinal logit model, namely:

$$\log\left\{ \frac{\Pr\left(Z \geq k|\mathbf{x}\right)}{\Pr\left(Z < k|\mathbf{x}\right)} \right\} = \alpha_k + \boldsymbol{\beta}^T \mathbf{x} \quad \text{for } k = 2, 3, \ldots, K. \tag{10.1}$$

Notice that if Z were converted into a binary variable B indicating whether $Z \geq k$, recorded as $B = 1$, or $Z < k$, recorded as $B = 0$, then under model (10.1), the binary variable would satisfy a binary logit model,

$$\log\left\{ \frac{\Pr\left(B = 1|\mathbf{x}\right)}{\Pr\left(B = 0|\mathbf{x}\right)} \right\} = \alpha_k + \boldsymbol{\beta}^T \mathbf{x},$$

and that as k changes, only the constant term α_k changes. See McCullagh (1980) for detailed discussion of the attractive features of this model. When model (10.1) correctly describes $\Pr\left(Z|\mathbf{x}\right)$, the scalar $\boldsymbol{\beta}^T \mathbf{x}$ has essentially the same properties as the propensity score for two groups. In particular, matching or stratifying on $\boldsymbol{\beta}^T \mathbf{x}$ tends to balance $\mathbf{x}$ in the K dose groups. The important issue is that the entire distribution $\Pr\left(Z|\mathbf{x}\right)$ is linked to $\mathbf{x}$ only through a scalar $\boldsymbol{\beta}^T \mathbf{x}$. Other models for $\Pr\left(Z|\mathbf{x}\right)$ with this property behave similarly; see Joffe and Rosenbaum (1999) for discussion. In practice, one estimates $\boldsymbol{\beta}$ by maximum likelihood, and matches on the estimated score, $\widehat{\boldsymbol{\beta}}^T \mathbf{x}$; see Lu, Zanutto, Hornik and Rosenbaum (2002) for an example.

The second generalization fits K separate propensity scores, $\Pr\left(Z = k|\mathbf{x}\right)$, $k = 1, 2, \ldots, K$. These K scores are used one at a time to adjust the responses of subjects who received dose $Z = k$ to yield an estimate of the distribution of responses that would have been observed had all subjects received dose $Z = k$. These K adjusted distributions are then compared. See Imbens (2000) and Rosenbaum (1987) for two aspects of this approach. By estimating $\Pr\left(Z = k|\mathbf{x}\right)$ separately for each k, it is not necessary to assume that $\Pr\left(Z|\mathbf{x}\right)$ follows a particular model, such as (10.1), so this second approach is somewhat more general. On the other hand, the K different propensity scores have little to do with one another. For instance, a person having $\Pr\left(Z = 1|\mathbf{x}\right) = 1/4$ has nothing in particular in common with a person having $\Pr\left(Z = 3|\mathbf{x}\right) = 1/4$, so there is no basis for matching these two people together, or placing them together in the same stratum. In other words, although the second approach is applicable in more circumstances, it preserves fewer of the convenient properties of the propensity score.

TABLE 10.2. Matching and Structure: Hypothetical Population with 2800 Subjects.

	$x = 0$	$x = 1$	$x = 2$
Treated	100	200	100
Control	2000	300	100

10.3 Optimal Strata

10.3.1 The Importance of Structure: Intuition

The structure of a matched or stratified comparison refers to constraints imposed on the number of controls that are matched to a treated subject. Pair matching, where each treated subject has its own single control, is one structure. Matching each treated subject to two controls is an alternative structure. Matching each treated subject to at least one and at most four controls is yet another structure, called variable matching. Full matching is the structure in which each matched set contains either one treated subject and one or more controls, or else one control and one or more treated subjects. In an unconstrained stratification, each stratum may have any nonnegative number of treated and control subjects, perhaps zero.

Statistical theory, simulations, and case-studies demonstrate that structure strongly affects the ability of matching and stratification to remove biases due to observed covariates, that is, to compare people who look comparable. Before considering the subject in detail, §§10.3.1 and 10.3.2 develop some intuition using hypothetical and actual examples.

Highly constrained forms of matching may fail to remove all of the bias due to observed covariates, whereas less constrained forms might succeed, and this remains true as the sample size increases. To see this, consider a simple hypothetical example. In the population before matching, there is one observed covariate at three levels, and its relationship to the treatment is given in the contingency Table 10.2. Notice that an exact pair matching of all 400 treated subjects is just barely possible here. However, if one matches each treated subject to two controls, obtaining 800 controls in total, then there are not enough controls with $x = 2$ or $x = 1$ to go around, and one might end up with 100 treated-control-control triples with covariate x triples $(2, 2, 1)$, 200 triples with $(1, 1, 0)$, and 100 triples with $(0, 0, 0)$. Many matched individuals have different values of the covariate x. Notice that the situation would not improve if all of the counts in Table 10.2 were doubled: a uniform increase in sample size does not make matching easier. On the other hand, if one matched each treated subject to at least one control and at most four controls, then one might have 100 treated-control pairs with covariate x pairs $(2, 2)$, 100 pairs with $(1, 1)$, 100 treated-control-control triples with covariate x triples $(1, 1, 1)$, and 100 matched sets of a treated subject and four controls with covariates x given by $(0, 0, 0, 0, 0)$, so each treated subject is compared to controls with the

TABLE 10.3. Matching and Structure: A Population in Which Even Pair Matching Is Not Possible.

	$x = 0$	$x = 1$	$x = 2$
Treated	100	200	100
Control	2000	300	50

same x, and there are still 800 controls in total. In this example, with a scalar x, the two matching procedures select the same controls but use them differently; however, with a vector $\mathbf{x}$, this would no longer be true.

The situation in Table 10.3 is more difficult. There is no way to exactly match the 100 treated subjects with $x = 2$ to 100 distinct controls; even pair matching is not possible here. Although there is no exact pair matching of 400 treated subjects to 400 controls, a full matching of this size is possible. It might contain 50 treated-treated-control triples with covariate x triples $(2, 2, 2)$, 150 treated-control pairs with $(1, 1)$, 100 pairs with $(0, 0)$, and 50 treated-control-control triples with $(1, 1, 1)$.

In principle, when observations are available without cost, full matching can use all available observations. In Table 10.3, one might select 50 treated-treated-control triples with covariates x triples $(2, 2, 2)$, 100 matched pairs with $(1, 1)$, 100 matched treated-control-control triples with $(1, 1, 1)$, and 100 matched sets with one treated subject and 20 controls all having covariate values 0. Because x takes just three values in Table 10.3, there would be no gain from subdividing the table into matched sets. However, in practical problems, $\mathbf{x}$ is a vector, often containing some continuous variables, and the matched sets would be more nearly homogeneous in $\mathbf{x}$ than a coarser stratification would allow; see §10.3.6.

Unconstrained stratifications are sometimes useful, but special methods are needed when some cells have zero frequencies; see Rosenbaum (1987) for detailed discussion. In the discussion of constructing strata and matched samples in this chapter, zero frequencies are avoided.

10.3.2   The Importance of Structure: Death After Surgery

To illustrate the importance of structure in a practical example, several matched samples with different structures are compared using pilot data for a case-referent study of the causes of death after surgery in the Medicare population. This section is based on an example in Ming and Rosenbaum (2000). The complete study examines more than 800 deaths randomly sampled from Medicare deaths following surgery at all Pennsylvania hospitals; see Silber, Rosenbaum, Trudeau, Even-Shoshan, Chen, Zhang, and Mosher (2001) and Rosenbaum and Silber (2001). The study attempts to compare patients who were in similar condition upon entering the hospital in the hope of learning about preventable causes of death during the hospital stay. Patients are matched on the basis of coarse computerized records

TABLE 10.4. Optimal Matching with Three Controls: Risk Scores for the First
Five Matched Sets. Source: Ming and Rosenbaum (2000).

Set	Death	Survivors
1	0.034	0.034, 0.034, 0.034
2	0.274	0.141, 0.114, 0.104
3	0.227	0.194, 0.158, 0.150
4	0.024	0.024, 0.024, 0.024
5	0.485	0.439, 0.198, 0.155

from Medicare. For matched patients, the hospital chart is abstracted in
detail, an extremely expensive process.

Using data from one hospital, the pilot data describe 38 eligible patients
who died following simple surgical procedures in 1993 and 1994, and 685
eligible patients who did not die. In the pilot, *using data from earlier
years*, a baseline risk of death was estimated using a logit model, and the
coefficients of this model were then used to score patient risk, and formed
the basis for the matching. A score of 0.1 means that the logit model
from earlier years estimated the risk of death upon admission to be 10%.
The final matching uses this risk score in a more complicated matching
algorithm described in Silber et al. (2001). Before matching, the 38 deaths
had a mean risk score of 0.152 and a standard deviation of 0.132, and the
685 survivors had a mean risk score of 0.037 and a standard deviation of
0.050, so the patients who went on to die appeared at much greater risk at
the time of admission.

Table 10.4 gives the risk scores for the first five of 38 matched sets when
each death is matched to three referents who did not die, 3×38 referents
in total. Although matched sets 1 and 4 are closely matched, many of the
other matches are quite poor. For example, in matched set 5, the death was
estimated to have a 0.485 chance of death upon admission, and one of the
matched referents had only a 0.155 estimated chance of death. The poor
matches are due to the unreasonable insistence that every death have three
matched referents, even though there are not enough high risk survivors to
go around.

Table 10.5 shows the same five deaths when each death is matched to at
least one and at most four referents, with 3×38 referents in total. Here,
matched patients have similar risk scores.

This example and associated theory are considered in detail in Ming
and Rosenbaum (2000). In particular, they show that: (i) the pattern for
the first five matched sets in Tables 10.4 and 10.5 applies generally to
the 38 matched sets; (ii) that matching with between 1 and 4 referents
removed 22% more bias than matching with three referents, and produced
only a slight increase in the standard error; and (iii) that these patterns
are consistent with general theory.

TABLE 10.5. Optimal Matching with One to Four Controls: Risk Scores for the First Five Matched Sets. Source: Ming and Rosenbaum (2000).

Set	Death	Survivors
1	.034	.034, .034, .034, .034
2	.274	.216
3	.227	.218
4	.024	.024, .024, .024, .024
5	.485	.439

The matchings described in this section are optimal in the sense described later in §10.4. This means: they minimized the total absolute difference in risk scores over all matchings with the given size and structure. The optimal matching was performed in SAS; see Ming and Rosenbaum (2001) and §10.6.

10.3.3 Evaluating Stratifications Based on the Average Remaining Distance

Section 10.3.6 characterizes the form of an optimal stratification. Optimality is defined in terms of a distance between each treated subject and each control, and the goal is a stratification that minimizes a weighted average of the distances within each stratum. Under mild conditions, the optimal stratification has a simple form developed in §10.3.6.

Initially, there are two sets of subjects: the treated subjects are in a set A and the controls are in a set B, with $A \cap B = \emptyset$. The initial number of treated subjects is $|A|$ and the number of controls is $|B|$, where $|\cdot|$ denotes the number of elements of a set. For each $a \in A$ and each $b \in B$, there is a distance, δ_{ab} with $0 \leq \delta_{ab} \leq \infty$. The distance measures the difference between a and b in terms of their observed covariates, say $\mathbf{x}_a$ and $\mathbf{x}_b$; however, it need not be a distance in the sense used to define a metric space, and it is not required to have any properties besides being nonnegative. An infinite distance, $\delta_{ab} = \infty$, indicates that $\mathbf{x}_a$ and $\mathbf{x}_b$ are so different that it is forbidden to place a and b in the same stratum.

The nature of the distance δ_{ab} is not important in characterizing the form of an optimal stratification. Many distances have been proposed. Cochran and Rubin (1973) consider the following distances:

(i) the categorical distance, if $\mathbf{x}_a = \mathbf{x}_b$, then $\delta_{ab} = 0$, otherwise $\delta_{ab} = \infty$;

(ii) caliper distance, if $|x_{aj} - x_{bj}| \leq c_j$ for each coordinate j of $\mathbf{x}$, then $\delta_{ab} = 0$, otherwise $\delta_{ab} = \infty$, where the c_j are given constants;

(iii) quadratic distances, $\delta_{ab} = (\mathbf{x}_a - \mathbf{x}_b)^{\mathrm{T}} \mathbf{D} (\mathbf{x}_a - \mathbf{x}_b)$ for some matrix $\mathbf{D}$, including the Mahalanobis distance in which $\mathbf{D}$ is the inverse of a sample variance–covariance matrix of the $\mathbf{x}$'s; and

(iv) the squared difference along a linear discriminant.

See also Carpenter (1977) and Rubin (1980) for discussion of the Mahalanobis distance. Smith, Kark, Cassel, and Spears (1977) standardized the coordinates of $\mathbf{x}$ and take δ_{ab} equal to the squared length of the standardized difference $\mathbf{x}_a - \mathbf{x}_b$; this is a quadratic distance with $\mathbf{D}$ equal to a diagonal matrix of reciprocals of sample variances. Similar measures have been defined with the coordinates of $\mathbf{x}$ replaced by their ranks. Rosenbaum and Rubin (1985) define three distances that use the propensity score and other coordinates of $\mathbf{x}$. Again, the nature of the distance is not important in §§10.3 and 10.4.

A *stratification* $(A_1, \ldots, A_S; B_1, \ldots, B_S)$ with S strata consists of S nonempty, disjoint subsets of A and S nonempty, disjoint subsets of B, so $|A_s| \geq 1$, $|B_s| \geq 1$, $A_s \cap A_{s'} = \emptyset$ for $s \neq s'$, $B_s \cap B_{s'} = \emptyset$ for $s \neq s'$, $A_1 \cup \cdots \cup A_S \subseteq A$, and $B_1 \cup \cdots \cup B_S \subseteq B$. For $s = 1, \ldots, S$, stratum s consists of the treated units from A_s, and the controls from B_s. Notice that a stratification may discard some units; that is, it may happen that $|A_1 \cup \cdots \cup A_S| < |A|$ or $|B_1 \cup \cdots \cup B_S| < |B|$. Write $\alpha = |A_1 \cup \cdots \cup A_S|$ and $\beta = |B_1 \cup \cdots \cup B_S|$, so the stratification includes α treated units and β controls, and call (α, β) the *size* of the stratification.

A *pair matching* is a stratification $(A_1, \ldots, A_S; B_1, \ldots, B_S)$ in which $|A_S| = |B_S| = 1$ for each s. A *matching with multiple controls* is a stratification $(A_1, \ldots, A_S; B_1, \ldots, B_S)$ in which $|A_s| = 1$ for each s. A *full matching* is a stratification $(A_1, \ldots, A_s; B_1, \ldots, B_S)$ in which $\min(|A_s|, |B_s|) = 1$ for each s, so a stratum consists of a single treated subject and one or more controls or else a single control and one or more treated subjects.

10.3.4   A Small Example: Nuclear Power Plants at New and Existing Sites

As an illustration, Table 10.6 describes 26 light water nuclear power plants built in the United States. The data are due to W. E. Mooz and are reported by Cox and Snell (1981). Seven of these plants were built on the site of an existing nuclear plant; they are plants $A = \{3, 5, 9, 18, 20, 22, 24\}$ and they form the columns of the table. The other nineteen plants were built at a new site; they are

$$B = \{1, 2, 4, 6, 7, 8, 10, 11, 12, 13, 14, 15, 16, 17, 19, 21, 23, 25, 26\}$$

and they form the rows of the table. The numerical labels in A and B correspond to those used by Cox and Snell (1981). Excluded are six "partial turnkey" plants whose costs may contain hidden subsidies. The comparison of interest is the cost of plants built at new or existing sites, adjusting for covariates related to the cost. The example is intended solely as a small illustration of matching and stratification.

TABLE 10.6. Covariate Distances Between Plants Built at a New Site (Rows) and Plants Built at an Existing Site (Columns). An Asterisk (*) Signifies a Plant Built in the Northeastern U. S.

ID#	3	*5	9	18	20	*22	24
*1	28	24	10	7	17	20	14
2	0	3	18	28	20	31	32
*4	3	0	14	24	16	28	29
*6	22	22	18	8	32	35	30
7	14	10	4	14	18	20	18
8	30	27	12	2	26	29	24
*10	17	14	5	10	20	22	17
11	28	26	11	6	18	20	16
*12	26	24	9	12	12	14	9
13	28	24	10	0	24	26	22
14	20	16	14	24	0	12	12
15	22	19	12	22	2	9	10
16	23	20	5	4	20	22	17
*17	26	23	14	24	6	5	6
19	21	18	22	32	7	15	16
21	18	16	10	20	4	12	14
23	34	31	16	18	14	9	4
25	40	37	22	16	20	11	8
26	28	25	28	38	14	12	17

The covariate $\mathbf{x}$ is two-dimensional and gives the year the construction permit was issued and the capacity of the power plant. The values in the table are distances δ_{ab} between power plants with $a \in A$ and $b \in B$. The distance is formed as follows. The two covariates in $\mathbf{x}$ were ranked separately from 1 to 26 with average ranks used for ties. Then δ_{ab} is the sum of the absolute values of the two differences in ranks for $\mathbf{x}_a$ and $\mathbf{x}_b$. For instance, $\delta_{3,2} = 0$ because plants 3 and 2 were started in the same year and had the same capacity, so their year covariates and their capacity covariates were each assigned the same tied rank, and the sum of the absolute differences in their tied ranks was zero.

The boxed distances describe a matching that is optimal among all matchings in which each plant in A is matched to two plants in B. For instance, plant 3 is matched to plant 2 and plant 21. Notice that plant 4 is better than plant 21 as a match for plant 3, but if plant 4 were matched to plant 3, then plant 5 would not receive its best match. The match described by the boxes is optimal in that it minimizes the total distance within pairs among all matchings with two controls. The construction of an optimal matching is discussed in §10.4.

To illustrate the notation in §10.3.3, the optimal match in Table 10.6 is the stratification of size $(\alpha, \beta) = (7, 14)$ with $S = 7$ strata $(A_1, \dots, A_7; B_1, \dots, B_7) = (\{3\}, \{5\}, \dots, \{24\}; \{2, 21\}, \{4, 7\}, \dots, \{23, 25\})$, so the first stratum consists of $A_1 = \{3\}$ together with $B_1 = \{2, 21\}$. The goal in §10.3 is to characterize optimal stratifications, and in §10.4 to construct them.

10.3.5 Evaluating Stratifications Based on the Average Distance Within Strata

A good stratification would place similar subjects in the same stratum. The distances between treated and control subjects in the same stratum would be small. Let $\delta(A_s, B_s)$ be the average of the $|A_s| \times |B_s|$ distances δ_{ab} with $a \in A_s$ and $b \in B_s$. For instance, in Table 10.6, $\delta(A_1, B_1) = \delta(\{3\}, \{2, 21\}) = (\delta_{3,2} + \delta_{3,21})/2 = (0 + 18)/2 = 9$. If $|A_s| = 2$ and $|B_s| = 5$ then $\delta(A_s, B_s)$ would be an average of 10 distances.

To find an optimal stratification, a numerical criterion is needed that evaluates a stratification $(A_1, \dots, A_S; B_1, \dots, B_S)$, combining the S distances $\delta(A_s, B_s), s = 1, \dots, S$ into a single number. For this purpose, introduce a weight function, $w(\cdot, \cdot)$. The distances within the S strata will be combined with weights $w(|A_s|, |B_s|)$, so the weights are a function of the sizes of the treated and control groups within each stratum. The weight function $w(\cdot, \cdot)$ is assumed to be strictly positive and finite, and it is defined for strictly positive integer arguments. Define the *distance* Δ for a stratification $(A_1, \ldots, A_S; B_1, \ldots, B_S)$ to be

$$\Delta = \sum_{s=1}^{S} w(|A_s|, |B_s|)\ \delta(A_s, B_s).$$

There are several natural choices of weight function. Let the stratification $(A_1, \ldots, A_S; B_1, \ldots, B_S)$ be of size (α, β). One weight function is $w(|A_s|, |B_s|) = |A_s|/\alpha$; that is, the weight is the proportion of the α treated subjects who fall in stratum s. In this case, Δ has a simple interpretation. From the stratification, pick one of the α treated subjects at random, and pick a control at random from the stratum containing the selected treated subject; then Δ is the expected distance between these two subjects. Another weight function is $w(|A_s|, |B_s|) = |B_s|/\beta$ with a similar interpretation. Still another weight function is $w(|A_s|, |B_s|) = (|A_s| + |B_s|)/(\alpha + \beta)$. With this weight function, pick one of the $\alpha + \beta$ subjects at random and pick a comparison subject at random in the same stratum but from the other treatment group; then Δ is the expected distance between these subjects. For these three weight functions, $1 = \sum w(|A_s|, |B_s|)$, so Δ is truly a weighted average of the $\delta(A_s, B_s)$. A different weight function is $w(|A_s|, |B_s|) = |A_S| \times |B_s|$; then Δ is the total of all distances within strata.

Distinguish three types of weight function. They are defined in terms of the impact of removing a pair of subjects from one stratum to form a new stratum comprised of just that pair. Does the total weight increase, decrease, or stay the same when a pair is separated? A weight function *favors large strata* if $w(p, q) < w(p - 1, q - 1) + w(1, 1)$ for all integers $p \geq 2$, $q \geq 2$, it *favors small strata* if $w(p, q) > w(p-1, q-1) + w(1, 1)$, and it is *neutral* if $w(p, q) = w(p - 1, q - 1) + w(1, 1)$. A weight function that favors large strata increases the total weight when a pair is separated, so there is a penalty for increasing the number of strata, increasing Δ even if the distances are not changed. Similarly, a weight function that favors small strata creates a reward for increasing the number of strata. A neutral weight function neither rewards nor penalizes the creation of additional strata. The three weight functions $|A_s|/\alpha$, $|B_s|/\beta$, and $(|A_s| + |B_s|)/(\alpha + \beta)$ are all neutral, while the weight function $|A_s| \times |B_s|$ favors small strata.

Notice that $\Delta = \infty$ if there is a stratum s and a pair of subjects $a \in A_s$ and $b \in B_s$ with $\delta_{ab} = \infty$. In words, if a comparison of a and b is forbidden, then placing a and b in the same stratum yields a stratification with an infinite distance Δ. If $\Delta = \infty$ call the stratification *unacceptable*, but if $\Delta < \infty$ call it *acceptable*.

10.3.6   The Structure of Optimal Strata

Consider a fixed data set, a fixed size (α, β) for the stratification, and a fixed weight function $w(\cdot, \cdot)$. A stratification of size (α, β) and distance Δ

is *optimal* if there is no other stratification of size (α, β) with a strictly smaller distance. It may happen that $\Delta = \infty$ for an optimal stratification, but in this case there is no acceptable stratification.

A *refinement* of a stratification $(A_1, \dots, A_S; B_1, \dots, B_S)$ is another stratification $(\tilde{A}_1, \dots, \tilde{A}_{\tilde{S}}; \tilde{B}_1, \dots, \tilde{B}_{\tilde{S}})$ of the same size with $\tilde{S} \geq S$ such that for each $\tilde{s}$, $\tilde{s} = 1, \dots, \tilde{S}$, there exists an s such that $\tilde{A}_{\tilde{s}} \subseteq A_s$ and $\tilde{B}_{\tilde{s}} \subseteq B_s$. In other words, a refinement subdivides strata.

The following proposition says that if there is no penalty for creating additional strata, then any stratification can be refined into a full matching which is as good or better. Recall that all of the weight functions discussed in §10.3.5 were either neutral or favored small strata. Proposition 30 is proved in the Problems in §10.7 and in Rosenbaum (1991).

Proposition 30 *Consider a weight function $w(\cdot, \cdot)$ that is either neutral or favors small strata. If the stratification $(A_1, \dots, A_S; B_1, \dots, B_S)$ has distance Δ_0 and is not a full matching, then it has a refinement which is a full matching and has distance no greater than Δ_0.*

For weight functions that are neutral or favor small strata, Proposition 30 has several consequences. First, if there is an acceptable stratification of size (α, β), then there is an acceptable full matching of size (α, β). Second, for each fixed size (α, β), there is a full matching that is an optimal stratification. In searching for an optimal stratification, it suffices to confine the search to full matchings because one of them is sure to be optimal. As will be seen in §10.3.7, neither statement is true for pair matching or for matching with multiple controls.

Under the conditions of Proposition 30, it can happen that the refinement is only as good as the original stratification, but no better. For this to happen, the distances δ_{ab} must satisfy a large number of linear equations. This will often happen when there are many distances δ_{ab} that are equal to zero. However, it may be shown that if the covariates $\mathbf{x}$ came from a multivariate normal distribution and the distance is the Mahalanobis distance, then with probability one the equations will not be satisfied, and the full matching will be strictly better than the original stratification. In other words, in the multivariate normal case, with probability one, a stratification that is not a full matching is not optimal. For proof, see Rosenbaum (1991, §4).

10.3.7 Pair Matching Is Not Optimal

Unlike an optimal full matching, an optimal pair matching is not generally an optimal stratification. The distance matrix in Table 10.7 demonstrates this. There are $|A| = 3$ treated subjects, $A = \{a, b, c\}$, $|B| = 3$ controls, $B = \{1, 2, 3\}$, and the distances take two values, $\omega > \varepsilon$.

TABLE 10.7. Hypothetical Distance Matrix.

		a	b	c
			A	
	1	ε	ε	ω
B	2	ω	ω	ε
	3	ω	ω	ε

For the neutral weight function $w(|A_s|, |B_s|) = (|A_s| + |B_s|)/(\alpha + \beta)$, the optimal stratification and optimal full matching is $(A_1, A_2; B_1, B_2) = (\{a, b\}, \{c\}; \{1\}, \{2, 3\})$ with distance $\Delta = (3\varepsilon + 3\varepsilon)/6 = \varepsilon$. The optimal pair matching is $(A_1, A_2, A_3; B_1, B_2, B_3) = (\{a\}, \{b\}, \{c\}; \{1\}, \{2\}, \{3\})$ with distance $(2\varepsilon + 2\omega + 2\varepsilon)/6 > \varepsilon$. By letting ω increase, the difference between optimal full matching and optimal pair matching may be made arbitrarily large.

In the same way, the optimal matching with multiple controls is not generally an optimal stratification.

10.4 Optimal Matching

10.4.1 Greedy Matching Versus Optimal Matching

In §10.4, the task is to build a matched sample from a matrix of distances δ_{ab} between treated subjects $a \in A$ and controls $b \in B$. Various types of matching are considered, including pair matching, matching with multiple controls, balanced matching, and full matching. Here, optimal refers to minimizing the total distance within matched sets.

The first impulse is to use what is known as a greedy algorithm. A greedy algorithm divides a large decision problem into a series of simpler decisions each of which is handled optimally, and makes those decisions one at a time without reconsidering early decisions as later ones are made. One greedy algorithm for matching finds the smallest distance δ_{ab}, for $a \in A$ and $b \in B$, calls this (a, b) the first matched pair, removes a from A and b from B, and repeats the process to find the next pair. Greedy algorithms do solve a small class of problems optimally, but the matching problem is not a member of that class.

In principle, greedy can perform very poorly as a matching algorithm. To see this, consider the distance matrix in Table 10.8 with two treated subjects and two controls, where ε is finite. Greedy pairs a with 1 at a cost of 0 and then is forced to pair b with 2 at a cost of ∞. The optimal match pairs a with 2 and b with 1 for a cost of 2ε. In principle, greedy can be arbitrarily poor compared to optimal matching.

The situation in Table 10.8 is not far-fetched when some pairings are forbidden through the use of infinite distances, as is true with category

TABLE 10.8. A Small Distance Matrix.

		A	
		a	b
B	1	0	ε
	2	ε	∞

matching and caliper matching. Suppose one were matching on age, requiring matched pairs to have ages that differ by less than five years. As with the caliper distance in §10.3.3, there is an infinite distance δ for any pair whose ages differ by more than five years; however, if the age difference is less than or equal to five years, the distance is the absolute difference in ages. If 1 and a are both 50 years old, 2 is 47 years old, and b is 53 years old, then Table 10.8 results with $\varepsilon = 3$. Greedy would pair the 50-year olds, and then it would have no acceptable match for the 53-year old. There are also theoretical results suggesting greedy can be very poor in large problems. See, for instance, Walkup (1979) and Snyder and Steele (1990). In §10.4.7, greedy and optimal matching are compared by simulation when covariates are multivariate normal.

If greedy were used to match each treated plant to two controls in Table 10.6, it would begin by pairing plant 3 to plant 2 at a cost of $\delta_{3,2} = 0$ units of distance, then plant 5 to 4 at a cost of $\delta_{5,4} = 0$, and so on. Table 10.9 compares greedy and optimal matching step by step.

In Table 10.9, greedy does well at the beginning when there is little competition for controls. In selecting the first 11 pairs, greedy and optimal matching have selected the same pairs and produced the same total distance. At Step 12, greedy misses a small opportunity because it never reconsiders previous decisions. Greedy adds the pair (22,26) at a cost of $\delta_{22,26} = 12$ units of distance. Instead, at Step 12, optimal matching deleted pair (20,15) and added pairs (20,21) and (22,15) at a cost of $-\delta_{20,15} + \delta_{20,21} + \delta_{22,15} = -2 + 4 + 9 = 11$. As the final matched pairs are selected and the competition for controls intensifies, greedy misses additional opportunities. In the end, the greedy match gives a total distance $(79 - 71)/71 = 11\%$ higher than necessary.

This figure, 11%, refers to this one example, and should not be anticipated in general. Greedy performs poorly when there is intense competition for controls and it performs well when there is little competition. There is, however, no reason to use a greedy algorithm. Fast optimal algorithms are widely available; see §10.6.

TABLE 10.9. Step-By-Step Comparison of Greedy and Optimal Matching.

	Greedy		Optimal		
Step	Add	Distance	Delete	Add	Distance
1	(3,2)	0		(3,2)	0
2	(5,4)	0		(5,4)	0
3	(18,13)	0		(18,13)	0
4	(20,14)	0		(20,14)	0
5	(18,8)	2		(18,8)	2
6	(20,15)	4		(20,15)	4
7	(9,7)	8		(9,7)	8
8	(24,23)	12		(24,23)	12
9	(22,17)	17		(22,17)	17
10	(9,10)	22		(9,10)	22
11	(24,25)	30		(24,25)	30
12	(22,26)	42	(20,15)	(20,21)	41
				(22,15)	
13	(5,21)	58	(9,7)	(9,16)	52
				(5,7)	
14	(3,19)	79	(22,15)	(22,26)	71
			(20,21)	(20,15)	
				(3,21)	

10.4.2   Optimal Matching and Minimum Cost Flow in a Network

Optimal matching is known to be equivalent to finding a minimum cost flow in a certain network, a problem that has been extensively studied and for which good algorithms exist. The current section sketches a few of the general ideas about network flow. These ideas are applied to matching problems in subsequent sections. One can, of course, use optimal matching without understanding network optimization, just as one can fit a logit model without understanding nonlinear optimization. An attractive, detailed, modern discussion of this subject is given by Bertsekas (1991); he also provides computer code in an Appendix and the code may be downloaded from a website. See also Ahuja, Magnanti, and Orlin (1993), Papadimitriou and Steiglitz (1982), Tarjan (1983), and Rockafellar (1984).

A network is a directed graph, that is, a set V of vertices and a set E of directed edges consisting of ordered pairs of elements of V. Later the vertices in the set V will include subjects available for matching, that is, V will contain $A \cup B$, and the set E of edges will include an edge (a, b) with $a \in A$ and $b \in B$ if a may be matched to b, that is, if $\delta_{ab} < \infty$. However, at times it will be convenient to include in V and E certain other vertices and edges, so for the moment V is any set of vertices and E is any set of ordered pairs of elements of V. A network (V, E) is depicted by drawing a dot for each vertex $v \in V$, and for each edge $e = (v_1, v_2) \in E$ an arrow from vertex v_1 to vertex v_2.

As an illustration, Figure 10.1 is a network for the matching problem in Table 10.8. The four vertices are $V = \{a, b, 1, 2\}$ and the three edges are $E = \{(a, 1), (a, 2), (b, 1)\}$. There is no edge $(b, 2)$ because $\delta_{b2} = \infty$, so b cannot be matched to 2.

$$
\begin{array}{ccc}
b & \rightarrow & 1 \\
 & \nearrow & \\
a & \rightarrow & 2
\end{array}
$$

Vertices $V = \{a, b, 1, 2\}$
Edges $E = \{(a, 1), (a, 2), (b, 1)\}$

FIGURE 10.1. Matching network for table 10.8.

Network flow theory was originally concerned with the movement of material from one vertex to another along the edges. One might think of a railroad network or a telephone network. A *flow* in a network assigns a number $\mathrm{FLOW}(i, j)$ to each edge $(i, j) \in E$ signifying that $\mathrm{FLOW}(i, j)$ units of material are to flow from vertex i to vertex j directly across edge (i, j). Network flow optimization finds the best flow for all $(i, j) \in E$ subject to various requirements. Each edge (i, j) must carry at least $\mathrm{MIN}(i, j)$ units of flow and at most $\mathrm{MAX}(i, j)$ units of flow, that is, $\mathrm{MIN}(i, j) \leq \mathrm{FLOW}(i, j) \leq \mathrm{MAX}(i, j)$ for all $(i, j) \in E$, where $\mathrm{MIN}(i, j)$ and $\mathrm{MAX}(i, j)$ are given numbers for each $(i, j) \in E$. For each vertex $i \in V$, the divergence is the total flow out from i minus the total flow into i; that is,

$$
\mathrm{DIV}(i) = \sum_{j:(i,j)\in E} \mathrm{FLOW}(i, j) - \sum_{k:(k,i)\in E} \mathrm{FLOW}(k, i).
$$

If there were a warehouse at vertex i, then $\mathrm{DIV}(i) > 0$ would signify that the stock in the warehouse is being depleted, $\mathrm{DIV}(i) < 0$ would signify that stock is building up, and $\mathrm{DIV}(i) = 0$ would signify that stock is arriving and leaving at the same rate. There are specified limits on the divergence at each vertex, $\mathrm{DMIN}(i) \leq \mathrm{DIV}(i) \leq \mathrm{DMAX}(i)$.

Shipping a unit of flow from i to j along (i, j) costs $\mathrm{COST}(i, j)$. The total cost of a flow is $\sum_{(i,j)\in E} \mathrm{FLOW}(i, j)\, \mathrm{COST}(i, j)$. The minimum cost flow problem is to find a flow, $\mathrm{FLOW}(i, j)$ for $(i, j) \in E$, which minimizes the total cost subject to the constraints, that is,

$$
\text{minimize} \sum_{(i,j)\in E} \mathrm{FLOW}(i, j)\, \mathrm{COST}(i, j), \tag{10.2}
$$

subject to

$$
\mathrm{MIN}(i, j) \leq \mathrm{FLOW}(i, j) < \mathrm{MAX}(i, j) \qquad \text{for all} \quad (i, j) \in E,
$$

and

$$\text{DMIN}(i) \leq \text{DIV}(i) \leq \text{DMAX}(i) \qquad \text{for} \quad i \in V.$$

See Bertsekas (1991, Exercise 1.6, p. 19) for discussion of the relationship between this definition of the minimum cost flow problem and other equivalent definitions.

An *integer flow* is a flow in which $\text{FLOW}(i, j)$ is an integer for each $(i, j) \in E$. Integer flows arise naturally when it is impossible to divide a unit in half for shipping. Also, in matching problems, only integer flows make sense because whole units must be matched. If all of the capacity constraints, $\text{MIN}(i, j)$, $\text{MAX}(i, j)$, $\text{DMIN}(i)$, and $\text{DMAX}(j)$, are integers, then whenever there is a minimum cost flow there is also an integer minimum cost flow (e.g., Tarjan (1983, p. 110)). Notice that the costs $\text{COST}(i, j)$ need not be integers. Throughout this chapter, $\text{MIN}(i, j)$, $\text{MAX}(i, j)$, $\text{DMIN}(i)$, and $\text{DMAX}(j)$ are integers, and the solution to the minimum cost flow problem is assumed to be one of the integer solutions. This happens automatically with commonly used algorithms.

Good algorithms exist for solving the minimum cost flow problem. In particular, there is an algorithm for optimal matching which requires computational effort that grows no faster than the cube of the number of subjects to be matched; see Papadimitriou and Steiglitz (1982, Theorem 11.1, p. 250) or Tarjan (1983, Theorem 8.13, p. 110). For comparison, the conventional way of multiplying two square matrices with one row per subject requires effort that grows as the cube of the number of subjects. In other words, the rate of growth in the difficulty of these two problems is similar. Bertsekas (1991, §5) compares the performance of several algorithms on a Macintosh Plus computer. In particular, with his favored algorithm for pair matching, he reports solving in less than six seconds optimal pair matching problems with 5000 treated subjects, 5000 controls, and 25,000 permitted matched pairs (i.e., $|A| = 5000$, $|B| = 5000$, $|E| = |\{(a, b) : a \in A, b \in B, \delta_{ab} < \infty\}| = 25,000$).

The remainder of §10.4 indicates how to set up various matching problems as minimum cost flow problems. Once cast as minimum cost flow problems, available algorithms provide optimal solutions.

10.4.3   Matching with a Fixed Number of Controls

In matching with a fixed number of controls, each of the treated subjects in A is to be matched with a fixed number, say $k \geq 1$, of controls from B to minimize the total of the $k|A|$ distances between treated subjects and their matched controls. This becomes a minimum cost flow problem with the following identifications. An alternative approach is discussed in Problem 9.

The vertices are the subjects, $V = A \cup B$. The edges link treated subjects to controls for which the distance is finite, $E = \{(a, b) : a \in$

$A, b \in B, \delta_{ab} < \infty\}$. See Figure 10.1. The cost is simply the distance, $\mathrm{COST}(a, b) = \delta_{ab}$ for all $(a, b) \in E$. The bounds on the flow are $\mathrm{MIN}(a, b) = 0$ and $\mathrm{MAX}(a, b) = 1$ for all $(a, b) \in E$. The bounds on the divergences are $\mathrm{DMIN}(a) = \mathrm{DMAX}(a) = k$ for all $a \in A$, and $\mathrm{DMIN}(b) = -1$, $\mathrm{DMAX}(b) = 0$ for all $b \in B$.

Suppose that there is an integer flow $\mathrm{FLOW}(a, b)$ that satisfies the constraints in (10.2). Since $\mathrm{MIN}(a, b) = 0$ and $\mathrm{MAX}(a, b) = 1$ and the $\mathrm{FLOW}(a, b)$ is an integer, it follows that $\mathrm{FLOW}(a, b) = 0$ or $\mathrm{FLOW}(a, b) = 1$ for all $(a, b) \in E$. Since $\mathrm{DMIN}(a) = \mathrm{DMAX}(a) = k$, it follows that for each $a \in A$, $\mathrm{FLOW}(a, b) = 1$ for exactly k subjects $b \in B$. Also, since $\mathrm{DMIN}(b) = -1$ and $\mathrm{DMAX}(b) = 0$, for each $b \in B$, there is at most one $a \in A$ such that $\mathrm{FLOW}(a, b) = 1$. In other words, a flow satisfies the constraints if and only if $\{(a, b): \mathrm{FLOW}(a, b) = 1\}$ is a matching of each treated subject to exactly k controls. A minimum cost flow is a matching that minimizes the total distance among all matchings that assign k controls to every treated unit. If no integer flow satisfies the constraints, then there is no acceptable matching with k controls, where an acceptable matching was defined in §10.3.5.

The boxed power plants in Table 10.6 are an optimal match with $k = 2$ controls; see also Problem 11. Notice that there is competition among treated plants for the same control. For instance, control 7 was matched to treated plant 5 at a distance of $\delta_{5,7} = 10$, though 7 is closer to plant 9, and so on. See also Table 10.9.

10.4.4 Matching with a Variable Number of Controls

An alternative to matching each treated subject $a \in A$ to k controls $b \in B$ is to require each treated subject to have at least $k_{\min}$ controls and at most $k_{\max}$ controls, with a total of h controls in the matched sample. In this case, the optimization chooses the best number of controls for each treated subject. To produce a matching, the parameters $k_{\min}$, $k_{\max}$, and h should satisfy $k_{\min} \geq 1$ and $k_{\max} \leq |B| - |A| + 1$, and $|A|k_{\min} \leq h \leq |A|k_{\max}$.

Optimal matching of h controls with at least $k_{\min}$ controls and at most $k_{\max}$ controls becomes a minimum cost flow problem in the following way. Add to the vertex set a new vertex called the SINK, so $V = A \cup B \cup \{\mathrm{SINK}\}$. Add an edge from each $b \in B$ to the sink, so $E = \{(a, b) : a \in A, b \in B, \delta_{ab} < \infty\} \cup \{(b, \mathrm{SINK}) : b \in B\}$. See Figure 10.2 which adds a SINK to Figure 10.1.

The SINK is needed to ensure that optimal matching does not reduce the cost by matching fewer than h controls. Set $\mathrm{DMIN}(a) = k_{\min}$, $\mathrm{DMAX}(a) = k_{\max}$ for all $a \in A$, $\mathrm{DMIN}(b) = \mathrm{DMAX}(b) = 0$ for all $b \in B$, and $\mathrm{DMIN}(\mathrm{SINK}) = \mathrm{DMAX}(\mathrm{SINK}) = -h$. Set $\mathrm{MIN}(b, \mathrm{SINK}) = 0$, $\mathrm{MAX}(b, \mathrm{SINK}) = 1$, and $\mathrm{COST}(b, \mathrm{SINK}) = 0$ for all $b \in B$. Define all other quantities as in §10.4.3. In an integer FLOW, each of h controls must send one unit of flow to the SINK, and each of these controls must receive its

TABLE 10.10. Three Matched Samples with a Variable Number of Controls.

Treated Unit	$k_{min} = 1$ $k_{max} = 13$	$k_{min} = 1$ $k_{max} = 4$	$k_{min} = 2$ $k_{max} = 3$
3	2	2	2, 6
5	4	4	4, 7
9	7, 10, 12	1, 7, 10, 16	1, 10, 16
18	1, 6, 8, 11, 13, 16	6, 8, 11, 13	8, 11, 13
20	14, 15, 19, 21	14, 15, 19, 21	14, 19, 21
22	17, 26	17, 26	15, 17, 26
24	23, 25	12, 23, 25	12, 23, 25
Total Distance	87	91	118

unit of flow from a different treated unit. An integer FLOW satisfies the constraints in (10.2) if and only if $\{(a, b) : a \in A, b \in B, \text{FLOW}(a, b) = 1\}$ is a matching with h controls in which each treated unit has at least k_{min} and at most k_{max} controls. A minimum cost flow is an optimal matching.

Table 10.10 gives optimal matched samples with a variable number of controls for three choices of k_{min} and k_{max}. The table identifies the controls matched to each treated plant. For instance, in the first match with $k_{min} = 1$ and $k_{max} = 13$, treated unit 9 is matched to controls 7, 10, and 12. Every control is matched to some treated unit, so $h = 19$, unlike the match in §10.4.3 where some controls were discarded. The first match in Table 10.10 has $k_{min} = 1$ and $k_{max} = 13$, so it requires only that every treated plant have at least one control. This first match assigns one control to treated unit 3 and six controls to treated unit 18. The last match has $k_{min} = 2$ and $k_{max} = 3$. Remember that there are $|B| = 19$ controls and $|A| = 7$ treated units. Since $19/7 = 2.71$, to match with $k_{min} = 2$ and $k_{max} = 3$ is to require the number of controls to be as nearly constant as possible.

Table 10.10 also gives the total distance within matched sets for each of the three matched samples. The match with $k_{min} = 1$ and $k_{max} = 4$ has a total distance only slightly greater than the match with $k_{min} = 1$ and $k_{max} = 13$, and both are much better than the match with $k_{min} = 2$ and $k_{max} = 3$. A substantial price was paid in giving more than one control to treated units 3 and 5.

$$b \rightarrow 1$$
$$\nearrow \qquad \searrow$$
$$a \rightarrow 2 \rightarrow SINK$$

Vertices $V = \{a, b, 1, 2, SINK\}$
Edges $E = \{(a, 1), (a, 2), (b, 1), (1, SINK), (2, SINK)\}$

FIGURE 10.2. A network with a sink.

An alternative method for matching with variable controls is discussed in Problem 10 and in Ming and Rosenbaum (2001).

10.4.5 Optimal Full Matching

Recall that in a full matching, each matched set contains a treated unit and one or more controls or a control unit and one or more treated units. This section finds a full matching of all subjects that minimizes the total distance. Assume at first that $\delta_{ab} > 0$ for each $a \in A$ and $b \in B$; this assumption can be removed, as indicated at the end of this section.

Optimal full matching uses the network in §10.4.3 without a SINK but with the following changes. Set $\text{DMIN}(a) = 1$ and $\text{DMAX}(a) = |B|$ for each $a \in A$ and $\text{DMIN}(b) = -|A|$ and $\text{DMAX}(b) = -1$ for each $b \in B$. Define other quantities as in §10.4.3. Then an integer FLOW that satisfies the constraints in (10.2) must have at least one unit of flow leaving each treated unit, since $\text{DMIN}(a) = 1$, and it must have at least one unit of flow entering each control, since $\text{DMAX}(b) = -1$, so every treated unit is connected to one or more controls and every control is connected to one or more treated units. However, the set $\{(a,b) : a \in A, b \in B, \text{FLOW}(a,b) = 1\}$ may not define a full matching because a treated unit a might be connected to several controls including b but b might be connected to several treated units. This is not a problem, however, if FLOW has minimum cost, as will now be demonstrated.

Find a minimum cost integer FLOW. It will now be shown that because FLOW has minimum cost, the set $\{(a,b) : a \in A, b \in B, \text{FLOW}(a,b) = 1\}$ does define a full matching. Suppose not, so there are a, $a^* \in A$ and b, $b^* \in B$ such that $\text{FLOW}(a,b) = 1$, $\text{FLOW}(a^*,b) = 1$, and $\text{FLOW}(a^*,b^*) = 1$. Define a new flow called BETTER such that $\text{BETTER}(a^*,b) = 0$ and $\text{BETTER}(i,j) = \text{FLOW}(i,j)$ for all $(i,j) \neq (a^*,b)$. Then BETTER satisfies the constraints and it has a cost that is lower by $\delta_{a^*,b} > 0$, so FLOW is not optimal, and there is a contradiction. Hence, a minimum cost integer FLOW is an optimal full matching.

It was assumed at the beginning of this section that $\delta_{ab} > 0$ for all $a \in A$ and $b \in B$. This assumption will now be removed. Suppose that $\delta_{ab} = 0$ for d pairs (a,b) with $a \in A$ and $b \in B$. Construct a new minimum cost flow problem identical to the original problem except that $\delta_{ab} = \varepsilon > 0$ for these d pairs. The minimum cost flow for the new problem is a full matching and the total cost is at most $d\varepsilon$ higher than the minimum cost of the original problem. Since ε was arbitrary, it follows that for sufficiently small ε, the minimum cost flow for the new problem is also a minimum cost flow for the original problem.

As it turns out, the first match in Table 10.10 is an optimal full matching.

10.4.6   Optimal Balanced Matching

In the nuclear power plant example in Table 10.6, two of the treated plants were constructed in the northeastern United States, specifically plants 5 and 22, and six of the control plants were constructed in the northeast. Construction costs, the outcome, are generally high in the northeast, so it would be desirable to match on this covariate in addition to the date of construction and the capacity of the plant. In the minimum distance match with two controls, $2/7 = 29\%$ of the treated plants were constructed in the northeast, but only $3/14 = 21\%$ of the controls were constructed there. The matched sample would be balanced if $4/14 = 29\%$ of the controls came from the northeast. This section finds the matched sample which has the minimum total distance among all balanced samples with two controls per treated unit.

In general, suppose that units are divided into C categories, $c = 1, \ldots, C$, that m_c treated subjects fall in category c, and that at least km_c controls are available from category c. The task is to match each treated unit with k controls so that a total of km_c controls fall in category c, and the total distance is as small as possible.

The following network is used. Introduce C new vertices, $\text{SINK}_1, \ldots,$ SINK_C. The vertices are the subjects together with the sinks, $V = A \cup B \cup \{\text{SINK}_1, \ldots, \text{SINK}_C\}$. The edges E are of two types. There is an edge linking each treated subject $a \in A$ to each control $b \in B$ for which the distance is finite, $\delta_{ab} < \infty$. These edges have $\text{COST}(a, b) = \delta_{ab}$. There is also an edge linking each control in category c to SINK_c, so each control is linked to exactly one sink. These edges connecting $b \in B$ to a sink have COST equal to zero. The set E of edges is the union of these two sets of edges. All edges $(i, j) \in E$ have $\text{MIN}(i, j) = 0$ and $\text{MAX}(i, j) = 1$. The bounds on the divergences are $\text{DMIN}(a) = \text{DMAX}(a) = k$ for all $a \in A$, $\text{DMIN}(b) = \text{DMAX}(b) = 0$ for all $b \in B$, and $\text{DMIN}(\text{SINK}_c) = \text{DMAX}(\text{SINK}_c) = -km_c$ for $c = 1, \ldots, C$.

Suppose that there is an integer flow $\text{FLOW}(a, b)$ that satisfies the constraints in (10.2). Then the set $\{(a, b) : a \in A, b \in B, \text{FLOW}(a, b) = 1\}$ defines a balanced matched sample with k controls for each treated subject. To see this, note the following. A total of k units of flow leave each treated unit $a \in A$ because $\text{DMIN}(a) = \text{DMAX}(a) = k$, so each treated unit is paired with k controls. No control $b \in B$ can be matched to more than one treated unit because $\text{DMIN}(b) = \text{DMAX}(b) = 0$ and $\text{MAX}(b, \text{SINK}_c) = 1$ for exactly one c, so unit b must pass on all the flow it receives and can pass on at most one unit of flow. The sample is balanced because $\text{DMIN}(\text{SINK}_c) = \text{DMAX}(\text{SINK}_c) = -km_c$, so there must be km_c controls from category c. A minimum cost integer flow therefore defines a balanced matched sample of minimum total distance.

Table 10.11 compares optimal balanced matching with greedy matching and optimal matching for the example in Table 10.6. All three matched

TABLE 10.11. Comparison of Greedy, Optimal and Balanced Matching.

Treated Unit	Greedy	Optimal	Optimal Balanced
3	2, 19	2, 21	2, 10*
*5	4*, 21	4*, 7	4*, 7
9	7, 10*	10*, 16	12*, 16
18	8, 13	8, 13	8, 13
20	14, 15	14, 15	14, 21
*22	17*, 26	17*, 26	15, 17*
24	23, 25	23, 25	23, 25
Total Distance	79	71	74.5
% Treated in Northeast	29	29	29
% Control in Northeast	21	21	29

samples have two controls for each treated plant. The greedy and optimal match are from §10.4.3. The optimal balanced match forces 29% of the control plants to come from the northeast because 29% of the treated plants are from the northeast.

In Table 10.11, optimal balanced matching corrects the imbalance in plants from the northeast with a distance $5\% = (74.5 - 71)/71$ higher than the unbalanced optimal match. The optimal balanced match is better than the greedy match both in being balanced and having a smaller total distance.

The method described in this section balanced a single categorical covariate and minimized distance among such balanced matched samples. It is possible to balance many variables at once, but this requires integer programming rather than network optimization. See Li, Propert, and Rosenbaum (2001) for discussion with an example.

10.4.7 Comparison of Matching Methods Using Simulation

In a simulation study, Gu and Rosenbaum (1993) compared the methods described in this chapter. This section outlines some of the findings, though the paper should be consulted for detailed discussion, numerical results, and specifics.

The simulation considered multivariate normal covariates $\mathbf{x}$ with common covariance matrix in the treated and control groups but different mean vectors. The number of covariates or dimension of $\mathbf{x}$ was 2, 5, or 20. In each matched sample, there were $|A| = 50$ treated subjects. The number of controls available for matching was $|B| = 50, 100, 150,$ or 300. For each matching situation considered, 50 different matched samples were constructed.

Three distances δ_{ab} were used. The first was the absolute difference in propensity scores estimated using a logit model. The second was the Mahalanobis distance. The third method, $M + P$, used a caliper on the propensity score, so $\delta_{ab} = \infty$ if a and b had propensity scores differing by more than 0.6 times the standard deviation of the propensity scores, and otherwise δ_{ab} was the Mahalanobis distance. This third distance was proposed by Rosenbaum and Rubin (1985).

Two types of measure were used to judge the quality of a matched sample. The first was the average distance within matched sets. The average distance is the total distance that was minimized by optimal matching divided by the number of distances summed to form the total. The second was the balance of the matched groups, that is, the difference in covariate mean vectors in treated and control groups after matching. When the matched sets have unequal sizes, as is true for some matching methods, balance is defined in terms of the average over matched sets of the average difference between treated and control subjects within matched sets. Different distances δ_{ab} may be compared in terms of balance but not in terms of total distance, since the difference in propensity scores is not commensurate with the Mahalanobis distance.

The main findings follow.

- When there were 20 covariates, matching on the propensity score produced greater covariate balance than matching using the Mahalanobis distance. When there were two covariates, the methods were similar with no consistent winner. The combined distance, $M + P$, often fell between the other methods but rarely performed poorly.

- When a fixed number $k|A|$ of controls were matched, full matching was found to be much better than matching k controls to each treated subject. That is, the method of §10.4.5 was much better than the method of §10.4.3 when the same total number of controls was matched. This was true for both the average distance and covariate balance.

- Optimal matching was sometimes much better than greedy matching and sometimes only marginally better. Optimal matching was much better than greedy at minimizing the average Mahalanobis distance when there were few controls to choose from, $|B| = 50$ or $|B| = 100$, and the number of covariates was 2 or 5. Optimal matching was only slightly better than greedy at minimizing the propensity distance. Optimal matching was no better than greedy at producing covariate balance.

Some caveats are needed. The comparisons of different distances refer only to covariate balance, because the distances themselves are not commensurate. The comparison of full matching and matching with a fixed number of controls refers only to average distance and balance, measures of bias. An optimal full matching may contain matched sets of extremely unequal sizes, and these may be unattractive for reasons of precision or aesthetics. In practice, with a given data set, a full matching, a matching with a fixed number of controls, and several compromises may be compared, as in §10.4.4.

10.5 Bibliographic Notes

The discussion in §10.2 of the balancing property of the propensity score is based on Rosenbaum and Rubin (1983, 1984). Three empirical comparisons of analyses using propensity scores are Rosenbaum and Rubin (1985), Smith (1997), and Dehejia and Wahba (1999). The discussion of optimal stratification in §10.3 is from Rosenbaum (1991), and the discussion in §10.3.2 is from Ming and Rosenbaum (2000). There is a vast literature on network optimization and its relationship to matching. A concise place to start is Bertsekas (2001). Some early references are Kuhn (1955) and Ford and Fulkerson (1962), and modern discussions are given by Papadimitriou and Steiglitz (1982), Tarjan (1983), Rockafellar (1984), Bertsekas (1991, 2001), and Ahuja, Magnanti, and Orlin (1993). The material and examples in §10.4 are from Rosenbaum (1989), except for §10.4.5 which is from Rosenbaum (1991). Section 10.4.7 summarizes the simulation study in Gu and Rosenbaum (1993). Other simulations of greedy matching with a single control are given in Rubin (1973, 1979, 1980). Li, Propert, and Rosenbaum (2001) discuss matching on longitudinal covariate histories when treatment is delayed for varied periods of time. Joffe and Rosenbaum (1999) and Lu, Zanutto, Hornik and Rosenbaum (2002) discuss matching with doses of treatment; see also Imbens (2000) for an alternative approach. Matching forms a useful link between quantitative and qualitative research; see Rosenbaum and Silber (2001).

The methods of this chapter are widely used, and a few selected illustrations follow. Propensity scores are used by: Lecher (2000), Normand, Landrum, Guadagnoli, et al. (2001), Petersen, Normand, Daley, and McNeil (2000), Phillips, Spry, Sloane, and Hawes (2000), Powers and Rock (1999), and Silber, Kennedy, Even-Shoshan, et al. (2000). Optimal matching is used by: Rephann and Isserman (1994), Nuako, Ahlquist, Sandborn, et al. (1998), Ghavamian, Bergstralh, Blute, et al. (1999), and Warner, Warner, Offord, et al. (1999).

10.6 Software

The availability of software changes rapidly. A brief discussion follows.

Propensity scores may be estimated using a logit model available in most statistical software packages. Propensity scores with several doses may be estimated using McCullagh's (1980) ordinal logit model, also available in most statistical software packages. Stratification on an estimated propensity score is straightforward.

When properly coded, algorithms for optimal matching are very fast, often faster than casually coded greedy algorithms. A brief, elementary, clear exposition of one of the simplest and fastest algorithms—the auction algorithm—for optimal pair matching is given by Bertsekas (2001). It is easy to translate Bertsekas' exposition of the auction algorithm into casual, brief S-Plus code; see Problem 11. Such an implementation is of value for learning about the algorithm in small problems, but a quick algorithm requires careful programming and cannot loop within S-Plus.

An algorithm for optimal pair matching can be tricked into performing optimal matching with a fixed number of controls as in §10.4.3 or with a variable number of controls as in §10.4.4. The trick involves introducing duplicate copies of treated subjects and phantom controls so that an optimal pair matching with the duplicates and phantoms equals an optimal matching with multiple controls. The easy details, an example, and SAS code are given by Ming and Rosenbaum (2001) and are discussed in Problems 8 through 11.

The SAS System has several procedures or PROCs that are useful in optimal matching, PROC ASSIGN and PROC NETFLOW in SAS/OR. Specifically, PROC ASSIGN implements the assignment algorithm, or optimal pair matching: it starts with a distance matrix, such as Table 10.6, and returns an optimal pair matching. Programming features in SAS may be used to construct the distance matrix. As just noted, PROC ASSIGN can be tricked into performing optimal matching with multiple controls. More complex matching problems, such as full matching, may be solved using PROC NETFLOW; here, however, the user must program the construction of a network. Bergstralh, Kosanke, and Jacobsen (1996) provide downloadable SAS macros that simplify optimal matching in SAS. In addition, the SAS Web site, http://www.sas.com, has a few basic technical reports on propensity scores and matching which may be located using their search engine.

Bertsekas (1991) provides downloadable subroutines for network optimization.

10.7   Problems

1. **Refinement without improvement.** Find a distance matrix δ_{ab} and a stratification $(A_1, \ldots, A_S; B_1, \ldots, B_S)$ such that the refinement in Proposition 30 is as good as but no better than the stratification.

2. **Full matching and the support of the distribution of covariates.** Let $x_1, \ldots, x_n$ be independent random variables with distribution $F(\cdot)$ and let $\tilde{x}_1, \ldots, \tilde{x}_n$ be independent random variables with distribution

$\tilde{F}(\cdot)$, where the x's are independent of the $\tilde{x}$'s. Suppose that we match x's to $\tilde{x}$'s to minimize the distance $\delta_{ab} = |x_a - \tilde{x}_b|$. Consider both a pair matching of each x to a single $\tilde{x}$ and a full matching. Within each matched set, calculate the average x and the average $\tilde{x}$ and difference these two averages. Each matched set produces one such difference; average the differences across matched sets and call the result B. Then B is a measure of covariate balance. Note that B is computed from certain $x_a - \tilde{x}_b$, not from $|x_a - \tilde{x}_b|$. Suppose that $F(\cdot)$ and $\tilde{F}(\cdot)$ are two beta distributions on the interval $[0, 1]$ with different means. How does B behave as $n \to \infty$ for pair matching and full matching? Suppose that $F(\cdot)$ is uniform on $[0, 1]$ and $\tilde{F}(\cdot)$ is uniform on $[1, 2]$. How does B behave as $n \to \infty$ for pair matching and full matching? (Hint: Does pair matching affect B?) What does this suggest about what full matching can and cannot do?

3. **Proof concerning full matching.** Problems 3 through 7 prove Proposition 30, and the notation of that proposition is used. A proof of Proposition 30 along these lines is given in Rosenbaum (1991, §4). Why can we assume, without loss of generality, that $\Delta_0 < \infty$? Why can we assume that there is some s such that $|A_s| \geq 2$ and $|B_s| \geq 2$? Fix that s.

4. **Proof concerning full matching, continued.** Pick any two units $a \in A_s$ and $b \in B_s$. Suppose that (a, b) are separated from stratum s to form a new stratum $S+1$ with $A_{S+1} = \{a\}$ and $B_{S+1} = \{b\}$. Let $\theta(a, b)$ be the difference between the distance Δ_0 for the original stratification and the distance for the new stratification in which (a, b) are in their own strata. Express $\theta(a, b)$ in terms of $w(|A_s|, |B_s|)$, $w(|A_s|-1, |B_s|-1)$, $w(1, 1)$, $\delta(A_s, B_s)$, $\delta(A_s - \{a\}, B_s - \{b\})$, and δ_{ab}. Here, $A_s - \{a\}$ means the set A_s with a removed.

5. **Proof concerning full matching, continued.** What is the average of δ_{ab} over all $a \in A_s$ and $b \in B_s$? What is the average of $\delta(A_s - \{a\}, B_s - \{b\})$ over $a \in A$, and $b \in B_s$?

6. **Proof concerning full matching, continued.** Show that the average of $\theta(a, b)$ over $a \in A_s$ and $b \in B_s$ is nonnegative. (Hint: Use Problem 5 and the assumption that the weight function is either neutral or favors small strata.)

7. **Proof concerning full matching, continued.** Conclude that there is some $a \in A_s$ and $b \in B_s$ such that $\theta(a, b) \geq 0$. Separate this (a, b) to form a new stratum. How does repeating the process prove Proposition 30?

8. **Pair matching and the assignment algorithm.** The assignment problem is optimal pair matching of m treated subjects to m controls, with an $m \times m$ distance matrix, **a**. Suppose you had an algorithm for

the assignment problem, but you had n treated subjects and $m > n$ controls, and you wanted an optimal pair matching. For instance, Table 10.6 is an $m \times n = 19 \times 7$ distance matrix. Pad the $m \times n$ distance matrix with zeros to make it a square $m \times m$ matrix; that is, add $m - n$ phantom columns. Show that an optimal assignment from the square $m \times m$ matrix yields an optimal pair matching of the n treated subjects once the $m - n$ phantoms and their matched controls are discarded.

9. Matching with two controls and the assignment algorithm. Continuing Problem 8, suppose $m \geq 2n$, and that you want to use the assignment algorithm to match each of the n treated subjects to two controls. Duplicate the $m \times n$ distance matrix, yielding an $m \times 2n$ matrix, and pad it with $m - 2n$ columns of zeros to yield an $m \times m$ matrix. Show that an optimal assignment from the square $m \times m$ matrix yields an optimal matching of the n treated subjects with two controls once the $m - n$ phantoms and their matched controls are discarded.

10. Matching with variable controls and the assignment algorithm. Continuing Problem 9, show how to use the assignment algorithm to match with a variable number of controls. (Ming and Rosenbaum 2001)

11. ToyAuction: Splus code for a toy version of the auction algorithm. Read Bertsekas (2001) and write up the auction algorithm for the assignment problem in S-Plus code. (Hint:

```
> ToyAuction
function(a)
{
# a is a square matrix of nonnegative integer distances
#
# Note carefully: a is square! Pad a nonsquare matrix
# with zeros (phantoms) to make it square & discard them
# later. See Ming & Rosenbaum (2001) J Comp Graph Stat
#
# b is a vector, where row i is matched to column b[i]
#
# Implements a simple version of the auction algorithm
# for the assignment problem. The code closely follows
# the description in Bertsekas (2001) Auction Algorithms
# in Encyclopedia of Optimization. The code is a toy --
# to learn about small examples. No effort has been made
# to make the code efficient. ToyAuction is slow, but
# the auction algorithm can be fast.
#
  a <- - a #change signs -- maximize negative distances
```

```
quit <- 10000 #loop limit
qcount <- 0 #loop counter
n <- dim(a)[1] #remember a is square!
eps <- 0.95/n #small eps is slow, but ensures optimal match
p <- rep(0, n) #initial prices
b <- 1:n #initial match: row i matched to b[i]
bid <- rep(0, n)
happy <- 0
go <- T
while(go) {
    for(i in 1:n) {
        gain <- a[i, ] - p
        o <- order(gain)
        if(gain[b[i]] >= gain[o[n]] - eps) {
            happy <- happy + 1
            }
        else {
            bid[o[n]] <- 1
            p[o[n]] <- p[o[n]] + gain[o[n]] + eps - gain[o[n -
1]] #update price
            k <- (1:n)[b == o[n]] #current owner of o[n]
            b[k] <- b[i] #give current owner i's object
            b[i] <- o[n] #give i best choice
                }
        }
    qcount <- qcount + 1
    if(happy == n) {
    go <- F
        }
    happy <- 0
    if(sum(bid) == n) {
    go <- F
        }
    if(qcount > quit) {
        b <- ''too many iterations''
        go <- F
        }
    }
    b
}
```

The distance matrix in Table 10.6 is given in the matrix `temp`.
```
> temp
        1  2  3  4  5  6  7
[1,] 28 24 10  7 17 20 14
[2,]  0  3 18 28 20 31 32
```

```
[3,]   3   0 14 24 16 28 29
[4,]  22 22 18  8 32 35 30
[5,]  14 10  4 14 18 20 18
[6,]  30 27 12  2 26 29 24
[7,]  17 14  5 10 20 22 17
[8,]  28 26 11  6 18 20 16
[9,]  26 24  9 12 12 14  9
[10,] 28 24 10  0 24 26 22
[11,] 20 16 14 24  0 12 12
[12,] 22 19 12 22  2  9 10
[13,] 23 20  5  4 20 22 17
[14,] 26 23 14 24  6  5  6
[15,] 21 18 22 32  7 15 16
[16,] 18 16 10 20  4 12 14
[17,] 34 31 16 18 14  9  4
[18,] 40 37 22 16 20 11  8
[19,] 28 25 28 38 14 12 17
```

The next step takes `temp` and converts it into a square 19×19 matrix using the idea in Problem 9. Specifically, the seven columns are duplicated and padded with zeros. Columns 1 and 8 refer to power plant 3, columns 2 and 9 refer to power plant 5, and so on. Columns 15 through 19 are phantoms, to be discarded after matching.

```
> temp2<-cbind(temp,temp,matrix(0,19,5))
```

Applying ToyAuction to `temp2` gives an optimal match with two controls.

```
> ToyAuction(temp2)
 [1] 19 8 9 16 2 4 3 15 14 11 5 12 10 13 18 1 7 6 17
```

For example, the first control in row #1 is matched to column #19, which is a phantom, so the first control is not used. The second control in row #2 is matched to column #8, which is the second copy of the first column; that is, the second control is matched to power plant 3 in the first column of Table 10.6. The third control, power plant 4 in Table 10.6, is matched to column 9, which is the second copy of column 2 in Table 10.6, that is, power plant 5. The optimal match found by ToyAuction has the same distance as the optimal match in Table 10.6, but they are not quite identical. The match in Table 10.6 included the power plant pairs $(25, 24)$ with a distance of 8 and $(26, 22)$ with a distance of 12, for a total of $8 + 12 = 20$. The match using ToyAuction replaced these two pairs by $(12, 24)$, or row 9 with column 14, with a distance of 9, and $(25, 22)$, or row 18 with column 6, for a distance of 11, for a total of $9 + 11 = 20$.)

10.8   References

Ahuja, R. K., Magnanti, T. L., and Orlin, J. B. (1993) *Network Flows: Theory, Algorithms, and Applications.* New York: Prentice-Hall.

Bergstralh, E. J., Kosanke, J. L., and Jacobsen, S. L. (1996) Software for optimal matching in observational studies. *Epidemiology,* **7,** 331–332. http://www.mayo.edu/hsr/sasmac.html.

Bertsekas, D. (1991) *Linear Network Optimization: Algorithms and Codes.* Cambridge, MA: MIT Press. http://www.mit.edu:8001//people/ dimitrib/home.html.

Bertsekas, D. (2001) Auction algorithms. *Encyclopedia of Optimization,* C. A. Foudas and P. M. Pardalos, eds., Kluwer.

Carpenter, R. (1977) Matching when covariables are normally distributed. *Biometrika,* **64,** 299–307.

Cochran, W. G. (1968) The effectiveness of adjustment by subclassification in removing bias in observational studies. *Biometrics,* **24,** 205–213.

Cochran, W. G. and Rubin, D. B. (1973) Controlling bias in observational studies: A review. *Sankya,* Series **A, 35,** 417–446.

Cox, D. R. (1970) *The Analysis of Binary Data.* London: Methuen.

Cox, D. R. and Snell, E. J. (1981) *Applied Statistics: Principles and Examples.* London: Chapman & Hall.

Dehejia, R. H. and Wahba, S. (1999) Causal effects in nonexperimental studies: Reevaluating the evaluation of training programs. *Journal of the American Statistical Association,* **94,** 1053–1062.

Ford, L. and Fulkerson, D. (1962) *Flows in Networks.* Princeton, NJ: Princeton University Press.

Ghavamian, R., Bergstralh, E. J., Blute, M. L., Slezak, J., and Zincke, H. (1999) Radical retropubic prostatectomy plus orchectomy versus orchietomy for pTxN+ prostate cancer: A matched comparison. *Journal of Urology,* **161,** 1223– 1227.

Gu, X. S. and Rosenbaum, P. R. (1993) Comparison of multivariate matching methods: Structures, distances and algorithms. *Journal of Computational and Graphical Statistics,* **2,** 405–420.

Imbens, G. W. (2000) The role of the propensity score in estimating dose-response functions. *Biometrika,* **87,** 706– 710.

Joffe, M. M. and Rosenbaum, P. R. (1999) Propensity scores. *American Journal of Epidemiology*, **150**, 327– 333.

Kuhn, H. W. (1955) The Hungarian method for the assignment problem. *Naval Research Logistics Quarterly*, **2**, 83–97.

Lechner, M. (2000) An evaluation of public-sector-sponsored continuous vocational training programs in East Germany. *Journal of Human Resources*, **35**, 347– 375.

Li, Y., Propert, K. J. and Rosenbaum, P. R. (2001) Balanced risk set matching. *Journal of the American Statistical Association*, **96**, September, to appear.

Lu, B., Zanutto, E., Hornik, R., and Rosenbaum, P. R. (2002) Matching with doses in an observational study of a media campaign against drug abuse. *Journal of the American Statistical Association*, to appear.

McCullagh, P. (1980) Regression models for ordinal data. *Journal of the Royal Statistical Society*, **B**, **42**, 109– 142.

Ming, K. and Rosenbaum, P. R. (2000) Substantial gains in bias reduction from matching with a variable number of controls. *Biometrics*, **56**, 118– 124.

Ming, K. and Rosenbaum, P. R. (2001) A note on optimal matching with variable controls using the assignment algorithm. *Journal of Computational and Graphical Statistics*, to appear.

Normand, S. T., Landrum, M. B., Guadagnoli, E., Ayanian, J. Z., Ryan, T. J., Cleary, P. D., and McNeil, B. J. (2001) Validating recommendations for coronary angiography following acute myocardial infarction in the elderly: A matched analysis using propensity scores. *Journal of Clinical Epidemiology*, **54**, 387– 398.

Nuako, K. W., Ahlquist, D. A., Sandborn, W. J., Mahoney, D. W., Siems, D. M., and Zinsmeister, A. R. (1998) Primary sclerosing cholangitis and colorectal carcinoma in patients with chronic ulcerative colitis: A case-control study. *Cancer*, **82**, 822– 826.

Papadimitriou, C. and Steiglitz, K. (1982) *Combinatorial Optimization*. Englewood Cliffs, NJ: Prentice-Hall.

Petersen, L. A., Normand, S., Daley, J., and McNeil, B. (2000) Outcome of myocardial infarction in Veterans Health Administration patients as compared with medicare patients. *New England Journal of Medicine*, 343, December 28, 2000, 1934– 1941.

Phillips C. D., Spry K. M., Sloane P. D., and Hawes, C. (2000) Use of physical restraints and psychotropic medications in Alzheimer special care units in nursing homes. *American Journal of Public Health*, **90**, 92– 96.

Powers, D. E. and Rock, D. A. (1999) Effects of coaching on SAT I: Reasoning test scores. *Journal of Educational Measurement*, **36**, 93– 118.

Rephann, T. and Isserman, A. (1994) New highways as economic development tools—An evaluation using quasi-experimental matching methods. *Regional Science and Urban Economics*, **24**, 723– 751.

Rockafellar, R. T. (1984) *Network Flows and Monotropic Optimization*. New York: Wiley.

Rosenbaum, P. R. (1984) Conditional permutation tests and the propensity score in observational studies. *Journal of the American Statistical Association*, **79**, 565–574.

Rosenbaum, P. R. (1987) Model-based direct adjustment. *Journal of the American Statistical Association*, **82**, 387–394.

Rosenbaum, P. R. (1989) Optimal matching for observational studies. *Journal of the American Statistical Association*, **84**, 1024–1032.

Rosenbaum, P. R. (1991) A characterization of optimal designs for observational studies. *Journal of the Royal Statistical Society*, Series **13**, **53**, 597–610.

Rosenbaum, P. R. and Rubin, D. B. (1983) The central role of the propensity score in observational studies for causal effects. *Biometrika*, **70**, 41–55.

Rosenbaum, P. R. and Rubin, D. B. (1984) Reducing bias in observational studies using subclassification on the propensity score. *Journal of the American Statistical Association*, **79**, 516–524.

Rosenbaum, P. R. and Rubin, D. B. (1985) Constructing a control group using multivariate matched sampling methods that incorporate the propensity score. *American Statistician*, **39**, 33–38.

Rosenbaum, P. R. and Silber, J. H. (2001) Matching and thick description in an observational study of mortality after surgery. *Biostatistics*, **2**, 217– 232.

Rubin, D. B. (1973) The use of matched sampling and regression adjustment to remove bias in observational studies. *Biometrics*, **29**, 185– 203.

Rubin, D. B. (1979) Using multivariate matched sampling and regression adjustment to control bias in observational studies. *Journal of the American Statistical Association*, **74**, 318–328.

Rubin, D. B. (1980) Bias reduction using Mahalanobis metric matching. *Biometrics*, **36**, 293–298.

Silber, J. H., Kennedy, S. K., Even-Shoshan, O., Chen, W., Koziol, L. F., Showan, A. M., Longnecker, D. E. (2000) Anesthesiologist direction and patient outcomes. *Anesthesiology*, **93**, 152–163.

Silber, J. H., Rosenbaum, P. R., Trudeau, M. E., Even-Shoshan, O., Chen, W., Zhang, X., and Mosher, R. E. (2001) Multivariate matching and bias reduction in the surgical outcomes study. *Medical Care*, to appear.

Smith, A., Kark, J., Cassel, J., and Spears, G. (1977) Analysis of prospective epidemiologic studies by minimum distance case–control matching. *American Journal of Epidemiology*, **105**, 567–574.

Smith, H. (1997) Matching with multiple controls to estimate treatment effects in observational studies. *Sociological Methodology*, **27**, 325–353.

Snyder, T. L. and Steele, J. M. (1990) Worst case greedy matching in the unit d-cube. *Networks*, **20**, 779–800.

Tarjan, R. (1983) *Data Structures and Network Algorithms*. Philadelphia: Society for Industrial and Applied Mathematics.

Walkup, D. (1979) On the expected value of a random assignment problem. *SIAM Journal on Computing*, **8**, 440–442.

Warner, D. O., Warner, M. A., Offord, K. P., Schroeder, D. R., Maxson, P., and Scanlon, P. D. (1999) Airway obstruction and perioperative complications in smokers undergoing abdominal surgery. *Anesthesiology*, **90**, 372–379.

11
Planning an Observational Study

11.1 Introduction and Some Ground Rules

11.1.1 Active Observation: Control and Choice

Many observational studies do not succeed in providing tangible, enduring, and convincing evidence about the effects caused by treatments, and those that do succeed often exhibit great care in their design. Particularly at the early stages of design, this care consists of choices that determine the circumstances of the study and the data to be collected. A convincing observational study is the result of active observation, an active search for those rare circumstances in which tangible evidence may be obtained to distinguish treatment effects from the most plausible biases. In an experiment, treatment effects are seen clearly because the environment is tightly controlled, whereas in a compelling observational study, control is, to a large extent, replaced by choice—the environment is carefully chosen.

Studies of samples that are representative of populations may be quite useful in describing those populations, but may be ill suited to inferences about treatment effects. For instance, describing early work in public program evaluation, Donald Campbell (1988, p. 324) wrote:

> There was gross overvaluing of, and financial investment in, external validity, in the sense of representative samples at the nationwide level. In contrast, the physical sciences are so provincial that they have established major discoveries like the hydrolysis of water... by a single water sample ....

Passive observation of a natural population followed by regression analysis is often unsuccessful as an approach to inference about treatment effects; see, in particular, Box (1966) and Freedman (1997). A report of the National Academy of Sciences (Meyer and Fienberg 1992, p106) concerning evaluation studies of bilingual education makes a similar point:

> In comparative studies, comparability of students in different programs is more important than having students who are representative of the nation as a whole. Elaborate analytic methods will not salvage poor design or implementation of a study.

Here, several principles are discussed which guide the choices made in the early planning and design of an observational study. The principles are illustrated using three observational studies that use choice effectively, one each from economics, clinical psychology and epidemiology. After discussing the quality of evidence, the examples are described, and then the subsequent discussion of principles makes reference to these examples.

11.1.2   The Quality of Evidence: Some Ground Rules

Most empirical disciplines make use of some form of mathematical reasoning. To discuss the quality of evidence provided by an empirical study one must first recognize that evidence is not proof, can never be proof, and is in no way inferior to proof. It is never reasonable to object that an empirical study has failed to prove its conclusions, though it may be reasonable, perhaps necessary, to raise doubts about the quality of its evidence. As expressed by Sir Karl Popper (1968, p. 50): "If you insist on strict proof (or strict disproof) in the empirical sciences, you will never benefit from experience and never learn from it how wrong you are."

One expects that a proof will be correct and that evidence will be candidly presented, despite occasional disappointments. One is often concerned about the relevance of a proof and the quality of evidence. If it is proved that a certain bridge can withstand certain forces, and yet the bridge collapses, it not the correctness but the relevance of the proof that is likely to be called into question.

The distinction between the mathematical correctness and relevance of a proof is familiar in most empirical sciences that use mathematical methods, for instance, economics, where Samuelson in his 1947 technical treatise, *The Foundations of Economic Analysis*, wrote:

> ... our theory is meaningless in the operational sense unless it does imply some restrictions upon empirically observable quantities, by which it could conceivably be refuted (p. 7) ... By a meaningful theorem I mean simply a hypothesis about empir-

ical data which could be refuted, if only under ideal conditions. A meaningful theorem may be false. (p. 4)

Evidence may refute a theorem, not the theorem's logic, but its relevance. A related point is made by Quine (1951).

Evidence, unlike proof, is both a matter of degree and multifaceted. Useful evidence may resolve or shed light on certain issues while leaving other equally important issues entirely unresolved. This is but one of many ways that evidence differs from proof.

Evidence, even extensive evidence, does not compel belief. Rather than being forced to a conclusion by evidence, a scientist is responsible and answerable for conclusions reached in light of evidence, responsible to his conscience and answerable to the community of scientists. Of this, Michael Polanyi (1964) writes:

> ...our decision what to accept as firmly established cannot be wholly derived from any explicit rules but must be taken in the light of our own personal judgment of the evidence.
>
> Nor am I saying that there are no rules to guide verification, but only that there are none which can be relied on in the last resort...
>
> We may conclude that just as there is no proof of a proposition in natural science which cannot conceivably turn out to be incomplete, so also there is no refutation which cannot conceivably turn out to have been unfounded. There is a residue of personal judgment required in deciding—as the scientist eventually must—what weight to attach to any particular set of evidence in regard to the validity of a particular proposition. (pp. 30-31)
>
> The scientist takes complete responsibility for every one of these actions... (p. 40) ... for the process as a whole—he will assume full responsibility before his own conscience. (p. 46)

Evidence may be convincing beyond reasonable doubt, yet being convinced or remaining skeptical is a judgment for which a scientist is responsible. The desire to be compelled rather than convinced by evidence is the desire to evade responsibility for judging the evidence.

Aesthetics matter in proof, but also in evidence. A beautiful proof is simple, illuminating its conclusion and rendering it obvious. The same aesthetic applies to evidence. In both cases, the simple and the obvious are the product of care, effort, and skill.

11.2   Three Observational Studies

11.2.1   Examples of Careful Planning

Three examples are used to illustrate the choice of circumstances in which an observational study is conducted. The studies employ choice in their design in an instructive manner which meaningfully strengthens the evidence, but of course does not prove, that the treatment caused its ostensible effects. Two of these studies, the lead exposure and minimum wage studies, were encountered more briefly in §§3.2.4, 4.3, and 5.4.6.

To emphasize, these studies are being used solely to illustrate careful planning and the use of choice in the design of observational studies. The conclusions of the minimum wage and bereavement examples remain controversial, and it is not my purpose here to suggest the controversy is warranted or unwarranted. The purpose is to illustrate how careful planning can improve the quality of evidence.

11.2.2   The Effects of Increasing the Minimum Wage

The most familiar of all arguments of economic analysis are those of comparative statics, as discussed, for example, in the early chapters of Samuelson's (1947) *Foundations of Economic Analysis*. Often motivated with the aid of diagrams in elementary economics courses, such an argument pictures the world as determined by an equilibrium of forces, and it describes how that equilibrium changes as the forces are changed. Arguments of this form (c.f. Samuelson's "Illustrative Market Case" on page 17) suggest that increasing the price of a commodity will decrease the demand for that commodity, other things "being equal" or *ceteris paribus*. Viewing labor as a commodity and applying such considerations leads many economists to expect that an increase in the minimum wage will cause a decline in employment among workers receiving the minimum wage. Developed in slightly different terms, such arguments suggest that an increase in the minimum wage shifts production to use less labor and more capital equipment, or to substitute goods produced outside the reach of the minimum wage. For this reason, many economists argue that increasing the minimum wage hurts the very individuals it is intended to benefit. Inspection of the details of such a theorem suggest that, as a theoretical argument removed from data, this expectation seems logical.

Writing in the *American Economic Review* in 1994, David Card and Alan Krueger (CK) attempted to estimate the effects of an increase in New Jersey's minimum wage that occurred on April 1, 1992. CK looked at employment in the fast-food industry (Burger King, Kentucky Fried Chicken, Wendy's, and Roy Rogers) in New Jersey before and after the increase in the minimum wage and compared this to a control group consisting of fast-food restaurants in adjacent eastern Pennsylvania. They found "no evidence that the rise in New Jersey's minimum wage reduced employment at fast-food restaurants in the state." Methodological aspects of CK are nicely discussed by Meyer (1995).

The CK study is highly controversial in its conclusions and somewhat unusual in its design. To shed light on its design, it is useful to describe a related study conducted in the more traditional manner, which reached very different conclusions. The traditional study used here for comparison actually came in response to Card and Krueger, and was conducted by Deere, Murphy, and Welch (1995), or DMW. The DMW study attempted to estimate the effects of the increases in the Federal minimum wage that took place from \$3.35 to \$3.80 on April 1, 1990 and from \$3.80 to \$4.25 on April 1, 1991 by applying regression to national monthly data from the Current Population Survey from 1985 to 1993. DMW conclude: "The regression estimates have no surprises. When the cost of employing low-wage laborers is increased, fewer low-wage laborers are employed." Without taking sides on the substantive conclusion, §11.4 at several places compares the designs of these two studies.

Before leaving the studies by CK and DMW, one should note that their conclusions do not, strictly speaking, contradict each other. First, CK discuss a particular change in a state minimum wage while DMW discuss a particular change in the Federal minimum wage, and it is possible in principle that these different interventions had different effects. Second, DMW examine changes in employment in certain demographic groups with varied percentages of minimum wage earners, while CK discuss employment in particular fast-food restaurants. Note, however, that the simple comparative statics argument above would suggest that these distinctions should not matter and that the same general pattern of effects should be seen in both cases.

11.2.3 Effects of Loss of a Spouse or Child in a Car Crash

Lehman, Wortman, and Williams (1987) (LWW) attempted to estimate the long-term psychological effects of a sudden and unexpected death of a spouse or a child in a motor vehicle crash. The study identified 39 individuals who had lost a spouse and 41 individuals who had lost a child in a motor vehicle crash four to seven years prior to the study. This group of exposed subjects was the result of a selection process that applied various criteria and sampling to a record of motor vehicle fatalities in Michigan between 1976 and 1979, with some nonresponse that is discussed in the LWW paper.

Noting that "many previous studies on the impact of bereavement have not included control or comparison groups...," so that these studies were "difficult to interpret," LWW constructed a control group in the following way. From a reservoir of 7581 individuals who came to renew their licenses, one control was matched to each exposed subject based on gender, age, family income in 1976 (i.e., before the crash), education level, number and ages of children. The outcomes included measures of depression and various psychiatric symptoms. Bereaved spouses and parents both were depressed substantially and significantly more than matched controls between four and seven years after the loss, and bereaved spouses exhibited higher rates of several other psychiatric symptoms. LWW concluded:

> The results presented here suggest that the current theoretical approaches to bereavement may need to be reexamined ... [The discussion goes on to contrast the study's results with the views of Bowlby and Freud, among others] (p. 228)   ... From 4 to 7 years after the sudden loss of a spouse or child, bereaved respondents showed significantly greater distress than did matched controls, suggesting little evidence of timely resolution. Contrary to what some early writers have suggested about the duration of the major symptoms of bereavement ... both spouses and parents in our study showed clear evidence of depression and lack of resolution at the time of the interview, which was 4 to 7 years after the loss occurred... The present study suggests that exposure to stress can trigger enduring changes in mental health and functioning (p. 229).

11.2.4 Lead in Children of Workers Subject to Occupational Exposures to Lead

Morton, Saah, Silberg, Owens, Roberts and Saah (1982) (MSSORS) examined lead levels in the blood of children whose parents worked in a factory that used lead in the manufacture of batteries. They suspected that parents brought lead home in their clothes and hair, thereby exposing their children. Thirty-three children from different families with a parent at the battery factory were matched to thirty-three unexposed control children, the matching being based on age, neighborhood, and exposure to traffic.

They found that exposed children had substantially higher levels of lead in their blood than did matched control children. The exposed children were also classified in two ways, namely, parental exposure to lead on the job (low, medium, or high), and parental hygiene upon leaving the factory (good, moderately good, poor). They found lower levels of lead in the blood of exposed children whose parents had lower exposures to lead on the job, and lower levels of lead among children whose parents had better hygiene. MSSORS (1982, p. 555) concluded: "...the data presented justify more stringent enforcement of lead containment practices..."

11.3   Choice of Research Hypothesis

11.3.1   Narrow, Focused, Controlled Examination of a Broad Theory

Consider, first, a laboratory experiment in the physical or biochemical sciences. It begins with a broad theory that makes assertions about the effects of a treatment. Specifically, such a broad theory makes innumerable predictions about what would be observed in the innumerable circumstances in which the particular treatment might be applied, now and in the future, and in different locations. A laboratory experiment makes no effort to draw up a frame comprising all circumstances, locations, and times when the treatment might be applied and to draw a representative sample of such circumstances. Rather, the laboratory experiment examines the theory under highly unrepresentative circumstances, namely, circumstances in which sensitive, calibrated measuring instruments are used in an environment carefully freed of forces that might intrude on the experiment, in which the treatment is delivered at doses sufficient to produce dramatic effects if the theory is correct or to produce an equally dramatic absence of effect if the theory is incorrect. In a way, it is part of the essence of a laboratory experiment that its circumstances are unrepresentative. In a well-conducted laboratory experiment, one of the rarest of things happens: the effects caused by treatments are seen with clarity.

Observational studies of the effects of treatments on human populations lack this level of control, but the goal is the same. Broad theories are examined in narrow, focused, controlled circumstances.

Broad theories are desired because they predict more, and in consequence are both more useful and can be more thoroughly scrutinized. Quoting Popper (1968) again:

> ... those theories should be given preference which can be most severely tested (p. 121) ... if the class of potential falsifiers of one theory is 'larger' than that of another, there will be more opportunities for the first theory to be refuted by experience [...and...] the first theory says more about the world of experience than the second theory, for it rules out a larger class of basic statements.... Thus it can be said that the amount of empirical information conveyed by a theory, or its empirical content, increases with its degree of falsifiability (p. 112–113) [which] explains why simplicity is so highly desirable... [Simple theories] are to be prized more highly than less simple ones because they tell us more; because their empirical content is greater; and because they are better testable. (p. 142) ... Theories are not verifiable, but they can be 'corroborated'. ...we should try to assess what tests, what trials, [the theory] has

withstood (p. 251) ... it is not so much the number of corroborating instances which determines the degree of corroboration as the severity of the various tests to which the hypothesis can be, and has been, subjected. But the severity of the tests, in its turn, depends upon the degree of testability, and thus upon the simplicity of the hypothesis: the hypothesis which is falsifiable in a higher degree, or the simpler hypothesis, is also the one which is corroborable in a higher degree (p. 267).

Similar claims about simplicity, falsification and corroboration are made by Milton Friedman (1953, pp. 8–9), who adds: "A hypothesis is important if it 'explains' much by little, that is, if it abstracts the common and crucial elements ... and permits valid predictions on the basis of them alone" (p. 14) and "...the only relevant test of the validity of a hypothesis is comparison of its predictions with experience" (p. 9). As noted by Putnam (1995, p. 71), a similar point had been made earlier by Charles Sanders Peirce (1903, p. 418–419):

> But if I had the choice between two hypotheses ... I should prefer ... [the one which] would predict more, and could be put more thoroughly to the test... It is a very grave mistake to attach much importance to the antecedent likelihood of hypotheses.... Every hypothesis should be put to the test by forcing it to make verifiable predictions.

The terms "falsify," "reject" or "refute," when applied to scientific theories, are not quite accurate. Lakatos (1981, p117) describes such a theory as "shelved," the implication being that the theory is stored along with the evidence against it. Under rare circumstances, a shelved theory might be reconsidered; see again the quote from Polanyi (1946, 1964).

Broad theories permit close scrutiny in numerous particular cases, and a research hypothesis is intended to focus attention on one such case in which close scrutiny—a severe test—is possible. Passing such a severe test corroborates but does not prove the theory. Platt (1964), Meehl (1978), and Dawes (1996) make similar points.

11.3.2 Choice of Research Hypothesis in the Examples

In the economics example, broad arguments from comparative statics predict that increases in the cost of labor will diminish employment. If true, this theory operates in innumerable instances in the US economy on a daily basis; however, in almost all such instances, it operates amid numerous other forces that obscure its effects. For instance, if wages rise faster in one company than in a second company making a different brand of a similar product, then the theory may be true even though employment does not decline in the first company, precisely because increasing demand for its brand of product may have led the first company to raise wages in an effort to increase output by expanding its workforce. Even if the theory is correct, it rarely operates in isolation. Card and Krueger attempted to find one of those rare instances in which the theory might be hoped to operate in relative isolation. Specific aspects of this isolation are discussed later, but the point here concerns their choice of research hypothesis. The effect of the increase in the New Jersey minimum wage on fast-food chains in 1992 is, in itself, at most a very minor footnote to economic history. However, as an opportunity to closely scrutinize and thereby possibly corroborate or refute the broad theory that forcible increases in the minimum wage cause declines in employment, the 1992 increase in New Jersey's minimum wage becomes much more important. In observational studies, one chooses a research hypothesis that permits a broad theory to operate dramatically and in relative isolation.

Some accounts in the popular press have emphasized, perhaps even slightly exaggerated, the breadth of the theory challenged by Card and Krueger's results. Writing in *Forbes*, the Stanford economist Thomas Sowell (1995) said:

> ... if true, these results challenge the very foundations of economics. If rises in the price of labor do not reduce employment, why should we expect that a rise in the price of anything else affects the quantity purchased? This is to economics what disproving the law of gravity would be to physics.

Whether or not this is what CK did, whether or not they were really operating on this scale, nonetheless, the spirit of Sowell's remark is right: one seeks broad and consequential theories exposed to decisive challenges in focused, clear circumstances.

In the psychology example, the belief that bereavement should have short-lived effects on mental functioning stems from a much broader theory, the dominant but not universally accepted theory in clinical psychology, which holds that the structure of mental functioning is largely shaped by a mixture of biology and experiences as a young child in relation to caretakers, usually parents. In particular, that theory suggests that bereavement should not greatly alter mental functioning over long periods of time. As in the economics example, if the broad theory is correct, then it is in constant operation in innumerable lives, but in almost all cases its operation is obscured by limited knowledge of the biology of mental functioning, by the difficulty in accurately measuring the experiences of early childhood, and by the difficulty in distinguishing what is cause and what is consequence in the mental life and experiences of an adult. Sudden deaths from car crashes are a situation in which the broad theory operates or fails to operate with clarity.

Is it really true that a broad theory will be refuted or shelved based on a decisive challenge in a single, focused circumstance? Probably not. As argued by Lakatos (1970), a single, focused challenge may challenge a research program, may quicken interest in competing research programs, may stimulate further challenges in other focused circumstances, but it is unlikely to cause the immediate abandonment of a research program that has enjoyed some success. And this is for the best. Empirical studies, particularly observational studies, are attended by various uncertainties and ambiguities, some of which are difficult to quantify or even to identify, so consistent results in several studies are typically needed to force changes in the direction of a research program. Nonetheless, each such influential study is likely to examine the same broad theory in different focused circumstances, attended by different uncertainties.

In short, the choice of a research hypothesis focuses a broad theory on a narrow instance in which the theory's operation may be viewed clearly. Often, this setting permits the theory to operate in relative isolation from other forces, or to operate on a dramatic scale due to concentrated exposures, or else the treatment is imposed suddenly, at a discrete moment, in a manner not influenced by the individuals under study.

11.4   A Control Group

In design, one must choose a situation in which a control group can be constructed. A control group consists of subjects or units that did not receive the treatment. The control groups used in the three examples were described previously.

Some observational studies do not have a control group. For instance, in the study by DMW, everyone covered by the Current Population Survey was in a region affected by the increase in the Federal minimum wage, so there is no control group—all subjects received the treatment. In contrast, the CK study of New Jersey's minimum wage had a control group consisting of branches of the same fast-food chains across the Delaware river in eastern Pennsylvania where the minimum wage had not been increased.

Lacking a control group, DMW estimate the effect of the increase in the Federal minimum wage using regression in the following way. The Federal minimum wage increases went into effect on April 1, 1990 and April 1, 1991, so DMW define years which begin on April 1 and end on March 31. They focus on two groups of individuals, namely, teenagers aged 15 to 19 and adult high school dropouts aged 20 to 54, reasoning that these groups contain a disproportionate number of individuals earning the minimum wage, so they should be most affected by changes in the minimum wage. These two groups are examined in parallel but separate analyses, but for brevity, only the analysis for teenagers is described here. Within these groups, men, women, and blacks are examined in similar but not exactly parallel analyses; to permit a brief discussion here, the focus is on the analysis for male teenagers. For each year from 1985 to 1992, for each state, there is an outcome measure, namely the log of the fraction of male teenagers in the US who were employed, as estimated by the current population survey. DMW regress this outcome on the following predictors: (i) the log of the fraction of 15 to 64-year old men who were employed in the same state and year, (ii) state indicator variables, and (iii) indicator variables identifying the level of the Federal minimum wage. Concerning male teenagers, they write: "Compared to the employment level projected from the movement in aggregate employment with the $3.35 minimum wage, teenage employment was 4.8 percent ... lower in 1990 ... and 7.3 percent ... lower in 1991–1992." In other words, employment among male teenagers in 1990 and 1991-2 fell more sharply than employment among all 15 to 64-year old men in the same period, adjusting for state-to-state differences that are constant through time. Qualitatively similar though numerically different results were obtained for females, blacks, and adult high school dropouts. Their conclusion from these regressions was quoted earlier.

Conclusions reached in the absence of a control group are not necessarily wrong, but they are typically open to plausible objections and legitimate skepticism of types that are inapplicable with a control group. The assumption implicit in DMW's method of estimation is that changes in the log employment fraction for male teenagers would have been linearly related to changes in the log employment fraction among all 15 to 64-year old men if the minimum wage had not been increased, and any departure from such a linear change is an effect caused by the change in the minimum wage. Is this assumption self-evidently true? Might not employment among teenagers and high school dropouts fall disproportionately more than employment in the general workforce during times of generally declining employment even without an increase in the minimum wage? Alternatively, a rise in the minimum wage might increase employment in certain demographic groups and decrease it in others because employers find that they can pull into the workforce better educated or more experienced workers who, at lower wages, would not work or who would work fewer hours. For instance, an anecdotal account in the popular press of changes in employment practices in response to the most recent increase in the Federal minimum wage describes an employer as introducing special hiring practices to avoid "wasting the extra 50 cents on unreliable help" (Duff 1996). Even a bit of nonlinearity on the log scale might be mistaken for an effect of changing the minimum wage. Objections of this sort may well be incorrect and unfounded, but what is important here is that they can reasonably be raised for a study which lacks a control group.

If raising the minimum wage decreases employment then, DMW reasoned, the larger decreases should occur in groups with more minimum wage earners. Although DMW focused on comparisons involving teenagers and adult dropouts, other groups with a disproportionate number of minimum wage earners were also examined briefly. For instance, low, medium and high wage states were compared, the first group being thought to be most affected by increases in the minimum wage. Men and women were compared, where women as a group receive lower wages. These comparisons pointed in the opposite direction from the comparisons discussed in earlier paragraphs; see DMW's Table 3. Employment in low wage states declined less than employment in high wage states and employment among women declined less than employment among men after increases in the minimum wage. Women in low wage states experienced no decline in employment. If increasing the minimum wage decreased employment, the reasoning DMW applied to teenagers and dropouts would have predicted larger employment declines in low wage states and among women. DMW (1995, p. 234) write: "The latter fact is easily dismissed based on long-standing trends." Evidently, certain interactions do exist in which different demographic groups experience larger or smaller declines in employment, although DMW believe they can distinguish which effects are caused by the minimum wage and which are irrelevant demographic trends, and perhaps others will agree with them. Nonetheless, looking at the methodology, inconsistencies of this sort can arise in studies in which treated and control groups are replaced by groups with higher and lower exposure to the treatment.

An interesting paper by Holland and Rubin (1983) looks at various "paradoxes" that arise when there is no control group and a model or a calculation is used to estimate the treatment effect. They interpret the paradoxes as consequences of diverging, uncheckable, and unarticulated assumptions about what the control group would look like.

If a study lacks a control group, a minimal requirement is the articulation of the assumptions about what the control group would look like, together with a discussion of the tangible evidence in support of those assumptions and the sensitivity of conclusions to violations of the assumptions. Articulation and evaluation of assumptions about the absent control group aid in judging the roles evidence and assumptions play in the study's conclusions.

In addition to a control group, baseline measures of the response are often useful. See §9.5, Cook, Campbell, and Peracchio (1990), Cook and Shadish (1994), Allison (1990), and Rosenbaum (2001a).

11.5   Defining Treated and Control Groups

11.5.1   Sharply Distinct Treatments That Could Happen to Anyone

In a randomized experiment, treated and control groups are defined by, first, defining the treatments themselves and, second, defining and implementing a mechanism for random assignment of treatments. Typically, in prolonged experiments with human populations, two or at most a few quite distinct treatments are compared (Peto, Pike, Armitage, Breslow, Cox, Howard, Mantel, McPherson, Peto, and Smith 1976).

The situation in observational studies is different. The investigator does not control the assignment of treatments to subjects, and so must define a treated and a control group using available subjects who have already received treatment or control. The goal in defining the treatment groups should be to produce a situation that resembles, to the extent possible, the situation in a randomized experiment: markedly distinct treatments that could happen to anyone. Consider the two parts separately.

In discussing the design of controlled trials, Peto, et al. (1976, p. 590) say:

> A positive result is more likely, and a null result is more informative, if the main comparison is of only 2 treatments, these being as different as possible. ... it is the mark of good trial design that a null result, if it occurs, will be of interest.

Treatment groups that are distinct in concept but not in actual implementation may not differ in their outcomes even if the conceived but unimplemented distinction would have an important effect on outcomes. For instance, this was a concern in a National Academy of Sciences report on studies of bilingual education which found that: "... Immersion and Early-exit Programs were in some instances indistinguishable from one another" (Meyer and Fienberg 1992, p. 102).

To say that the distinct treatments "could happen to anyone" is shorthand. One wishes to define the treated and control groups in such a way that the treatment could easily have happened to the controls, and the treated subjects could easily have been spared the treatment, the actual assignment of subjects to treatments being determined by haphazard or ostensibly irrelevant circumstances. Haphazard is not random, and haphazard treatment assignments can produce severe, consequential, and undetected biases that would not be present with random assignment of treatments. Still, haphazard or ostensibly irrelevant assignments are to be preferred to assignments which are known to be biased in ways that cannot be measured and removed analytically.

11.5.2 Examples of Defining Treated and Control Groups

An example of careful defining of treated and control groups comes from the study by LWW of the effects of a traumatic loss of a spouse or child. First, the treated and control conditions are markedly distinct. The loss is confined to a spouse or a child less than 18 years of age living at home, and the loss was produced suddenly by a car crash. One could study the effects of a loss of other relatives, such as an adult's parent or an adult's adult sibling, or gradual losses due to chronic disease, but these might have smaller psychological effects. As in experiments, effects should be demonstrated for markedly distinct treatments before refined studies of smaller effects are undertaken.

Second, LWW took care to define the treated group so it "could happen to anyone". In particular, they used published, relatively objective criteria to appraise probable responsibility or fault in the car crashes, and then insisted that the treated group consist of individuals from cars that were not at fault. For instance, if one car crossed a center dividing line and collided with an oncoming car, the occupants of the first car would be ineligible for the study while the occupants of the second car would be eligible. Their reasoning was that fault in car crashes is related to alcohol and drug use and to certain forms of psychopathology, all of which would be studied later as outcomes for survivors. In contrast, a car crash for which one is not responsible could happen to any driver. No matter what the particulars, car crashes are a far cry from random numbers, but car crashes for which one is not responsible are plausibly a limited but meaningful step closer to random.

In studying the effect of class size on academic achievement, Angrist and Lavy (1999) give another interesting example of sharply distinct treatments that could happen to anyone. In the US, a class of size 40 will often be located in a very different school district than a class of size 20, so it is difficult to distinguish the effects of class size from the other consequences of the differences between school districts. In contrast, Israeli public schools implement the following version of a rule due to Maimonides: when class size exceeds 40 students, the class must be divided. In this case, a class of size 41 becomes two much smaller classes. In that setting, one might compare certain classes of size near 40 to split classes of size near 20, knowing that these are very different class sizes, and yet the rather minor event of enrolling one or two more students determined this dramatic change in class size. Again, this is not a random assignment of class sizes to students, but it is a meaningful step closer to random. Angrist and Krueger (1999) discuss additional examples.

Still another example of sharply distinct treatments that could happen to anyone is found in Bronars and Grogger's (1994) study of the economic consequences of unwed motherhood. Here, too, as a group, women who bear children prior to marriage differ from unmarried women who do not bear children, and it is important to avoid mistaking these differences from economic effects of an additional unplanned child. Instead of comparing these two groups, Bronars and Grogger compared unwed mothers who had twins to unwed mothers who had singletons, reasoning that having twins rather than a single child is a comparatively haphazard event, one that could happen to anyone. See also Rosenzweig and Wolpin (1980, 2000).

11.6 Competing Theories, Not Null and Alternative Hypotheses

In his essay, "How to be a good empiricist—a plea for tolerance in matters epistemological," Paul Feyerabend (1968) writes:

> You can be a good empiricist only if you are prepared to work with many alternative theories rather than with a single point of view and 'experience'. This plurality of theories must not be regarded as a preliminary stage of knowledge which will at some time in the future be replaced by the One True Theory. Theoretical pluralism is assumed to be an essential feature of all knowledge that claims to be objective.... The function of such concrete alternatives is, however, this: They provide means of criticizing the accepted theory in a manner which goes beyond the criticism provided by a comparison of that theory 'with the facts'.... This, then, is the methodological justification of a plurality of theories: Such a plurality allows for a much sharper criticism of accepted ideas than does the comparison with a domain of 'facts' which are supposed to sit there independently of theoretical considerations. (p. 14-15)

Elsewhere, Feyerabend (1975) argues that alternative theories are needed to unearth new facts, advising one to:

> ...introduce and elaborate hypotheses which are inconsistent with well-established theories and/or well-established facts. ... [T]he evidence that might refute a theory can often be unearthed only with the help of an incompatible alternative (p. 29) ... [M]any facts become available only with the help of alternatives, [so] the refusal to consider [alternative theories] will result in the elimination of refuting facts as well (p. 42).

Popper (1965, p. 112) makes a closely related point:

> A theory is tested not merely by applying it, or trying it out, but by applying it to very special cases—cases for which it yields results different from those we would have expected without that theory, or in the light of other theories. In other words we try to select for our tests those crucial cases in which we should expect the theory to fail if it is not true. Such cases are 'crucial' in Bacon's sense; they indicate the cross-roads between two (or more) theories.

Lakatos (1981, p. 114-5) distinguishes the "internal" testing of theories from the "external" competition between theories, suggesting the latter is an aid to the former:

... facts are only noticed if they conflict with some previous expectation. [This is a] cornerstone of Popper's psychology of discovery. Feyerabend developed another interesting psychological thesis of Popper's, namely, that proliferation of theories may—externally—speed up internal Popperian falsification (pp. 114–115).

In his paper of 1890, T. C. Chamberlin (1965, p756) argued for the "method of multiple working hypotheses":

> The value of a working hypothesis lies largely in its suggestiveness of lines of inquiry that might otherwise be overlooked. Facts that are trivial in themselves are brought into significance by their bearings upon the hypothesis, and by their causal indications ... In the use of the multiple method, the re-action of one hypothesis upon another tends to amplify the recognized scope of each, and their mutual conflicts whet the discriminative edge of each.

Endorsing Chamberlin's "method of multiple working hypotheses," John Platt (1964, p. 347) argued:

> ... rapidly moving fields are fields where a particular method of doing scientific research is schematically used and taught, an accumulative method of inductive inference that is so effective that I think it should be given the name of 'strong inference.' ... Strong inference consists of applying the following steps to every problem in science, formally and explicitly and regularly:
>   1) Devising alternative hypotheses;
>   2) Devising a crucial experiment (or several of them), with alternative possible outcomes, each of which will, as nearly as possible, exclude one or more of the hypotheses;
>   3) Carrying out the experiment so as to get a clean result;
>   1*) Recycling the procedure ...
> Any conclusion that is not an exclusion is insecure ...

A scientific theory that has been fairly successful—a theory discussed in journals and textbooks, explained to undergraduates, offered to graduate students as an area for research, and so on—is likely to agree with empirical observations at many points and to seem helpful in interpreting those observations. Certainly, sufficiently large, externally imposed increases in the price of a commodity *may* cause demand to decrease, and both biology and the events of early childhood *may* have enduring effects on mental functioning. Both of these theories offer coherent interpretations of frequent observations. A successful theory is likely to work in many situations; this is an aspect of its success. Unaided by a competing theory, an empirical investigation may do no more than rediscover why the successful theory achieved success in the first place. Such an investigation may not place the successful theory at much risk of refutation, and since it was never at much risk, failing to refute the theory provides little corroborating evidence in support of the successful theory.

A competing theory focuses attention on certain observable events about which the successful theory and its competitor make very different predictions. The competing theory directs attention to places where the successful theory might fail, placing the successful theory at severe risk of refutation. The competing theory anticipates a particular refutation of the successful theory, making refutation of the successful theory more likely and more decisive, and providing stronger corroboration of the successful theory if its predictions turn out to be correct. Card and Krueger quote the following remark of Paul Samuelson: "In economics it takes a theory to kill a theory; facts can only dent a theorist's hide" (Card and Krueger 1995, p. 355).

The conventional argument anticipates that an increase in the minimum wage in New Jersey drives up the cost of some labor in New Jersey, with two consequences for employment: first, an increase in prices of final products resulting in reduced demand, and second a tendency to substitute capital equipment, whose cost has not increased, for labor, whose cost has increased. The increase in the minimum wage affects all New Jersey firms, so the price increases may have smaller effects on a single firm's business than would be the case if that one firm raised prices while other firms did not. Card and Krueger (1995, p. 359) express this same idea more formally as follows:

> For purposes of modeling the effect of an industry-wide wage increase, however, the relevant product-demand elasticity is one that takes into consideration simultaneous price adjustments at all firms. This elasticity will tend to be smaller (in absolute value) than the elasticity of demand for a firm's output with respect to its own price. In the case of the restaurant industry, for example, any individual restaurant presumably faces a relatively elastic demand for its product, holding constant prices at nearby restaurants. When the minimum wage increases, however, prices will tend to rise at all restaurants, resulting in a smaller net reduction in demand at any particular firm.

In other words, if Roy Rogers alone raised the price of hamburgers, it might face a substantial decline in business as customers switched to Burger King, but if all restaurants raise prices at the same time, the decline in business might be much smaller. Notice that this is particularly true of the fast-food industry, because it is not practical for a New Jersey restaurant to import cooked food from, say, Hong Kong or Pennsylvania, to escape the effects of the minimum wage increase in New Jersey. If the conventional argument were incorrect, if increases in the minimum wage had only slight effects when simultaneous price adjustments occur in all firms, then the fast-food industry, though unrepresentative of all industries, is one place to see this. Both theories accept that increases in the minimum wage might result in higher prices for fast-food, but only one theory predicts a dramatic decline in business and employment in the fast-food industry. In fact, Card and Krueger (1994, §§3E and 5) found no association between increases in the minimum wage and (i) the number of hours a restaurant is open on a weekday, (ii) the number of cash registers, (iii) the number of cash registers in operation at 11:00am, but they did find some evidence of slightly higher increases in prices in New Jersey than in Pennsylvania during this period. During the period of the minimum wage increases, CK (1984, Table 2) found that the average price of a specifically defined "full meal" changed from $3.04 to $3.03 in the Pennsylvania restaurants and from $3.35 to $3.41 in the New Jersey restaurants. In other words, the structure of a competing theory suggests where to look to test a standard theory, for instance, which industry to study and which outcome measures to examine.

Card and Krueger (1995) write: "We suspect that the standard model ... does correctly predict the effect of the minimum wage on some firms," (p. 355) but they say their results are "...inconsistent with the proposition that the standard model is always correct" (p. 383). Whether or not one agrees with CK about the minimum wage, the methodological issue is clear: A study of a narrow and unrepresentative corner of a population cannot, by itself, be the sole basis for policy for the whole population, but it may provide a severe test of a theory that purports to apply throughout the population. Failing such a test raises doubts about using that theory as the sole basis for policy, whereas passing such a severe test corroborates the theory, and tends to strengthen the case for using it as a basis for policy.

Similar, if more direct, considerations apply in the LWW study. A theory which suggests that mental functioning is largely shaped by biology and early childhood experiences might reasonably be contrasted with a theory which holds that events later in life shape mental functioning over long periods. Sudden deaths of close relatives in car crashes are instances in which these theories make markedly different predictions.

In short, scrutiny of a theory is aided by a competing theory. The competition between theories suggests circumstances in which the theories make sharply different predictions about particular observable quantities.

11.7   Internal Replication: Multiple Treatment Assignment Mechanisms

Replication is important in observational studies, as it is in experiments. Susser (1987) writes:

> The epidemiologist's alternative to exact replication is the consistency of a result in a variety of repeated tests... Consistency is present if the result is not dislodged in the face of diversity in times, places, circumstances, and people, as well as of research design (p. 88).

In part, biases that are peculiar to the circumstances of one study may not replicate, whereas an effective treatment is expected to produce similar results in studies of varied circumstance and design.

Some observational studies incorporate a form of internal replication. These studies replicate the treatment assignment mechanism, so that essentially the same treatment is assigned to subjects by more than one process. As discussed in a moment, this is true of two of the examples. If two treatment assignment mechanisms produce a pattern of associations consistent with an actual treatment effect, then to explain the pattern as a hidden bias, one must attribute a bias to both assignment mechanisms, and moreover a bias yielding the pattern anticipated from an actual effect. In Popper's terms, each mechanism provides a check on the theory that the treatment is the cause of its ostensible effects, so if several assignment mechanisms produce compatible estimates of effect, then this theory receives greater corroboration.

In the MSSORS study, children were exposed to low levels of lead in three different ways. First, the parents of control children did not work in the battery factory. Second, among exposed children whose parents did work in the battery factory, some were believed to have received lower doses of lead because their parents worked in jobs in the factory that provided little exposure to lead. Third, some parents exposed to high levels of lead practiced good hygiene. In fact, no matter which device assigned a child to low exposures to lead, the results were similar: children with lower exposures tended to have less lead in their blood. This pattern could, conceivably, be the result of hidden bias, but it is somewhat more difficult, though of course not impossible, to imagine biases that would produce all three associations.

In the CK study of the increase in the New Jersey minimum wage, a fast-food restaurant could escape a legal requirement to increase wages in either of two ways. Restaurants in Pennsylvania were not required to increase wages. Restaurants in New Jersey whose lowest wage was above the new minimum wage were not required to increase wages. In connection with their Table 3, CK (1994, p. 778) write: "Within New Jersey, employment expanded at the low-wage stores ... and contracted at the high-wage stores... Indeed, the average change in employment at the high-wage stores... is almost identical to the change among Pennsylvania stores..." In other words, restaurants placed under no new legal requirement by the minimum wage saw similar employment changes, whether they were Pennsylvania restaurants or high-wage New Jersey restaurants.

In short, the central concern in an observational study is that treatments may be assigned to subjects in a biased manner. The choice of a setting in which essentially the same treatment is assigned to subjects by several very different processes provides a partial check of this central concern. More precisely, the theory that the treatment is the cause of its ostensible effects predicts similar effects no matter which mechanism delivered the treatment, and this prediction offers an opportunity to refute or corroborate the theory.

11.8   Nondose Nonresponse: An Absent Association

A causal theory not only predicts the presence of certain associations, but also the absence of certain others. In some studies, it is possible to determine the dose of treatment that a control would have received had this control received the treatment. Call this the potential dose. While it is often reasonable to expect higher responses from treated subjects who received higher doses (see §8.5), the same pattern is not expected among comparable controls: higher potential doses that were never received should not predict higher responses if higher responses are being caused by the treatment.

The minimum wage study contains an example. Card and Krueger (1994) define a measure, GAP_i, of the impact of the minimum wage legislation on restaurant i, as the percentage increase in the starting wage in restaurant i needed to achieve the new New Jersey minimum wage. A restaurant that already paid more that the new minimum would have $GAP_i = 0$. A restaurant that paid the old minimum wage would have $GAP_i = 18.8\%$. This variable GAP_i resembles a dose of treatment in that the law has greater impact on the starting wage when GAP_i is larger. A concern, however, is that GAP_i is not only a dose of treatment, but also a variable describing local labor market conditions. Presumably, some restaurants must pay higher starting wages than others to attract employees, so GAP_i is confounded with labor market conditions. For instance, the labor market in the poor city of Camden is different from the labor market in the relatively affluent suburb of Princeton. Still, it is not unreasonable to think that this confounding operates in a similar way in New Jersey and Pennsylvania. For the control restaurants in Pennsylvania, the undelivered potential dose of treatment is the percentage change in the starting wage needed to achieve New Jersey's new minimum wage. If these undelivered potential doses predicted employment changes in Pennsylvania, then that could not be an effect of New Jersey's minimum wage legislation, and would strongly suggest confounding. Card and Krueger (1994, p. 784) write:

> ...we exclude stores in New Jersey and (incorrectly) define the variable for Pennsylvania stores as the proportional increase in wages necessary to raise the wage to $5.05 per hour. In principle the size of the wage gap for stores in Pennsylvania should have no systematic relation with employment growth. In practice, this is the case. There is no indication that the wage gap is spuriously related to employment growth.

In short, there is often the concern that not only the assignment to treatment or control but also the dose of treatment is confounded with unobserved covariates. When potential but undelivered doses are known for controls, the theory that the treatment is the cause of its ostensible effects predicts that delivered doses in the treated group should be related to the magnitudes of responses, but, at each fixed value x of observed covariates, undelivered doses in the control group should be unrelated to responses. This prediction provides an additional check of the theory that the treatment is the cause of its ostensible effects.

11.9 Stability Analyses, and Minimizing the Need for Stability Analyses

A complex analysis involves numerous implementation or analytical decisions. The audience for such an analysis typically wishes to be assured that conclusions are not artifacts of such decisions, but rather are stable over analyses that differ in apparently innocuous ways. All three examples include stability analyses for certain decisions, as described below.

A sensitivity analysis asks to what extent plausible changes in assumptions change conclusions. In contrast, a stability analysis asks how ostensibly innocuous changes in analytical decisions change conclusions. A sensitivity analysis typically examines a continuous family of departures from a critical assumption, in which the magnitude of the departure and the magnitude of the change in conclusions are the focus of attention. In contrast, a stability analysis typically examines a discrete decision, and the hope and expectation is that the conclusions are largely unaltered by changing this decision. Stability analyses are necessary in most complex analyses; however, the extent to which they are needed varies markedly from study to study.

As an example of a stability analysis, consider the MSSORS study of lead exposures, which needed to address the possibility that workers exposed to lead were more likely to have lead-related hobbies. MSSORS (1982, pp. 552-553) write:

...11 pairs were found in which the study children had potential for lead exposure other than that due to the father's occupation, while their matched controls had no such exposures. Exogenous sources of lead found in the children's environment were automobile body painting, casting of lead and playing with spent gun shell casings in the home. Six children in the study group had fathers whose hobby was casting lead into fish sinkers; none of the control children's fathers did this. It was speculated that those who work with lead on the job are more accustomed to handling lead, thereby promoting its use in the home environment. ... [A]ny study/control matched pair in which one of these hobbies was present for the study child and not present for the control was eliminated from the analysis. ... When these 11 children and their controls were eliminated, and the remaining 22 pairs were analyzed, the study and control groups continued to be statistically different ($p < .001$).

MSSORS go on to show that the estimates of lead levels do not change much when the 11 pairs are excluded. Notice that a discrete decision—whether or not to include the 11 pairs—is investigated by carrying out the analysis both ways and comparing the results, a process that is very different from sensitivity analysis.

Similarly, in the LWW study of the effects of the death of a spouse or child in a car crash, there was concern that the death of a spouse might alter family income and, possibly, that it was this sustained loss of income and not the death itself that has psychological effects. This was investigated by adjusting for income by regression, concluding that most of the findings remained stable (LWW 1987, p. 226). The minimum wage study included several stability analyses, for instance, various ways of handling temporarily closed restaurants; see CK (1984, Table 5).

The examples mentioned above have a common feature that arises frequently in observational studies. Because it is clear that one wishes to compare treated and control subjects who were comparable prior to treatment, it is clear that visible pretreatment differences need to be removed by adjustments of one kind or another. It may happen, however, that treated and control groups are noted to differ after the start of treatment. In this case, the posttreatment difference may reflect unobserved pretreatment differences or effects caused by the treatment or the effects of other treatments occurring at the same time. Lead hobbies are another treatment coexisting with occupational lead exposure. Income loss is an outcome affected by the loss of a spouse. The closing of restaurants may be related to business conditions, the level of the minimum wage being one such condition. Adjustments for posttreatment differences may remove part of the actual treatment effect, and they may either remove bias or introduce bias into comparisons (Rosenbaum 1984). Stability analyses are common in this setting, and they seek to demonstrate that results are stable whether or not adjustments are made for a posttreatment difference. An alternative approach is to be explicit about the effect of the treatment on the posttreatment variable, replacing the discrete choice of a stability analysis by the continuous variation in assumptions of a sensitivity analysis; see Rosenbaum (1984, §§4.3, 4.4) for detailed discussion of this alternative.

As another example with a different result, the minimum wage debate between Neumark and Wascher (1992) and Card, Katz, and Krueger (1994) turns in part on a stability analysis. Neumark and Wascher (1994) had conducted a panel study of changes in state minimum wages between 1973 and 1989 in relation to unemployment among teenagers and young adults, reaching the conclusion that increases in minimum wages depress employment. Card, Katz, and Krueger then commented that the results were unstable in the following sense. Newmark and Wascher had made adjustments for the "proportion of the age group enrolled in school". Card, Katz, and Krueger argued that, first, the conclusions about the minimum wage change dramatically if adjustments are not made for this variable and, second, that the definition of this variable is such that it is "mechanically" related to the response variable, namely employment, because anyone working even part-time was counted as not enrolled in school. In their response, Neumark and Wascher disagreed.

The purpose here is to examine methodology and not the minimum wage debate. From a methodological view, the study by Newmark and Wascher (1992) relies heavily on analytical models in comparisons, whereas the study by Card and Krueger (1994) relies more on the design of the study and the choice of circumstances in which fairly comparable units are compared, here restaurants from the same chains in adjacent states. A study that relies heavily on analytical models and adjustments will make many more implementation decisions in analysis, and so will need to conduct more extensive stability analyses to be convincing. This is an argument in favor of simple study designs that compare ostensibly comparable units under alternative treatments.

11.10   The Role of Time: Abrupt, Short-Lived Treatments

The concept that a treatment must precede its effects influences the design of an observational study in many ways. A treatment may be delivered over a prolonged period of time, so the distinction between what precedes a treatment and what follows it may not be sharp. Subjects may switch from one treatment to another, possibly in part in response to an earlier failure of the first treatment. Other extraneous treatments may intervene, and these interventions may themselves be, in part, effects stimulated by the treatments under study. Subjects may know about the treatment before they receive it, and part of the ultimate effect of the treatment may begin to materialize before the treatment is delivered. For instance, the increases in New Jersey's minimum wage were public legislation well before the increases actually took place. For discussion of various aspects of the role of time in intervention studies, see Campbell and Stanley (1963, p. 5-6), Diggle, Liang and Zeger (1994), Holland (1993), Joffe et al. (1998), Li, Propert, and Rosenbaum (2001), Peto et al. (1976), Robins (1989, 1992, 1998), Robins, Rotnitzky and Zhao (1995), Rosenbaum (1984, 2001a), Rubin (1991, §5.2), Schafer (1996), Sobel (1995), and Susser (1987, p. 86).

In research design, given the choice, one would prefer a single, abrupt, unexpected, short-lived treatment of dramatic proportions. With such a treatment, the line between what precedes treatment and what follows it is sharply drawn. This is true of only one of the three examples, namely the LWW study of the psychological effects of the death of a spouse or a child in a car crash.

Abrupt, unexpected, short-lived treatments of dramatic proportions are sometimes informally associated with the term "exogenous" as it is used in econometrics. However, formal discussions of "exogeneity" actually define the matter rather differently (e.g., Engle, Hendry, and Richard 1983). In sociology, Giddens (1979, p. 127–128) makes a related point about the role of "critical situations":

> We can learn a good deal about day-to-day life in routine settings from analysing circumstances in which those settings are radically disturbed. . . . By a critical situation I mean a set of circumstances which—for whatever reason—radically disrupts accustomed routines of daily life.

Another example of an abrupt, short-lived treatment is found in a study of the effects of immigration on labor markets. This is generally a difficult topic because immigration occurs gradually and immigrants may favor labor markets where jobs are available and attractive. It is usually difficult to disentangle the effects of immigration on labor markets from the effects of labor markets on immigration. Card (1990) exploited a rare exception, in which immigration was abrupt, short-lived, unexpected, and of dramatic proportions. In the Mariel Boatlift, about 125,000 Cubans immigrated to Miami between May and September, 1980, increasing Miami's labor force by seven percent. Card compared changes in Miami's labor market following the Mariel Boatlift to the concurrent changes in four unaffected cities, Atlanta, Houston, Los Angeles, and Tampa–St. Petersburg.

Many interesting treatments are not short-lived, but rather, by their nature, are prolonged or chronic. For such treatments, given the choice, one would prefer an abrupt, comparatively haphazard start of a chronic treatment that, once started, is inescapable. In discussing the effects of prolonged or chronic stress as a cause of depression, Kessler (1997, p. 197) discusses this clearly:

> ... a major problem in interpret[ation] ... is that both chronic role-related stresses and the chronic depression by definition have occured for so long that deciding unambiguously which came first is difficult ... The researcher, however, may focus on stresses that can be assumed to have occurred randomly with respect to other risk factors of depression and to be inescapable, in which case matched comparison can be used to make causal inferences about long-term stress effects. A good example is the matched comparison of the parents of children having cancer, diabetes, or some other serious childhood physical disorder with the parents of healthy children. Disorders of this sort are quite common and occur, in most cases, for reasons that are unrelated to other risk factors for parental psychiatric disorder. The small amount of research shows that these childhood physical disorders have significant psychiatric effects on the family.

11.11   Natural Blocks

Another opportunity to use choice in place of control in the design of an observational study involves natural blocks. Whereas matching is used to pair unrelated individuals having similar values of measured covariates, natural blocks create pairs or groups of individuals who are related in ways judged to be important but difficult to measure explicitly. Twins, siblings, neighbors, and schools are familiar examples of natural blocks.

Twins, for example, resemble one another in terms of genetics and childhood environment in many ways that cannot practically be described in measured covariates. The LWW study of the psychological effects of the loss of a spouse or of a child could not adjust for genetic differences between exposed subjects and controls. In related work, Lichtenstein, Gatz, Pedersen, Berg, and McClearn (1996) examined the psychological effects of widowhood by comparing twins, one bereaved and one still married. In partial corroboration of the LWW study, they also found long term psychological effects of the loss of a spouse, suggesting that genetic differences are not a likely explanation of the psychological outcomes.

Behrman, Rosenzweig, and Taubman (1996) use twins in an economic observational study of the effect of college quality on subsequent earnings. Ashenfelter and Krueger (1994) make a similar use of twins, while Altonji and Dunn (1996) use siblings instead.

It is possible to combine matching for covariates and pairing using natural blocks. In the MSSORS study of lead exposure, exposed and control children were compared to a neighbor's child of about the same age. Here, age is a covariate whereas neighborhood is a block. Similarly, in Rosenbaum (1986), high school dropouts were matched to students with similar grades, test scores, and behavior who remained in the same high school. Here, the high school is the block. In both cases, matching controlled for blocks and began the adjustment for covariates.

11.12   Refute Several Alternative Explanations

The common, perhaps inevitable, criticism of an observational study is that individuals who appeared comparable before treatment in terms of observed covariates were, in fact, not comparable, and that differing outcomes after treatment reflect this lack of comparability, not an effect caused by the treatment. In this context, Campbell (1957) insisted on a certain logic, a certain standard for criticism, namely that objections to an observational study be expressed as specific, credible threats to validity, or "grounds for doubt" in Wittgenstein's phrase (1972, p. 18); see also Bross (1960), Gastwirth, Krieger and Rosenbaum (1997), and Shadish and Cook (1999). Anticipating specific grounds for doubt, Campbell (1957) argued for improving the design to address these specific issues. Instead of assuming that hidden biases are absent, that what isn't visible is equal—the so-called *ceteris paribus* clause—one assumes specific biases may be present and investigates them. This is similar to advice offered by Lakatos (1970, p. 110):

> How can one test a *ceteris paribus* clause severely? By assuming that there *are* other influencing factors, by specifying such factors, and by testing these specific assumptions. If many of them are refuted, the *ceteris paribus* clause will be regarded as well-corroborated.

Strengthen evidence by eliminating weaknesses, one weakness at a time.

11.13   Bibliographic Notes

This chapter is adapted from Rosenbaum (1999). The paper discusses several issues in greater depth, and is accompanied by interesting discussion by Manski (1999), Robins (1999), and Shadish and Cook (1999), together with a rejoinder. Good, recent general discussions of the early planning of observational studies are given by Cook, Campbell, and Peracchio (1990), Cook and Shadish (1994), Meyer (1995), Angrist and Krueger (1999), Rosenzweig and Wolpin (2000), and Shadish, Cook and Campbell (2002). The design of replications of observational studies is discussed in Rosenbaum (2001b).

11.14   Problems

1. **Environmental influences on eating and physical activity.** Successful planning of an observational study requires a genuine puzzle in a context that is familiar. The context needs to be familiar so one can think concretely about plausible sources of bias, grounds for doubt, and what to do about them. In advanced scientific work, the context may be familiar only to the relatively few people who conduct studies in that area, but for a textbook problem, the context needs to be generally familiar. There is evidence that obesity is increasingly common in the United States; see Kuczmarski, Flegal, Campbell, and Johnson (1994) for evidence from the National Health and Nutrition Examination Survey. The question is: Why? Presumably it reflects changes in either eating patterns or physical activity or both, but the question is: What caused those changes? A thoughtful survey of possibilities is given by French, Story, and Jeffery (2001). They write: "Documenting the environmental influences on population physical activity and eating behaviors has posed an even greater challenge than documenting individual behaviors because such influences are difficult to define, measure, and study experimentally." Over roughly 25 years, they document: (i) dramatic increases in the consumption of cheese, up 146%, and soft drinks, up 131%, but a 3% increase in grams of fat consumed per day, (ii) dramatic increases in the number and use of restaurants, especially fast-food restaurants, and the commercial failure of several efforts by fast-food restaurants to market lower-fat options, and (iii) growing portion sizes, for instance, for soft drinks. They discuss convenience foods, pizza, "take-out," women working outside the home, the large amounts spent on food and restaurant advertising, and the relative prices of different foods. Concerning physical activity, they discuss television, VCR's, computer games, and health clubs. Many possible hypotheses, and a context that is, well, familiar. As one specific suggestion: consider the hypothesis that food prepared by fast-food restaurants plays a large role, versus the alternative that its role is minor. Can you think of a control group, that is, a group that rarely or never eats in fast-food restaurants? Preferably, the control group would be formed not by food preferences, but rather by something ostensibly irrelevant. Can you think of more than one control group of this sort? Are these control groups representative of the US population? If not, can you find an exposed group or several exposed groups, perhaps also unrepresentative, but quite similar to the controls? Can you identify instrumental variables, that is, haphazard limitations on access to fast-food restaurants, limitations that can affect diet and physical activity only through limitations on access? What outcomes would you measure? What would constitute a coherent treatment effect?

Will you frame the hypothesis to have implications for meals eaten at home?

2. **Read ethnographic accounts.** The previous problem suggested that successful planning of an observational study requires familiarity with the context in which the study will be conducted. Familiarity with context is easily attained for a few topics, but for many others, familiarity is attained only through years of effort, perhaps accompanied by personal risk. For such difficult topics, some familiarity can be attained by reading a few ethnographic accounts before planning an observational study. The best way to cultivate the habit of reading ethnographic accounts before planning an observational study is to read one now. Some good choices are Becker (1972), Bosk (1981), Estroff (1985), and Anderson (2000).

11.15 References

Allison, P. D. (1990) Change scores as dependent variables in regression analyses. In: *Sociological Methodology*, C. C. Clogg, ed., Oxford: Basil Blackwell, pp. 93–114.

Altonji, J. G. and Dunn, T. A. (1996) Using siblings to estimate the effect of school quality on wages. *Review of Economics and Statistics*, **77**, 665–671.

Anderson, E. (2000) *Code of the Streets: Decency, Violence and the Moral Life of the Inner City.* New York: Norton.

Angrist, J. D. and Krueger, A. B. (1999) Empirical strategies in labor economics. In: *Handbook of Labor Economics, III,* New York: Elsevier, Chapter 23.

Angrist, J. D. and Lavy, V. (1999) Using Maimonides' rule to estimate the effect of class size on scholastic achievement. *Quarterly Journal of Economics*, 533–575.

Ashenfelter, O. A. and Krueger, A. B. (1994) Estimates of the economic return to schooling from a new sample of twins. *American Economic Review*, **84**, 1157–1173.

Becker, H. S. (1972) *Outsiders: Studies in the Sociology of Deviance.* New York: The Free Press.

Behrman, J., Rosenzweig, M., and Taubman, P. (1996) College choice and wages: Estimates using data on female twins. *The Review of Economics and Statistics*, 672–685.

Bosk, C. L. (1981) *Forgive and Remember: Managing Medical Failure.* Chicago: University of Chicago Press.

Box, G. E. P. (1966) The use and abuse of regression. *Technometrics,* **8,** 625–629.

Bronars, S. G. and Grogger, J. (1994) The economic consequences of unwed motherhood: Using twin births as a natural experiment. *American Economic Review,* **84,** 1141–1156.

Bross, I. D. J. (1960). Statistical criticism. *Cancer,* **13,** 394–400. Reprinted in: *The Quantitative Analysis of Social Problems,* E. Tufte, ed., Reading, MA: Addison-Wesley, pp. 97–108.

Campbell, D. T. (1957) Factors relevant to the validity of experiments in social settings. *Psychological Bulletin,* **54,** 297–312.

Campbell, D. T. (1984, 1988) Can we be scientific in applied social science? *Evaluation Studies Review Annual,* **9,** 26–48. Reprinted in: D. Campbell, *Methodology and Epistemology for Social Science: Selected Papers.* Chicago: University of Chicago Press, pp. 315–333.

Campbell, D. and Stanley, R. (1963) *Experimental and Quasi-Experimental Designs for Research.* Chicago: Rand McNally.

Card, D. (1990) The impact of the Mariel Boatlift on the Miami labor market. *Industrial and Labor Relations Review,* **43,** 245–257.

Card, D., Katz, L., and Krueger, A. (1994) Comment. *Industrial and Labor Relations Review,* **48,** 487–496.

Card, D. and Krueger, A. (1994) Minimum wages and employment: A case study of the fast-food industry in New Jersey and Pennsylvania. *American Economic Review,* **84,** 772–793.

Card, D. and Krueger, A. (1995) *Myth and Measurement: The New Economics of the Minimum Wage.* Princeton, NJ: Princeton University Press.

Card, D. and Krueger, A. (2000) Minimum wages and employment: A case study of the fast-food industry in New Jersey and Pennsylvania: Reply. *American Economic Review,* **90,** 1397–1420.

Chamberlin, T. C. (1890, 1965) The method of multiple working hypotheses. Originally in *Science* 1890, **15,** 92. Reprinted in *Science* 1965, **148,** 754–759.

Cook, T. D., Campbell, D. T. and Peracchio, L. (1990) Quasi-experimentation. In: *Handbook of Industrial and Organizational Psychology*, M. Dunnette and L. Hough, Palo Alto, eds., CA: Consulting Psychologists Press, Chapter 9, pp. 491–576.

Cook, T. D. and Shadish, W. R. (1994) Social experiments: Some developments over the past fifteen years. *Annual Review of Psychology*, **45**, 545–580.

Dawes, R. (1996) The purpose of experiments: Ecological validity versus comparing hypotheses. *Behavioral and Brain Sciences*, **19**, 20.

Deere, D., Murphy, K., and Welch, F. (1995) Employment and the 1990–1991 minimum-wage hike. *American Economic Review*, **85**, 232–237.

Diggle, P. J., Liang, K.Y., and Zeger, S. L. (1994) *Analysis of Longitudinal Data*. New York: Oxford University Press.

Duff, C. (1996) New minimum wage makes few waves: Employers offset 50-cent raise with minor shifts. *Wall Street Journal*, 20 November 1996, pp. 2–4.

Engle, R., Hendry, D., and Richard, J. (1983) Exogeneity. *Econometrica*, **51**, 277–304.

Estroff, S. E. (1985) *Making It Crazy: An Ethnography of Psychiatric Clients in an American Community*. Berkeley, CA: University of California Press.

Feyerabend, P. (1968) How to be a good empiricist—a plea for tolerance in matters epistemological. In: *The Philosophy of Science* (Oxford Readings in Philosophy), P. H. Nidditch, ed., New York: Oxford University Press, pp. 12–39.

Feyerabend, P. (1975) *Against Method*. London: Verso.

Fisher, R. A. (1935) *The Design of Experiments*. Edinburgh: Oliver and Boyd.

Freedman, D. (1997) From association to causation via regression. *Advances in Applied Mathematics*, **18**, 59–110.

French, S. A., Story, M., and Jeffery, R. W. (2001) Environmental influences on eating and physical activity. *Annual Review of Public Health*, **22**, 309–335.

Friedman, M. (1953) The methodology of positive economics. In: *Essays in Positive Economics*, Chicago: University of Chicago Press, pp. 3–43.

Gastwirth, J. L., Krieger, A. M., and Rosenbaum, P. R. (1997) Hypotheticals and hypotheses. *American Statistician*, **51**, 120–121.

Giddens, A. (1979) *Central Problems in Social Theory*. Berkeley: University of California Press.

Holland, P. (1993) Which comes first, cause or effect? In: *A Handbook for Data Analysis in the Behavioral Sciences: Methodological Issues*, G. Keren and C. Lewis, eds., Hillsdale, NJ: Lawrence Erlbaum, pp. 273–282.

Holland, P. & Rubin, D. (1983) On Lord's paradox. In: *Principles of Psychological Measurement: A Festschrift for Frederic Lord*, H. Wainer and S. Messick, eds., Hillsdale, NJ: Lawrence Erlbaum, pp. 3–25.

Joffe, M., Hoover, D., Jacobson, L., Kingsley, L., Chmiel, J., Visscher, B., and Robins, J. (1998) Estimating the effect of zidovudine on Kaposi's sarcoma from observational data using a rank preserving structural failure-time model. *Statistics in Medicine*, **17**, 1073-1102.

Kessler, R. C. (1997) The effects of stressful life events on depression. *Annual Review of Psychology*, **48**, 191–214.

Kuczmarski, R. J., Flegal, K. M., Campbell, S. M., and Johnson, C. L. (1994) Increasing prevalence of overweight among US adults. The National Health and Nutrition Examination Surveys, 1960 to 1991. *Journal of the American Medical Association*, **272**, 205–211

Lakatos, I. (1970) Falsification and the methodology of scientific research programs. In: *Criticism and the Growth of Knowledge*, I. Lakatos and A. Musgrave, eds., New York: Cambridge University Press, pp. 91–196. Reprinted in: I. Lakatos (1978) *Philosophical Papers*, Volume 1, New York, Cambridge University Press, pp. 8–101.

Lakatos, I. (1981) History of science and its rational reconstructions. In: *Scientific Revolutions*, I. Hacking, ed., New York: Oxford University Press, pp. 107–127. Reprinted from *Boston Studies in the Philosophy of Science*, VIII, 1970.

Lehman, D., Wortman, C., and Williams, A. (1987) Long-term effects of losing a spouse or a child in a motor vehicle crash. *Journal of Personality and Social Psychology*, **52**, 218–231.

Li, Y., Propert, K. J., and Rosenbaum, P. R. (2001) Balanced risk set matching. *Journal of the American Statistical Association*, **96**, September, to appear..

Lichtenstein, P., Gatz, M., Pedersen, N., Berg, S., and McClearn, G. (1996) A co-twin-control study of response to widowhood. *Journal of Gerontology: Psychological Sciences*, **51**B, 279–289.

Manski, C. (1999) Comment. *Statistical Science*, **14**, 279–281.

Meehl, P. (1978) Theoretical risks and tabular asterisks: Sir Karl, Sir Ronald, and the slow progress of soft psychology. *Journal of Consulting and Clinical Psychology*, **46**, 806–834. Reprinted in: P. Meehl, *Selected Philosophical and Methodological Papers*, Minneapolis: University of Minnesota Press, 1991, pp. 1–42.

Meyer, B. D. (1995) Natural and quasi-experiments in economics. *Journal of Business and Economic Statistics*, **13**, 151–161.

Meyer, M. and Fienberg, S, eds. (1992) *Assessing Evaluation Studies: The Case of Bilingual Education Strategies*. Washington, DC: National Academy Press.

Morton, D., Saah, A., Silberg, S., Owens, W., Roberts, M., and Saah, M. (1982) Lead absorption in children of employees in a lead-related industry. *American Journal of Epidemiology*, **115**, 549–555.

Neumark, D. and Wascher, W. (1992) Employment effects of minimum and subminimum wages: Panel data on state minimum wage laws. *Industrial and Labor Relations Review*, **46**, 55–81.

Neumark, D. and Wascher, W. (1994) Reply. *Industrial and Labor Relations Review*, **48**, 497–512.

Peirce, C. S. (1903) On selecting hypotheses. In: *Collected Papers of Charles Sanders Peirce*, C. Hartshorne and P. Weiss, eds., Volume 5, Cambridge, MA: Harvard University Press, 1960, pp. 413–422.

Peto, R., Pike, M., Armitage, P., Breslow, N., Cox, D., Howard, S., Mantel, N., McPherson, K., Peto, J., and Smith, P. (1976) Design and analysis of randomized clinical trials requiring prolonged observation of each patient, I: Introduction and design. *British Journal of Cancer*, **34**, 585–612.

Platt, J. (1964) Strong inference. *Science*, **146**, 347–352.

Polanyi, M. (1946, 1964) *Science, Faith and Society*. New York: Oxford University Press. Reprinted, Chicago: University of Chicago Press.

Popper, K. (1965) *Conjectures and Refutations*. New York: Harper and Row.

Popper, K. (1968) *The Logic of Scientific Discovery*. New York: Harper and Row. (English translation of Popper's 1934 *Logik der Forschung*.)

Putnam, H. (1995) *Pragmatism.* Oxford: Blackwell.

Quine, W. (1951) Two dogmas of empiricism. *Philosophical Review.* Reprinted in: W. Quine (1980) *From a Logical Point of View*, Cambridge, MA: Harvard University Press, pp. 20–46.

Robins, J. (1989) The control of confounding by intermediate variables. *Statistics in Medicine*, **8**, 679–701.

Robins, J. (1992) Estimation of the time-dependent accelerated failure time model in the presence of confounding factors. *Biometrika*, **79**, 321–334.

Robins, J. (1998) Correction for non-compliance in equivalence trials. *Statistics in Medicine*, **17**, 269–302.

Robins, J. (1999) Comment. *Statistical Science*, **14**, 281–293.

Robins, J., Rotnitzky, A. and Zhao, L. (1995) Analysis of semiparametric regression models for repeated outcomes in the presence of missing data. *Journal of the American Statistical Association*, **90**,106–121.

Rosenbaum, P. R. (1984) The consequences of adjustment for a concomitant variable that has been affected by the treatment. *Journal of the Royal Statistical Society*, Series **A**, **147**, 656–666.

Rosenbaum, P. R. (1986) Dropping out of high school in the United States: An observational study. *Journal of Educational Statistics*, **11**, 207–224.

Rosenbaum, P. R. (1999) Choice as alternative to control in observational studies (with discussion). *Statistical Science*, **14**, 259–304.

Rosenbaum, P. R. (2001a) Stability in the absence of treatment. *Journal of the American Statistical Association*, **96**, 210–219.

Rosenbaum, P. R. (2001b) Replicating effects and biases. *American Statistician*, **55**, 223–227.

Rosenzweig, M. and Wolpin, K. (1980) Testing the quantity-quality fertility model: The use of twins as a natural experiment. *Econometrica*, **48**, 227–240.

Rosenzweig, M. R. and Wolpin, K. I. (2000) Natural "natural experiments" in economics. *Journal of Economic Literature*, **38**, 827–874.

Rubin, D. B. (1991) Practical implications of modes of statistical inference for causal inference and the critical role of the assignment mechanism. *Biometrics*, **47**, 1213–1234.

Samuelson, P. (1947, 1983) *Foundations of Economic Analysis*. Cambridge, MA: Harvard University Press.

Shadish, W. R. and Cook, T. D. (1999) Comment—Design rules: More steps toward a complete theory of quasi-experimentation. *Statistical Science*, **14**, 294–300.

Shadish, W. R., Cook, T. D., and Campbell, D. T. (2002) *Experimental and Quasi-Experimental Designs for Generalized Causal Inference*. Boston: Houghton-Mifflin.

Shafer, G. (1996) *The Art of Causal Conjecture*. Cambridge, MA: MIT Press.

Sobel, M. (1995) Causal inference in the social and behavioral sciences. In: *Handbook of Statistical Modelling for the Social and Behavioral Sciences*, G. Arminger, C. Clogg, and M. Sobel, eds., New York: Plenum, pp. 1–38.

Sowell, T. (1995) Repealing the law of gravity. *Forbes*, 22 May, p. 82.

Susser, M. (1987) Falsification, verification and causal inference in epidemiology: Reconsideration in the light of Sir Karl Popper's philosophy. In: M. Susser, *Epidemiology, Health and Society: Selected Papers*, New York: Oxford, pp. 82–93.

12
Some Strategic Issues

12.1   What Are Strategic Issues?

By and large, the discipline of statistics is concerned with the development of correct and effective research designs and analytical methods, together with supporting theory. Here, correct and effective refer to formal properties of the designs and methods. The first eleven chapters discussed issues of this sort. In contrast, a strategic issue concerns the impact that an empiric investigation has on its intended audience. Often, the audience is not focused on statistical technique and theory, may have limited training in statistics, and may be comprised of laymen, that is, laymen with respect to their knowledge of statistics. When this is so, strategic issues may arise in which formal properties are weighed against impact on the intended audience.

Should strategic issues be considered at all? Why, after all, should laymen manage the conduct of science? Why should the decisions of laymen govern the interpretation of empiric research? Why should laymen sit as judge and jury on statistical analyses?

The answer, of course, is that public and corporate managers, governors, judges, and juries are, typically, not professional statisticians. The person responsible for a decision hopes and expects to find the evidence somewhat convincing, not merely to hear that an expert finds the evidence convincing. Often, these hopes and expectations are reasonable, and often they determine the impact of an empiric investigation.

12.2 Some Specific Suggestions

This brief, final chapter discusses a few strategic issues in the design and conduct of observational studies. Motivation and elaboration of these issues is found in the chapters or sections indicated within parentheses.

- *Strategic issues are more than communication and presentation.* Strategic issues affect the choice of research design and analytical methods.

- *Design observational studies.* A recent report of the National Academy of Sciences (Meyer and Fienberg 1992, p. 106) concludes: "Care in design and implementation will be rewarded with useful and clear study conclusions.... Elaborate analytical methods will not salvage poor design or implementation of a study." Exert as much experimental control as is possible. Use the same measurement techniques in treated and control groups (§1.2). Carefully consider the process that selects individuals into the study—selection can introduce or eliminate biases (§§7.1.3, 7.3). Anticipate hidden biases that pose the greatest threat to the study. Actively collect data that can reveal those biases if they are present (Chapters 6, through 8). Employ strategies that can reduce sensitivity to hidden bias (§5.3.7, Chapters 9 and 11).

- *Focus on simple comparisons.* Tukey (1986) writes: "increase impact of results on consumers ... by focusing on meaningful results (e.g., simple comparisons)." Cox (1958, p. 11) lists "simplicity" as one of the five "requirements for a good experiment" and writes: "This is a very important matter which must be constantly borne in mind ...." Peto, Pike, Armitage, Breslow, Cox, Howard, Mantel, McPherson, Peto, and Smith (1976, §4, p. 590) make a similar point. Simplicity is of greater importance in observational studies. Comparisons are often subject to genuine ambiguity or credible challenge. Issues of this kind are far more likely to be resolved if they are not compounded by unnecessary complications. A complex analysis can often be divided into several simple analyses, each of which can be challenged and debated separately. Writing within economics, Blaug (1992, p. 245) offers good advice for all fields in saying that empirical work should be judged "on the basis of the likely validity of the results reported and not on the technical sophistication of the techniques employed."

- *Compare subjects who looked comparable prior to treatment.* The most direct, most compelling way to address overt biases is to compare treated and control groups that looked comparable prior to treatment in terms of observed covariates. Susser (1973, §7) calls this "simplifying the conditions of observation." The matching and stratification methods in Chapter 10 can often produce matched pairs or sets or strata that balance many observed covariates.

- *Use sensitivity analyses to inform discussions of hidden biases due to unobserved covariates.* Even with the greatest care, undetected hidden bias is a legitimate concern in an observational study. However, claims about hidden biases do not become credible merely because the covariates involved were not observed. The issue is explored through sensitivity analyses (Chapter 4). Sensitivity analyses for unobserved covariates are likely to have greatest impact if they build upon standard statistical techniques applied to simple comparisons of treated and control groups that appear comparable in terms of observed covariates.

12.3   References

Blaug, M. (1992) *The Methodology of Economics* (second edition). New York: Cambridge University Press.

Cox, D. R. (1958) *The Planning of Experiments*. New York: Wiley.

Meyer, M. and Fienberg, S., eds. (1992) *Assessing Evaluation Studies: The Case of Bilingual Education Strategies*. Washington, DC: National Academy Press.

Peto, R., Pike, M., Armitage, P., Breslow, N., Cox, D., Howard, S., Mantel, N., McPherson, K., Peto, J., and Smith, P. (1976) Design and analysis of randomized clinical trials requiring prolonged observation of each patient, I: Introduction and design. *British Journal of Cancer*, **34**, 585–612.

Susser, M. (1973) *Causal Thinking in the Health Sciences*. New York: Oxford University Press.

Tukey, J. (1986) Sunset salvo. *The American Statistician*, **40**, 72–76.

Index

Springer Series in Statistics *(continued from p. ii)*

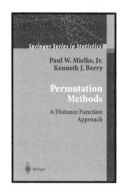